The Politics of Sustainability in the Arctic

The Politics of Sustainability in the Arctic argues that sustainability is a political concept because it defines and shapes competing visions of the future. In current Arctic affairs, prominent stakeholders agree that development needs to be sustainable, but there is no agreement over what it is that needs to be sustained. In original conservationist discourse, the environment was the sole referent object of sustainability; however, as sustainability discourses have expanded, the concept has been linked to an increasing number of referent objects, such as society, economy, culture, and identity.

This book sets out a theoretical framework for understanding and analysing sustainability as a political concept, and provides a comprehensive empirical investigation of Arctic sustainability discourses. Presenting a range of case studies from Greenland, Norway, Canada, Russia, Iceland, and Alaska, the chapters in this volume analyse the concept of sustainability and how actors are employing and contesting this concept in specific regions within the Arctic. In doing so, the book demonstrates how sustainability is being given new meanings in the postcolonial Arctic and what the political implications are for postcoloniality, nature, and development more broadly.

Beyond those interested in the Arctic, this book will also be of great value to students and scholars of sustainability, sustainable development, and identity and environmental politics.

Ulrik Pram Gad is Associate Professor of Arctic Culture and Politics at Aalborg University, Denmark.

Jeppe Strandsbjerg is Editor-in-Chief for Social Science at Djøf Publishing. Previously, he was Associate Professor at the Department of Business and Politics, Copenhagen Business School, Denmark.

Routledge Studies in Sustainability

Sustainable Modernity
The Nordic Model and Beyond
Edited by Nina Witoszek and Atle Midttun

Towards the Ethics of a Green Future
The Theory and Practice of Human Rights for Future People
Edited by Marcus Düwell, Gerhard Bos and Naomi van Steenbergen

Sustainability Transitions in South Africa
Edited by Najma Mohamed

Strongly Sustainable Societies
Organising Human Activities on a Hot and Full Earth
Edited by Karl Johan Bonnedahl and Pasi Heikkurinen

Managing Complexity, Earth Systems and Strategies for the Future
Walter R. Erdelen and Jacques G. Richardson

Measuring Intangible Values
Rethinking How to Evaluate Socially Beneficial Actions
Marie Harder and Gemma Burford

Education for Sustainable Human and Environmental Systems
From Theory to Practice
Edited by Will Focht, Michael A. Reiter, Paul A. Barresi and Richard C. Smardon

The Politics of Sustainability in the Arctic
Reconfiguring Identity, Space, and Time
Edited by Ulrik Pram Gad and Jeppe Strandsbjerg

Global Planning Innovations for Urban Sustainability
Edited by Sébastien Darchen and Glen Searle

www.routledge.com/Routledge-Studies-in-Sustainability/book-series/RSSTY

The Politics of Sustainability in the Arctic

Reconfiguring Identity, Space, and Time

Edited by Ulrik Pram Gad and
Jeppe Strandsbjerg

Routledge
Taylor & Francis Group
LONDON AND NEW YORK

earthscan
from Routledge

First published 2019
by Routledge
2 Park Square, Milton Park, Abingdon, Oxon OX14 4RN

and by Routledge
711 Third Avenue, New York, NY 10017

Routledge is an imprint of the Taylor & Francis Group, an informa business

British Library Cataloguing-in-Publication Data
A catalogue record for this book is available from the British Library

Library of Congress Cataloging-in-Publication Data
A catalog record has been requested for this book

ISBN: 978-1-138-49183-0 (hbk)
ISBN: 978-1-351-03198-1 (ebk)

Typeset in Goudy
by Wearset Ltd, Boldon, Tyne and Wear

Contents

Illustrations

Figures

Table

Contributors

Rikke Becker Jacobsen holds a PhD in Planning and Development (2014) with a focus on power and participation in Greenlandic fisheries governance. Rikke has conducted ethnographic research on fisheries governance at Aalborg University since 2008 and worked as an affiliated researcher at the Greenland Climate Research Centre from 2010 to 2015.

Lill Rastad Bjørst is Associate Professor at Aalborg University and holds a PhD in Arctic Studies from the Department of Cross-Cultural and Regional Studies at the University of Copenhagen. Her recent research focuses on the climate change debate in Greenland and the political debate about Greenland's uranium. She is head of the research centre CIRCLA and academic coordinator for Arctic Studies, a Master's programme specialization at Aalborg University. Her research interests include Inuit culture and society; climate change and sustainability; mining and industrialization; and postcolonialism and tourism.

Johanne Bruun recently completed her doctorate in the Department of Geography at Durham University, UK. Her research sits at the intersection between scientific practice, material politics, and spatialized governance, and her particular emphasis is on questions of terrain and territory. Her current work explores the role of science in constructing territory across a range of geologic volumes in Cold War Greenland.

Klaus Dodds is Professor of Geopolitics at Royal Holloway University of London and a Fellow of the Academy of Social Sciences. He is co-author (with Mark Nuttall) of *The Scramble for the Poles* (2016) and *Ice: Nature and Culture* (2018). In 2014–15, he served as specialist adviser to the House of Lords Select Committee on the Arctic.

Ulrik Pram Gad is Associate Professor of Arctic Culture and Politics at Aalborg University, Denmark. His main research interests are Greenlandic and Danish nationalism, postcoloniality, and Othering. Publications on the Arctic include *National Identity Politics and Postcolonial Sovereignty Games* (2016), and articles in *Cooperation and Conflict*, the *Journal of Language and Politics*, *Nordiques*, and *Politik*.

Hannes Gerhardt is Professor of Geography at the University of West Georgia. His research focuses on how geographic imaginaries shape political power, capitalism, and alternatives to capitalism. He is a co-author of the book *Contesting the Arctic: Politics and Imaginaries in the Circumpolar North.*

Naja Dyrendom Graugaard is a PhD Fellow at Aalborg University. Inspired by her own mixed background in Denmark and Greenland, Naja's research involves Inuit sealing, Indigenous knowledge, alternative narratives, coloniality, and decolonization in the Arctic. She holds an MES from York University and a BA from Trent University (Canada). Naja is also a mother, script writer, and theatre artist.

Victoria Herrmann is the Managing Director of the Arctic Institute and a Gates Scholar at the Scott Polar Research Institute, University of Cambridge. Her research focuses on climate change adaptation, mitigation, and cultural heritage in the circumpolar north, with a particular focus on coastal communities. Most recently, she authored 'America's first climate change refugees: victimization, distancing, and disempowerment in journalistic storytelling' in *Energy Research & Social Science* and was a co-editor of the *Politik* journal special issue 'Arctic international relations in a widened security perspective' (with Marc Jacobsen).

Marc Jacobsen is a PhD Candidate at the Department of Political Science, University of Copenhagen, where he investigates how Denmark and Greenland make discursive use of global Arctic interests to enhance and expand their international relations – collectively as well as separately. Most recently, he has co-edited a special issue of the journal *Politik*, 'Arctic international relations in a widened security perspective' (with Victoria Herrmann), and co-authored a book chapter (2017) on Greenlandic foreign policy narratives (with Ulrik Pram Gad).

Kathrin Keil is Scientific Project Leader at the Institute for Advanced Sustainability Studies (IASS) in Potsdam, Germany. Her research focuses on Arctic politics and governance, and interrelations between global and Arctic change processes. She is co-editor of the book *Governing Arctic Change: Global Perspectives*, published in 2017.

Berit Kristoffersen is Associate Professor and Political Geographer at the Department of Social Sciences, UiT – the Arctic University of Norway. Her research contributions include work on state space, ocean governance, and critical geopolitics, focusing on geographies of energy production, climate change, and environmental security.

Ingrid A. Medby is a Lecturer in Political Geography at Oxford Brookes University, UK. Her research focuses on the relationship between Arctic identities, statehood, and geopolitics. She has published on Arctic state identity and political relations in journals such as *Political Geography*, *Geography Compass*, and *Polar Geography*.

Mark Nuttall is Professor and Henry Marshall Tory Chair of Anthropology at the University of Alberta and Fellow of the Royal Society of Canada. He is also visiting Professor of Climate and Society at the University of Greenland and Greenland Climate Research Centre. He is co-author (with Klaus Dodds) of *The Scramble for the Poles* (2016), author of *Climate, Society and Subsurface Politics in Greenland: Under the Great Ice* (2017), and co-editor of the *Routledge Handbook of the Polar Regions*.

Frank Sejersen's research focus is on the relation between environmental use in Greenland and its importance for the negotiation of identities and societal development. He addresses historical as well as contemporary debates and issues. Questions of hunting, environmental management, urbanization, climate change, indigenous rights, and mining are approached as cross-cultural fields of negotiation and analysed from a postcolonial perspective, often departing in political ecology and anthropology. Publications include *Rethinking Greenland and the Arctic in the Era of Climate Change* (2015).

Philip Steinberg is Professor of Political Geography at Durham University, where he directs IBRU, the Centre for Borders Research, the Durham Arctic Research Centre for Training and Interdisciplinary Collaboration (Durham-ARCTIC), and the ICE LAW Project. His publications include *The Social Construction of the Ocean* (2001), *Contesting the Arctic* (2015), and *Territory beyond Terra* (2018).

Jeppe Strandsbjerg is currently Editor-in-Chief at Djøf Publishing and was previously Associate Professor at the Copenhagen Business School. His research has explored the relationship between political order, power, cartography of space theoretically, historically (globality and state formation) as well as in current politics (maritime boundaries). He has published books with Palgrave and Cambridge University Press.

Kirsti Stuvøy is Associate Professor at the Department of International Environment and Development Studies, Noragric, at the Norwegian University of Life Sciences. Stuvøy teaches International Relations and her research interests are in security theory, feminist and practice approaches; state–society relations in post-communist Russia; and urban violence in a global perspective. Among her publications are 'Power and public chambers in the development of civil society in Russia' and (with Gunhild Hoogensen) 'Gender, resistance and human security'.

Kirsten Thisted is an Associate Professor at Copenhagen University, Institute of Cross-Cultural and Regional Studies, Minority Studies Section. Her research areas include minority–majority relations, cultural and linguistic encounters, cultural translation, and postcolonial relations. She has published several books and a large number of articles about Greenlandic oral traditions, modern Greenlandic literature and film, Arctic explorers, and Scandinavia seen in a postcolonial perspective. She is currently leading the project 'Denmark and the New North Atlantic', and is a project participant in the 'Centre of Excellence for Resources, Extractive Industries and Sustainable Arctic Communities (REXSAC)'.

Elana Wilson Rowe is a Research Professor at the Norwegian Institute of International Affairs (NUPI, Oslo). Her recent publications include *Russian Climate Politics: When Science Meets Policy* (2013) and *Arctic Governance: Power in Cross-Border Cooperation* (2018).

Preface

The research project materializing in this volume can probably be traced back to cosy conversations in Ulrik's then office in the Center for Advanced Security Theory (CAST) at the University of Copenhagen in 2012, where Jeppe spent time as a visiting scholar. We both shared an interest in Arctic politics but came to it from very different angles. During our conversations it became clear, however, that the concept of sustainability – playing such a key role in the ongoing debates over the Arctic – brought together our interests in geopolitics and space (Jeppe) and identity and postcolonialism (Ulrik). The more we talked, the more excited we became, and gradually the present project developed during conversations, meetings, and white board diagramming. We hope that this book can spark and fuel discussions about the politics of sustainability and maybe change some hard-rooted perceptions – it certainly did for us in the making.

The project has survived tectonic shifts in our work-lives. Ulrik moved from the University of Copenhagen to Aalborg University; Jeppe left Copenhagen Business School to take up a position in the publishing sector. We are grateful to the people who made it all possible. Conducting PhD research as part of the overall project, Naja Graugaard made irreplaceable contributions to our discussions and workshops along with Marc Jacobsen (who also co-authored the introductory chapter). As a research assistant, Nikoline Schriver contributed research to initial versions of the introductory chapter but left the project to seek a career outside academia. Our chapter contributors congregated for two workshops and endless discussions and helped us sharpen our ideas. Among the contributors, research coordinator of AAU Arctic, Lill Bjørst, doubled as a close companion when the tectonics got tough. Justiina Dahl, Uffe Jacobsen, Kristian Søby Kristensen, Birger Poppel, and Jørgen S. Søndergaard also contributed to our discussions in and outside the workshops. Draft versions of the introductory and concluding texts were discussed at a 2015 University of Aarhus Matchpoint conference on the Arctic; in special panel sessions at the ICASS IX in Umeå and the 59th ISA Convention in San Francisco; at the University of Copenhagen's Centre for Advanced Security Theory; and in the Arctic Politics seminar series co-convened by Aalborg University and the University of Copenhagen. The University of Greenland charitably hosted our concluding workshop. Research assistants Stine Brix and Mikkel Østergaard, as well as the AAU Language and Communication Services, helped process various parts of the manuscript. Ulla Boss Henrichsen and Ulla Langballe were invaluable in helping us get the most out of the budget, which the VELUX Foundation generously granted us to make it all happen.

1 Introduction

Sustainability as a political concept in the Arctic

Ulrik Pram Gad, Marc Jacobsen, and Jeppe Strandsbjerg

In 2013, Greenland's legislature (*Inatsisartut*) overturned a 1988 ban on the mining of radioactive materials. While the critics of this controversial decision highlighted the environmental hazards involved in the mining process, as well as ethical problems, the proponents argued that lifting the ban would contribute to the sustainable development of Greenland. Sustainable in this context means that the Greenlandic society would be able to sustain itself economically. The logic of this argument flies in the face of one of the most common assumptions about sustainability: that it is about protecting nature from adverse effects from human activity. Moreover, the argument sits uneasily with another understanding prevalent in the Arctic, namely that Indigenous ways of living are also worth sustaining. However, it makes sense within a national logic according to which it is neither nature nor culture but a particular community – in this case the modern, postcolonial Greenlandic one – that needs to be sustained. But unsustainable global levels of CO_2 emission destroy the natural habitat of the polar bear and make seal hunting difficult. So producing energy from uranium rather than oil may also contribute to sustaining certain Arctic ecosystems and cultural practices. The decision to lift the ban clearly exhibits the political character of the concept of sustainability.

The Greenlandic controversy is just one example of how debates over sustainability in the Arctic often come across as conflicting questions of life and death answered in slow motion. Listening to people talking and reading what academics write, sustainability appears to be at the centre of politics. For the presence in the Arctic of any activity or body – individual or collective – to be legitimate, it must present itself as sustainable or at least on track to becoming so. It was not always so. In that sense, sustainability has become a precondition for life in the Arctic. At the same time, it seems that 'sustainability' is able to serve any purpose. Sustainability as a concept entails radically different futures depending on what it is that should be sustained. The difficulties involved in prioritizing or combining the sustainability of a community, of Indigenous ways of life, of the global climate, and of a prospective nation state highlights the political character of the concept of sustainability and also why it is worth analysing.

The purpose of this book, then, is to investigate what it means to think of sustainability as a political concept. The way we do that is by answering the overall research question: *How are struggles over rights and resources in the Arctic reconfigured by the concept of sustainability?* To answer this question, it is necessary to engage the question of what sustainability *does*. What are the consequences of sustainability becoming an 'obligatory concept'? And when we talk about consequences, we are not thinking about what sustainability does to the environment or to development, but rather what it does to political discourse. In response, this volume aims to posit sustainability as a political concept, suggest a framework for studying sustainability as a political concept, and set out a trajectory indicating the political and analytical purchase of such an approach. We want to be able to analyse and understand how, when the concept of sustainability is introduced, struggles over rights and resources are reconfigured: e.g. what difference it makes that Greenlanders and Nunavummiut – along with foreign investors and Danish and Canadian authorities – debate mining in terms of sustainability. How does 'sustainability' facilitate some and impede the promotion of other identities, projects, and scales?

Despite the fairly obvious political content of the concept, Krueger and Gibbs' decade-old observation that '[e]ngaging the politics of sustainability represents a gap in the current sustainability literature' (2007:2) still holds. Sustainability is a political concept because it defines and shapes different discourses about future developments; that is, competing visions of the future. Across the Arctic, sustainability plays a central role in almost every development programme. Aspirations of economic exploitation, business strategies, and social planning are defined in terms of sustainability. But so are local and Indigenous efforts to maintain a community or a particular way of life. Sometimes sustainability appears in conceptual majestic solitude in which case it signifies the urge, desire, or need to simply maintain something – or find a way to make everything form a synthesis.

The basic idea of sustainability has long historical roots. However, the articulation in the work of the Brundtland Commission (WCED 1987) became a defining moment: it combined caring for the natural environment with 'economic development'. When 'development' is added to the concept, 'sustainability' emerges as a more obviously political concept. The combination of a desire to change while keeping something stable fuels the political character of the concept. It raises the questions of what it is that should be preserved in the future while we at the same time undergo change? When? How? And who should be responsible? After the wedding of 'sustainability' and 'development', it was clear from the wider discourses involved that it was societies that should develop – both to become more equitable but also to allow the natural environment to be preserved (Redclift 1987). Soon, however, human collectives were featured at the 'stable' side of the equation: under the banner of 'sustainable development', advocates and analysts promoted communities, cultures, groups, livelihoods, and cultural diversity as worthy of being sustained (Jacobs 1999:37; Kates *et al.* 2005:11). Our contention is that this tendency has continued:

a wider and wider array of entities and phenomena appears as objects of sustainability.

It makes little sense to study sustainability in a vacuum. Concepts always carry with them a baggage of meaning conveyed by other concepts accompanying them – and when sustainability is introduced in a new context, it inevitably articulates pre-existing meaning structures. This will be obvious to anyone studying sustainability in the Arctic: here, changes to the climate, global power balances, demands for natural resources, and aspirations for self-determination set the stage for new political struggles. Central to the struggles is the notion of the Arctic as a special place characterized by a nature at once hostile and fragile. Moreover, sustainability has entered an Arctic political reality, which may be characterized as postcolonial: Indigenous peoples hold a prominent place and have comparatively strong organizations in the Arctic (Jacobsen 2015; Strandsbjerg 2014). Their relations to the respective states involve a variety of autonomy arrangements designed to distance the present from histories of colonialism, paternalism, and exploitation. Legitimizing Indigenous people's claim to a stake in Arctic governance is not just the fact that they were there first, but also that they managed to sustain themselves on Arctic resources. Hence, 'sustainability' has become a pivotal concept in struggles over rights and resources in the Arctic: it increasingly organizes the way Arctic nature and Indigenous identities are presented; it shapes what strategies for the future organization of postcoloniality and that future extractive projects are deemed viable and legitimate.

In order to analyse sustainability as a political concept, we commence by a historical and conceptual positioning of sustainability. In the following, we proceed by outlining a brief history of sustainability as a concept: we identify the marriage between sustainability and development as crucial for the way it plays out as a political concept; we characterize how prevailing images of the Arctic articulate sustainability; and we introduce the postcolonial and Indigenous question as an important vector in the politics of sustainability in this region. We then proceed with a theoretical suggestion on how to approach sustainability as a political concept. By dissecting 'sustainability' from 'development', we explain how our approach is discursive, but with special emphasis on the role of concepts in structuring discourse: We want to investigate the alterations in meaning structures and struggles for possible futures when 'sustainability' is introduced into the grammar of development. In other words, we want to know how identity, space, and time in the Arctic are reconfigured by sustainability.

Problems of sustainability

Concerns with human dependency on limited resources and particular ecosystems can be found throughout history.[1] Central, however, for the present debate on sustainability is the intellectual trajectory that can be traced back to eighteenth-century forest management and political economy (Warde

2011:153). Within this literature, a genre developed advising the head of the household (the *Hausvaterliteratur*) to cut wood in a durable (*nachhaltende*) way (Du Pisani 2006:85; Petrov *et al.* 2017:3; Warde 2011).

From household to globe

Writing before fossil fuel could be utilized, a shortage of timber was predicted (Grober 2007:7), threatening the existence of both states (in need of timber for ships) and households (in need of wood for fire) (Warde 2011:159). Sustainability would be achieved by ensuring that the harvest of timber was made to balance the growth of new forest (Brander 2007:8). Connecting the local harvest with the interest of state, and planning longer than the normal year-to-year horizon, were the first steps towards establishing the resource management literature (Warde 2011). At the same time, this literature wrote the state in as a central institution/actor for nature preservation. The ideas of managing limited resources and connecting the future of the state with resource use remain core elements of the concept of sustainability today (Brander 2007; Lumley and Armstrong 2004; Warde 2011).

However, the defining moment for sustainability as a political concept was the work of the Brundtland Commission (Kirkby *et al.* 1996:1). With the work of the Brundtland Commission, the concept of sustainability emerged as a global concern in a way that was politically programmatic before it was academic, and the most cited publications remain commissioned reports. Brundtland's definition of 'sustainable development' as 'meeting the needs of the present without compromising the ability of future generations to meet their own needs' (WCED 1987), has framed both environmentalism and developmental interests ever since (Quental *et al.* 2011:20). The 2002 World Summit on Sustainable Development reformulated the task ahead by identifying three 'interdependent and mutually reinforcing pillars' or dimensions of sustainable development: economic, social, and environmental (Kates *et al.* 2005:12). Within this framework, sustainability generally relates humanity to the global ecosystem in a way that prescribes socioeconomic development to be shaped in particular ways, rather than delimited (but see Kristoffersen and Langhelle 2017:28).

When social sciences have engaged sustainability debates, what have been called mainstream voices (Krueger and Gibbs 2007:2) has joined the normative commitment to perfecting the concepts of sustainability and sustainable development (e.g. Sen 2009), identifying problems in terms of lack of sustainability (e.g. Parker 2014), and developing and implementing solutions in the form of sustainability (e.g. Edwards 2005). Partly in response to this, a critical tradition including postcolonial, Marxist and political ecology voices have insisted on the political effects of sustainable development (Bryant 1991); some have, for example, found sustainability to be yet another neo-colonial way for the West to dominate the rest by imposing standards limiting prospects for development (e.g. Banerjee 2003; Sachs 1990). While there is intellectual merit and political purchase to both constructive and critical perspectives, the binary choice appears premature: neither loyal implementation nor wholesale rejection of the sustainability agenda help our understanding of the diverse political effects of the concept. Hence, we take our cue from a distinct strand of scholarship pointing to how the concept's contingent meanings may vary depending on inclusion and exclusion of actors, and the use of different indicators and time scales (Beckerman 1994:239; Lélé 1991:179). In this context, the contribution of this volume is to investigate systematically – within a particular region – exactly to what political uses the concept of sustainability is put and what practices it facilitates. It is to this end we theorize sustainability as a political concept and operationalize our theory as a tool for empirical analysis. For our immediate purpose – to investigate the Arctic – but also with the wider aim of contributing to a generally reproducible analytical strategy applicable to parallel projects in other regions across the globe or to studies focusing on, e.g. a particular socio-economic sector.

The problems of Arctic sustainability

Sustainability takes on new characteristics when moving from global to regional scales. Prevailing images of both the Arctic population and the Arctic geography set them apart from the logic of global sustainability. Both scholarship and public imagination has long agreed that the Arctic is a special place; even if a variety of imaginaries differ over what makes the region special (Steinberg *et al.* 2015). Global sustainability discourse lists the Arctic among a few iconic biotopes – along with, among others, the rainforest and coral reefs (Gillespie 2009) – but the particularity of the Arctic goes beyond ecology. First and foremost, the Arctic has been defined as a forbidding space facilitating only fragile ecosystems and, consequently, home only to fragile human communities (Lorentzen *et al.* 1999:5) that were only relatively recently brought the joys and perils of modernity and substantial statehood. However, as time allowed white people to develop technologies to navigate this forbidding space, the Arctic is increasingly presented as a new resource frontier waiting to be exploited (Howard 2009) and, related, as a matter of global (military) security concern (Kraska 2011). Scholarly writing on sustainability in the Arctic was from the outset concerned with the fragile ecosystems, echoing the feeble intergovernmental institutionalization – the Arctic Environmental Protection Strategy – defining the Arctic in international politics in the early post-Cold War years (Tennberg and Keskitalo 2002). As this strategy was given a firmer organizational basis with the formation of the broader mandated Arctic Council in 1996 (Tennberg 2000:120), academics followed suit, branched out, and placed human communities at the centre of a wider approach to sustainable development, including studies of, among others, whaling (Caulfield 1997) and hydrocarbon extraction (Mikkelsen and Langhelle 2008).

Most of the literature on Arctic sustainability relies on two very different but related storylines:[2] an image of the past in which vulnerable Indigenous communities were challenged by the forbidding Arctic environment; and an image of the present in which modern industrialized extraction, production, and consumption unsettle global climate, Arctic ecosystems, Indigenous cultures, and local communities. When pollution threatens fragile Arctic ecosystems and changes the global climate in ways that spur further regional changes, scholarship often focuses on how particular local ways of living are gilded or threatened (Anisimov *et al.* 2007:672; Berman *et al.* 2004; Buckler *et al.* 2009). Indigenous people's experiences with surviving in the Arctic for centuries without undermining their own livelihood endows them a certain legitimacy in discussions about sustainability (compare Petrov *et al.* 2017:13 with Thisted in this volume).[3] However, the normative decision on whether change is sustainable or not relies on how Inuit culture is defined – and often it is defined by Western stereotypes of the Indigenous Other (Bjørst 2008:119; Fienup-Riordan 1995; Ryall *et al.* 2010): if Inuit are seen as essentially 'traditional' (e.g. Ford and Smit 2004), change is necessarily exogenous and potentially damaging (Cameron 2012). If Inuit are defined as part of an ecosystem, Inuit voices advocating socio-economic change are silenced (Bravo 2008).[4] Moreover, the Arctic long ago ceased to be an environment in which Indigenous peoples were the sole actors or even sole inhabitants (Wenzel 2009:94 in Sejersen 2015:183).

So the prominent role of Indigenous peoples in the Arctic does not have one straightforward implication for the politics of sustainability. Quite to the contrary, it gives rise to a series of complications and variations. Questions of how to organize postcolonial sovereignty and statehood are crucial for the political struggles currently unfolding in the Arctic. It is well documented how the Arctic plays a particular role for established national identities of some of the states laying claim to parts of the region (Hønneland 2014; Medby 2014; Williams 2011). Making Northern territories a periphery of Southern states obviously opens up sustainability projects at the national scale. When first contemplating how to implement the Brundtland agenda in the Arctic, one prominent group of academics and diplomats had so little confidence in the state that they advocated philanthropic foundations bypassing Southern capitals by directly sponsoring Indigenous communities – because 'we' should learn new directions from them – pursuing their own sustainable development (Griffiths and Young 1989). In the case of Greenland, the simultaneous presence of formal Danish sovereignty and visions of future Greenlandic independence (Gad 2014) invites a separate set of struggles over how to scale sustainability. Whereas the creation of new industries is a circumpolar ambition, it is crucial to the Greenlandic debate on how to create a sustainable economy and, hence, make the postcolonial future take the form of a separate state. Whether the state is Southern or Arctic, it needs to legitimize itself by telling stories about how sustainability is secured locally. These stories are intimately entangled with images of what constitutes a legitimate Arctic community: who belongs? Settlers or only Indigenous peoples? How should they live? What combinations of tradition and modernity are deemed sustainable? Such questions reappear in repeated calls for 'cultural sustainability' in the Arctic (Gad *et al.* 2017:18; Petrov *et al.* 2017:20; Søndergaard 2017).

In order to capture the political significance of all these variations, the following sections first separate the concept of sustainability from the discourse of sustainable development, then explain how we understand and want to analyse 'sustainability' as a political concept.

Between environmental and developmental discourse

As the silver anniversary of the Brundtland Report has come and gone, you could build an entire library of texts on the true meaning of sustainable development (Redclift 2006). It seems well established that getting the concept of sustainability right is important. Both for getting development right and for understanding how it, nevertheless, goes wrong. So, why another volume on sustainability? Because, we argue, despite its growth, the library on sustainability still struggles to capture the political effects of sustainability beyond an affirmative or critical normative stance. Neither rationalistic conceptual analysis nor constructivist discourse analysis accounts adequately for its political effects. As laid out above, either sustainability is seen as a desirable goal, or the concept is identified as playing a negative role. This binary fails to describe the more fundamental political nature of the concept. In making this argument, it is important to assert that the concept of sustainability is not the same as the discourse of sustainable development; it has a distinct effect.

Within the broader literature, a tradition has been established that analyses both environmental and developmental policy as discourse (della Faille 2011; Dryzek 2013; Hajer and Versteege 2005) Discourse analysts, particularly developing the Foucauldian approach, have done a convincing job in focusing attention on the importance of language for the formation of environmental and developmental policies (Hajer and Versteege 2005). Discourses are basic packages of meaning, more or less tightly sealed, which allow actors to say and do things that make sense to each other.[5] Talking about objects, subjects, relations, and trajectories in a specific way makes them meaningful – and it makes them real. Concepts are one type of a wider range of rhetorical devices, which establishes the relations and trajectories of a discourse. For discourse analysts, concepts are little nuclei of meaning, imposing a certain structure on the discourses articulating them. Therefore, it is important to get them right; just as important as getting the discursive construction of objects and subjects right.

Rationalist conceptual analysts agree that getting the concepts right is important. That is what conceptual analysis is about: distilling the precise meaning of a concept, so that it can inform rather than obscure scientific and public debates and policy-making. However, conceptual analysts and discourse analysts disagree on the procedure. For rationalist conceptual analysts, getting a concept right is at heart a logical exercise establishing the correct meaning. For discourse analysts, it is an empirical exercise establishing how the concept is used; meaning what; to what effect.[6] On the one hand, ours is fundamentally a discursive project. As other constructivists, we find that language has important effects on the formation of policy and on the creation of reality as we know it.

On the other hand, we find the role awarded to concepts in most discourse analyses on sustainability, Arctic and elsewhere, has been somewhat muted. Granted, it has a huge effect on whether a certain object is 'allowed' existence by a discourse: Understanding a fish as part of an ecosystem is something radically different from thinking of it as part of a stock ready for exploitation. Obviously, it has huge political effect whether you are offered the subject position of a savage tribe in need of enlightenment, of an Indigenous people liable for preservation, as a partner for developing politics, or of a nation working towards independence. However, if you think through these examples, you will find that these subject positions and these categories of objects receive much of their meaning from concepts establishing relations and trajectories: enlighten, preserve, independence, system, exploit. In other words, these concepts do things on their own. Over time, they may ultimately be malleable, but immediately when introduced into a sentence or a discourse, concepts carry with them meaning. Some of the most important concepts constitute very basic ideas about how objects and subjects can and do relate, and how relations can and do unfold over time. From this perspective, the composite concept of 'sustainable development' is an oxymoron, combining dynamism and stasis (Lempinen 2017:37).

Development focuses attention on change. It draws on the notion of progress; a basic idea that humans may – individually and collectively – over time, turn into better versions of themselves (Du Pisani 2006:85). As established by critical geographers, modern development discourse relies on a certain assumption about space as being fixed and operating as a stable background for social practice. Classic development theories such as 'modernisation' and 'stage theory' (Struck *et al.* 2011) operate within a unified and uniform global space without paying historical or theoretical attention to different spatial histories across the globe. Such assumed fixity and stability has in turn allowed academic engagement to ignore the social production of space (Lefebvre 1991). Related to this problematique, the clear demarcation of a social, human sphere distinct from a natural environment has been thoroughly deconstructed (Latour 2005; Lempinen 2017:52–55; Swyngedouw 2007). The absence of space has not only been questioned by academics, but also, for example, by non-state actors. Many Indigenous peoples do not apprehend their identity in a way that allows for a separate concept of a natural environment in the same way as Western thought (Bielawski 2003:318–320; Leduc 2013:109). Furthermore, we like to entertain the idea that the environment has called this absence into question. Understood as such, sustainability is indeed a concept that draws space in as a necessary dimension of political and economic theory.

If development is relational, this quality is secondary – relationality only appears by comparing entities as to their state of development. Sustainability, on the contrary, focuses attention on what should *not* change – and as a concept, it is relational before anything else. It claims that an object, a subject, or a process relies on its environment, so that both need to be taken care of if the existences of either are to continue.[7] Possibly, the paradoxes and detours of politics of 'sustainable development' – but also its proliferation and success in framing debates, projects, and policies – comes from the paradoxical combination of change and constancy in this discourse. On its own, 'sustainability' is a question of 'preserving something in its relation to something else'. Combining with development compromises, supplements, or complicates the temporality of sustainability.

In this volume, we focus on what the concept of sustainability does. First and foremost, what it does in and to the discourse of 'sustainable development', because this is where the concept is most explicitly highlighted these days – whether it is allowed to do its work or it is in effect submitted to development. But also what the concept of sustainability does when appearing 'on its own'. Because, even if this work is less conspicuous, the very basic idea of continued mutually dependent co-existence between an object and its environment (natural or social) may be found way beyond the sustainable development discourse. And it is this quality of being a well-known rhetorical figure that makes it possible for 'sustainability' to be attached to surprising relations – and which, in turn, produces some of the surprising twists and turns of the composite discourse of 'sustainable development'.

Sustainability as a political concept

When we suggest analysing sustainability as a political concept, we make the claim that the concept intervenes in discursive struggles over the future allocation of rights and resources. This draws on an understanding of politics as a struggle between competing visions for the future (Laclau and Mouffe 1985; Palonen 2006; Skinner 2002). Concepts employed to implicitly or explicitly prognosticate and prescribe the future are central means and goals in these struggles (Koselleck 1985:21). And because, in brief '[s]ustainability is always about maintaining something' (McKenzie in Jacobsen and Delaney 2014), it prioritizes the preservation of a particular dimension of life even in the context of an effort of overall change to something better. To unravel the political effects of speaking in sustainability terms, we need to ask specific questions: What is to be sustained? In relation to what? How? As these questions indicate, sustainability is a concept that facilitates and structures a diversity of – partially conflicting, and therefore political – narratives of the future at a series of scales.[8]

The question remains, how do we know sustainability when we see it? Concepts are generally names for things and ideas (Bartelson 2001:25). However, words take on different meanings in different contexts, and similar meanings may be expressed in a variety of words. In principle, this conundrum leaves two diametrically opposed approaches open for the analyst: an onomasiological approach would begin with a specific word ('sustainability') and map its different meanings in different contexts, whereas a semasiological approach would begin with a specified meaning and map how this meaning is expressed in various contexts (see Elden 2013:18). To illustrate the difference between these two approaches, we can draw attention to how the US Arctic Strategy does not use the word 'sustainability' to describe the relationship between development and the environment. However, through a semasiological analysis focused on the meaning of sustainability, we can show how the word 'conservation' is related to other words (like nature, culture, community, development) in ways that may (or may not) convey the meaning of the concept of sustainability. A purely onomasiological approach – registering only the use of the word 'sustainability' – would not produce a nuanced account of what narratives the concept or idea of sustainability facilitates: the meaning of sustainability can be produced without speaking the word.

When deciding the meaning of sustainability, which we search for in texts, we have to strike a balance between the ways in which meaning is produced both in synchronic and diachronic relations (Saussure 1916): when a concept is used in one context rather than another, it conveys a different meaning, as the concept is related to new things and ideas. So parallel synchronic analyses would show how sustainability does not mean exactly the same thing in, say, UN debates on development aid and in an Arctic business proposal. Meanwhile, a diachronic analysis will show how a concept moving into a new context necessarily carries with it some baggage of meaning, as its relations 'backwards' cannot be entirely erased from social memory.

The more central a concept becomes within a certain discourse the more likely it is that it is either taken for granted or implicitly invoked (Bartelson 2001:10). Moreover, a certain emptying of semantic content appears to be a precondition for a concept becoming central: an 'empty signifier' may articulate more different meanings (Laclau 1996). This does not mean, however, that it is futile to define what sustainability means because, even if the discourse of sustainable development may facilitate the promotion of a variety of mutually contradictory projects and programmes, the concept of sustainability nevertheless plays a similar role in all these narratives, albeit articulating different objects, subjects, and environments. To pin down such a moving target for structured analysis – beyond mere description of how words, meanings, concepts, and contexts are all in flux – the analyst needs to fixate the most important nucleus of meaning of a concept as a criterion delimiting what part of reality should come in focus (Andersen 1999). Hence, the delimitation of a core meaning of the concept of sustainability is pivotal. However, the less the semantic content of a concept is specified in advance, the more open it is to historical inquiry (Bartelson 2001:17), so we want to keep our definition as parsimonious as possible, to capture what is core.

The concept of sustainability refers to a relationship between (1) identity, (2) space, and (3) time. Global discourses on sustainable development link humanity at large (or a particular society), that is to develop, with its natural environment that should stay the same (Jacobs 1999:26; Lélé 1991). Understood in this manner, sustainability represents a specific way of temporally mediating the relation between society and nature. In its most common articulation with development, sustainability maintains the distinction between identity and nature, claiming that one can develop while protecting the other. When sustainability refers to society as the thing that should be maintained, it does not necessarily entail a shift from nature to identity, but rather a rearticulating of the relationship between the two. Combining sustainability and development invites more complicated stories like 'changing something progressively over time while at the same time preserving something else' or 'changing progressively over time to arrive at a state where this or that can then be preserved'.

As a mediating concept, sustainability points out a referent object – something valuable enough to sustain (Lempinen 2017:52) – and relates it to time and to specific environments (whether conceived as natural or social). At the highest level of abstraction, we define sustainability as the narrative positioning of (1) a given entity and (2) a specified environment in (3) a relation characterized by interdependence successfully sustained over time. Or in plain words: When someone claims that *x* and *y* are interacting in a way which may continue without terminating the existence of neither *x* nor *y*, their relation is described as sustainable.

Most empirical uses postpone sustainability to the future: They are formulated as 'sustainability narratives', i.e. visions, plans, and programmes for how to achieve sustainability. As part of a discourse, sustainability makes specific agents responsible within a specified space, organizes other concepts in coherent narratives, and inscribes specific forms of knowledge with authority. Hence, narratives built around sustainability rather than around another concept ascribe legitimacy to some claims to rights and resources rather than others. In other words, sustainability narratives constitute and empower certain types of actors. However, actors will, in turn, select and seek to manipulate narratives in ways that they find in accordance with their identities, given their perception of competing identities and narratives. The future is never given; rather, a plurality of narratives co-exist, making it a political question which future should unfold and how.

In the attempts to frame and discuss sustainability in the Arctic, we employ the concept of scale.[9] The analytical advantage of scalar analysis is that it highlights the variety of interconnections between different scalar materializations of sociopolitical power (Brenner 1999). In practice, scales are historically contingent, in principle fluid and malleable, and the very construction of scales is an important part of the politics of sustainability. However, for the most part in the present project, we employ scale in a more heuristic taken-for-granted sense that allows us to discuss the different political effects of sustainability in the encounter between different scales. Actors at one level may be made responsible for making a referent object at a different scale sustainable in relation to an environment at a third scale. On a global scale, sustainability often means something very different than what it does at a state or local scale, and any specific development project (such as mining in, say, Kotzebue, Nuuk, or Lac de Gras) will most likely have consequences (benefits, threats, risks, externalities) on many scales from the global to the local. Moreover, the construction and prioritization of scales play out surprisingly differently across the Arctic, not least due to differences in how the (post)colonial relations between Southern centres and Northern populations are organized.

Analysing sustainability politics

As set out in the opening of this chapter, the overall purpose of this volume is to investigate how struggles over rights and resources in the Arctic are being reconfigured by the concept of sustainability. As discussed in the preceding sections, the concept of sustainability seeks to impose a specific configuration of identity, space, and time on discourse. To distil how this imposition works, the research questions guiding our overall approach to the politics of sustainability in the Arctic are:

- What is it that should be sustained? In other words: what is the *identity* of the referent object of sustainability?
- In relation to what environment should this referent object be sustained? What *space* is the referent object dependent on?
- How should sustainability come about? What *temporality* is produced when sustainability is combined with development (and/or other concepts)?

However, our focus is on the implications of the promotion of conflicting sustainability narratives; conflicting visions of the future, each structured by the concept of sustainability. In other words, our focus is on the interplay between competing claims about how *x* and *y* should (adapt to) sustain their mutually dependent existence. Hence, when each chapter approaches a body of empirical material asking 'What should be sustained? In relation to what? How?', the chapter is likely to identify more than one narrative. Moreover, approaching texts and practices with this reading strategy allows us to identify when the core meaning of sustainability is shaping them even without explicit use of any of the derivative forms of the word 'sustain'.

Apart from applying the basic reading strategy, the chapters go about their task in very different ways. We have organized them according to how they weigh their implementation of our analytical strategy. Nevertheless, whether the focus of the chapter's analysis is initially put on referent object, environment, responsibility, or the conditions for new interventions, in the end they each provide insights related to what sustainability does to identity, space, and time. Hence, they contribute to the overview, established in the concluding chapter, of how the politics of sustainability plays out in some of the most important issues across the Arctic.

Four chapters drive their analysis from a focus on what, according to the Arctic sustainability narratives, should be sustained. *Rikke Becker Jacobsen* lays out how three competing referent objects for sustainability – stocks, communities, and the public purse – complicates the governance of Greenlandic fisheries. *Kathrin Keil* surveys interventions on Arctic shipping and finds a surprising array of referent objects for sustainability. *Marc Jacobsen* compares how minerals extraction in Nunavut and Greenland is meant to serve different purposes – the sustainability of local communities and of the national economy of a nascent state – spurring different sovereignty dynamics. *Naja Dyrendom*

Graugaard investigates how postcolonial sustainability narratives surrounding Inuit seal hunting both depend on and seek to escape colonial ideas of indigeneity.

Three chapters focus on mechanisms for making distinct environments (ir)relevant as part of Arctic sustainability narratives: *Frank Sejersen* investigates how projects are phased and the social world continuously rescaled to produce sustainability in Greenlandic authorities' strategy to transform society by inviting in large-scale industries. *Elana Wilson Rowe* reads Russian policy documents along with political statements of the Kremlin and the RAIPON Indigenous peoples' organization to find how space is carved up to allow simultaneous protection of some natural environments and development made sustainable with reference to Indigenous social environments. *Lill Bjørst* traces how Greenland's own CO_2 emissions – and, hence, its contribution to global climate change severely impacting the Arctic – have been excluded as relevant for sustainable development of the island.

Four chapters focus their analysis on different ways of claiming responsibility and authority in the Arctic in relation to the sustainability of communities and ecologies. *Berit Kristoffersen and Philip Steinberg* analyse how Norway's comprehensive Blue Economy initiative appropriates the Arctic Ocean to sustain the legitimacy of the managerial state. *Hannes Gerhardt, Berit Kristoffersen, and Kirsti Stuvøy* compare how Russia, Greenland, and Norway each re-assert their version of state authority to protect hydrocarbon extraction in response to Greenpeace's vision of a transnational, networked solution to the global problem of ecosystem and climate sustainability. *Ingrid A. Medby* explores how Norway, Iceland, and Canada draw on discourses of sustainability when performing legitimacy for their Arctic identity. *Kirsten Thisted* distils how – even if indigeneity has been a potent signifier in discourses of sustainability – the way in which the Government of Greenland works with the sustainability concept affirms modernity and nation, rather than tradition and indigeneity.

Three chapters, in each their own way, focus attention on the conditions of possibility for new sustainability narratives: *Johanne Bruun* brings to light some of the challenges associated with constructing the Kuanersuit mountain in Southern Greenland as a uranium resource on which current popular and political narratives of economic sustainability in Greenland rely as natural fact. *Victoria Herrmann* surveys how remote Alaska Native communities seized the transition from diesel fuel to renewable energy as a way to build wider community sustainability. *Klaus Dodds and Mark Nuttall* trace how a variety of materiality, objects, and networks of knowledge create multiple contexts for ideas about sustainability to emerge, circulate, play out, and make themselves felt, in ways that make sustainability narratives stretch Greenland both geophysically and geopolitically.

Notes

1 Du Pisani (2006) mentions how Egyptians, Mesopotamians, Greek, and Romans discussed agricultural/environmental problems in sustainability-like terms. By invoking the imagery of humans 'roaming the earth ... noticing the negative effects of overuse [of natural resources]', Petrov *et al.* (2017:3) extend 'sustainability thinking' to hunter/gatherer societies.
2 A wider set of storylines is lined up by Kristoffersen and Langhelle (2017), Steinberg *et al.* (2015), and Wilson (2007).
3 Petrov *et al.* (2016:170) condense a general tendency of extending this legitimacy into *research* on sustainability in the Arctic in highlighting how 'a special role belongs to the indigenous researcher'. The same group of authors, on the one hand, advocate '[b]ringing forward the important role of indigenous conceptualizations of sustainability [to] ensure that sustainability is not a vehicle of (neo)colonialism' (2016:171), while, on the other hand, assuring that 'the principle and meanings of IQ [i.e. traditional Inuit knowledge] are largely consistent with sustainable notions of human–environment relations' (2017:13).
4 In one perspective, the way that Indigenous peoples are systematically awarded some voice in the Arctic may give the impression of a comparatively strong position – in contrast to the erasure of Indigenous peoples in the sustainability discourse further south on the American continent, documented, among others, by Shapiro (2005). In another perspective, the subject positions available are products of colonization (see Graugaard, this volume; Spivak 1988).
5 Hajer (1995) points to the role of basic 'story lines' in keeping even conflicting positions together within one discourse. Hence, discourses may be simultaneously interwoven and competing, leaving even central concepts contested (Feindt and Oelse 2005:162).
6 In this respect, conceptual historians like Koselleck and Skinner join discourse analysts. Hence, Bartelson (2007) describes two approaches to the study of concepts: the historical approach studying the diachrone development in the empirical use of a concept over time; and the philosophical approach establishing the logical meaning of a concept within one synchronous slice of discourse.
7 Dryzek (2013:150) notes how sustainability concerns were first introduced in development scholarship – with reference to the natural environment – but primarily with a view to making development sustainable in the sense that (local) people should be able to sustain development by themselves and find the resulting society liveable. Sneddon (2000) advocates an analytical and theoretical focus on sustainability rather than sustainable development, to recover its critical edge towards capitalism.
8 It should be noted that we are not alone in seeking answers to the reference of sustainability discourse. Sneddon (2000:525), Hattingh (2009:64), and Lempinen (2017:52) have all developed related sets of questions to serve their analytical purposes.
9 Hence, we join a recent tendency towards focusing on scale-making in the study of environmental discourse (Feindt and Oels 2005:168) and particularly of Arctic development discourse (Petrov *et al.* 2017:62; Sejersen 2014, 2015; Tennberg *et al.* 2014).

References

Andersen, N.Å. (1999): *Diskursive analysestrategier. Foucault, Koselleck, Laclau, Luhmann*, Frederiksberg: Nyt fra Samfundsvidenskaberne.

Anisimov, O.A., Vaughan, D.G., Callaghan, T.V., Furgal, C., Marchant, H., Prowse, T.D., Vilhjálmsson, H., and Walsh, J.E. (2007): Polar regions (Arctic and Antarctic). Climate change 2007: Impacts, adaptation and vulnerability, contribution of Working Group II. In: Parry, M.L., Vaughan, D.G., Callaghan, T.V., Furgal, C., Marchant, H., Prowse, T.D., Vilhjálmsson, H., and Walsh, J.E. (eds), *The fourth assessment report of the Intergovernmental Panel on Climate Change*, Cambridge: Cambridge University Press, pp. 653–685.

Banerjee, S.B (2003): Who sustain whose development? sustainable development and the reinvention of nature. *Organization Studies* 24(1):143–180.

Bartelson, J. (2001): *The critique of the state*. Cambridge: Cambridge University Press.

Bartelson, J (2007): Philosophy and history in the study of political thought. *Journal of the Philosophy of History* 1(1):101–124.

Beckerman, W. (2006 [1994]): 'Sustainable development': is it a useful concept?. In: Redclift, M. (ed.), *Sustainability, critical concepts in the social sciences*, volume 2: *Sustainable development*. London: Routledge, pp. 236–255. Original source: *Environmental Values*, 3, 1994:191–209.

Berman, M., Nicolson, C., Kofinas, G., Tetlichi, J., and Martin, S. (2004): Adaptation and sustainability in a small Arctic community: results of an agent-based simulation model. *Arctic* 57(4):401–414.

Bielawski, E. (2003): 'Nature doesn't come as clean as we can think it': Dene, Inuit, scientists, nature and environment in the Canadian north. In: Selin, H (ed.), *Nature across cultures: views of nature and the environment in non-Western cultures*. Doordrecht: Kluwer, pp. 311–328.

Bjørst, L.R. (2008): *En anden verden. Fordomme og stereotyper om Grønland og Arktis*. Copenhagen: Forlaget BIOS.

Brander, J.A. (2007): Sustainability: Malthus revisited? *Canadian Journal of Economics* 40(1):1–38.

Bravo, M. (2008): Voices from the sea ice: the reception of climate impact narratives. *Journal of Historical Geography* 35(2):256–278.

Brenner, N. (1999): Beyond state-centrism? Space, territoriality, and geographical scale in globalization studies. *Theory and Society* 28:39–78.

Bryant, R.L. (1991): Putting politics first: the political ecology of sustainable development. *Global Ecology and Biogeography Letters* 1(6):164–166.

Buckler, C., Wright, L., Normand, L. *et al.* (2009): Securing a sustainable future in the Arctic: engaging and training the next generation of northern leaders. Winnipeg, International Institute for Sustainable Development. Available at: www.iisd.org/library/securing-sustainable-future-arctic-engaging-and-training-next-generation-northern-leaders (accessed 13 September 2018).

Cameron, E.S. (2012): Securing indigenous politics: a critique of the vulnerability and adaptation approach to the human dimensions of climate change in the Canadian Arctic. *Global Environmental Change* 22:103–114.

Caulfield, R.A. (1997): *Greenlanders, whales, and whaling: Sustainability and self-determination in the Arctic*. Hanover, NH: University Press of New England.

della Faille, D. (2011): Discourse analysis in international development studies: mapping some contemporary contributions. *Journal of Multicultural Discourses* 6(3):215–235.

Dryzek, J. (2013): *The politics of the earth: environmental discourses*, 3rd edn. Oxford: Oxford University Press.

Du Pisani, A. (2006): Sustainable development: historical roots of the concept. *Environmental Sciences* 3(2):83–96.

Edwards, A.R. (2005): *The sustainability revolution: portrait of a paradigm shift*. Gabriola Islands: New Society Publishers.

Elden, S. (2013): *The birth of territory*. Chicago, IL: University of Chicago Press.

Feindt, P.H. and Oels, A. (2005): Does discourse matter? Discourse analysis in environmental policy making. *Journal of Environmental Policy & Planning* 7(3):161–173.

Fienup-Riordan, A. (1995): *Freeze frame: Alaska Eskimos in the movies*. Seattle, WA: University of Washington Press.

Ford, J. and Smit, B. (2004): A framework for assessing the vulnerability of communities in the Canadian Arctic to risks associated with climate change. *Arctic* 57(4):389–400.

Gad, U.P. (2014): Greenland: a post-Danish sovereign nation state in the making. *Cooperation and Conflict* 49(1):98–118.

Gad U.P., Jakobsen U., and Strandsbjerg, J. (2017): Politics of sustainability in the Arctic: a research agenda. In: Fondahl, G. and Wilson, G.N. (eds), *Northern sustainabilities: understanding and addressing change in the circumpolar world*. New York: Springer, pp. 13–24.

Gillespie, A. (2009): From the Galapagos to Tongariro: recognizing and saving the most important places in the world. In: *Resource Management Theory & Practice*. Auckland: Resource Management Law Association of New Zealand, pp. 115–155.

Griffiths, F. and Young, O.R. (1989): *Sustainable development and the Arctic: impressions of the co-chairs, second session, Ilulissat and Nuuk, Greenland 20–24 April*. Working Group on Arctic International Relations, reports and papers 1989-1.

Grober, U. (2007): Deep roots: a conceptual history of 'sustainable development' (Nachhaltigkeit) [pdf]. Discussion Papers, Presidential Department P 2007-002. Social Science Research Center Berlin (WZB). Available at: https://bibliothek.wzb.eu/pdf/2007/p07-002.pdf (accessed 10 October 2017).

Hajer, M. (1995): *The politics of environmental discourse: ecological modernization and the policy process*. Oxford: Oxford University Press.

Hajer, M. and Versteeg, W. (2005): A decade of discourse analysis of environmental politics: achievements, challenges, perspectives. *Journal of Environmental Policy & Planning* 7(3):175–184.

Hattingh, J. (2009): Sustainable development in the Arctic: a view from environmental ethics. In: UNESCO, *Climate change and Arctic sustainable development: scientific, social, cultural and educational challenges*. Paris: UNESCO.

Hønneland, G. (2014): *Arctic politics, the law of the sea and Russian identity: the Barents Sea Delimitation Agreement in Russian public debate*. Basingstoke: Palgrave Macmillan.

Howard, R. (2009): *The Arctic gold rush: the new race for tomorrow's natural resources*. London: Continuum.

Jacobs, M. (1999): Sustainable development as a contested concept. In: Dobson, A. (ed.), *Fairness and futurity: essays on environmental sustainability and social justice*. Oxford: Oxford University Press, pp. 21–45.

Jacobsen, M. (2015): The power of collective identity narration: Greenland's way to a more autonomous foreign policy. *Arctic Yearbook* 2015:102–118.

Jacobsen, R.B. and Delaney, A.E. (2014): When social sustainability becomes politics: perspectives from Greenlandic fisheries governance. *Maritime Studies* 13(6).

Kates, R.W., Parris, T.M., and Leiserowitz, A.A. (2005): What is sustainable development? Goals, indicators, values and practice. *Environment: Science and Policy for Sustainable Development* 47(3):8–2.

Kirkby, J., O'Keefe, P, and Timberlake, L. (1996): Sustainable development: an introduction. In: Kirkby, J., O'Keefe, P., and Timberlake, L. (eds), *The Earthscan reader in sustainable development*. London: Earthscan.

Koselleck, R. (1985): *Futures past. On the Semantics of Historical Time*, trans. Trib, K., New York: Columbia University Press.

Kraska, J. (2011): *Arctic security in an age of climate change*. Cambridge, Cambridge University Press.

Kristoffersen, B. and Langhelle, O. (2017): Sustainable development as a global-Arctic matter: imaginaries and controversies. In: Keil, K. and Knecht, S. (eds), *Governing Arctic change: global perspectives*. London: Palgrave Macmillan, pp. 21–42.
Krueger, R. and Gibbs, D. (2007): Introduction: problematizing the politics of sustainability. In: Krueger, R. and Gibbs, D. (eds), *The sustainable development paradox*. New York: The Guilford Press.
Laclau, E. (1996): Why do empty signifiers matter to politics? In: *Emancipation(s)*. London: Verso.
Laclau, E. and Mouffe, C. (1985): *Hegemony and socialist strategy*. London: Verso.
Latour, B. (2005): *Reassembling the social: an introduction to actor–network theory*. Oxford: Oxford University Press.
Leduc, T.B. (2013): Ancestral climate wisdom: return to a thoughtful etiquette. In: Schönefeld, M. (ed.), *Global ethics on climate change: the planetary crisis and philosophical alternatives*. Abingdon: Routledge, pp. 107–120.
Lefebvre, H. (1991): *The production of space*. Oxford: Blackwell.
Lélé, S.M. (2006 [1991]): Sustainable development: a critical review. In: Redclift, M. (ed.), *Sustainability, critical concepts in the social sciences*, Volume II: *Sustainable development*. London: Routledge, pp. 165–190. Original source: *World Development*, 19(6), 1991:607–621.
Lempinen, H. (2017): *The elusive social: remapping the soci(et)al in the Arctic energyscape*. University of Lapland.
Lorentzen, J., Jensen, E.L., and Gulløv, H.C. (1999): *Inuit, Kultur og Samfund. En grundbog i eskimologi*. Århus: Systime.
Lumley, S. and Armstrong, P. (2004): Some of the nineteenth century origins of the sustainability concept. *Environment, Development and Sustainability* 6(3):367–378.
Medby, I.A. (2014): Arctic state, Arctic nation? Arctic national identity among the post-Cold War generation in Norway. *Polar Geography* 37(3).
Mikkelsen, A. and Langhelle, O. (2008): *Arctic oil and gas: sustainability at risk?* New York: Routledge.
Palonen, K. (2006): Two concepts of politics: conceptual history and present controversies. *Distinktion* 12:11–25.
Parker, J. (2014): *Critiquing sustainability, changing philosophy*. New York: Routledge.
Petrov, A.N., BurnSilver, S., Chapin III, F.S., Fondahl, G., Graybill, J., Keil, K., Nilsson, A.E., Riedlsperger, R. and Schweitzer, P. (2016): Arctic sustainability research: toward a new agenda. *Polar Geography* 39(3):165–178.
Petrov, A.N., BurnSilver, S., Chapin III, F.S., Fondahl, G., Graybill, J., Keil, K., Nilsson, A.E., Riedlsperger, R. and Schweitzer, P. (2017): *Arctic sustainability research: past, present and future*. London: Routledge.
Quental, N., Lourenço, J., and Silva, F. (2011): Sustainable development policy: goals, targets and political cycles. *Sustainable Development* 19:15–29.
Redclift, M. (1987): *Sustainable development: exploring the contradictions*. London: Methuen & Co. Ltd.
Redclift, M. (ed.) (2006): *Sustainability, critical concepts in the social sciences*, Volumes I–III. London: Routledge.
Ryall, A., Schimanski, J., and Wærp, H.H. (eds) (2010): *Arctic discourses*. Newcastle-upon-Tyne: Cambridge Scholars Publishing.
Sachs, W. (2006 [1990]): On the archaeology of the development idea. In: Redclift, M. (ed.), *Sustainability, critical concepts in the social sciences*, Volume II: *Sustainable development*. London: Routledge, pp. 328–353. Original source: *The Ecologist*, 20(2), 1990:42–43.

Saussure, F.E. (1983 [1916]): *Course in general linguistics*, ed. Charles Bally and Albert Sechehaye, trans. Roy Harris. La Salle, IL: Open Court.

Sejersen, F. (2014): Klimatilpasning og skaleringspraksisser. In: Sørensen, M. and Eskjær, M.F. (eds), *Klima og mennesker: Humanistiske perspektiver på klimaforandringer*. Copenhagen: Museum Tusculanums Forlag.

Sejersen, F. (2015): *Rethinking Greenland and the Arctic in the era of climate change*. London: Routledge.

Sen, A. (2009): Sustainable development and our responsibilities. *Notizie di Politeia* 26 (98):129–136.

Shapiro, M. (2005): The discursive spaces of global politics. *Journal of Environmental Policy & Planning* 7(3):227–238.

Skinner, Q. (2002): *Visions of politics*. Cambridge: Cambridge University Press.

Sneddon, C.S. (2000): Sustainability in ecological economics, ecology and livelihoods: a review. *Progress in Human Geography* 24(4):521–549.

Søndergaard, J.S. (2017): Noget om aktuelle muligheder for anvendelse af begrebet 'bæredygtig udvikling,' når det gælder udviklingen af en 'Bæredygtig udvikling for et arktisk velfærdssamfund'. *Grønland* 65(2):118–124.

Spivak, G. (1988): Can the subaltern speak? In: Nelson, C. and Grossberg, L. (eds), *Marxism and the interpretation of culture*. Urbana, IL: University of Illinois.

Steinberg, P.E., Tasch, J., and Gerhardt, H. (2015): *Contesting the Arctic: rethinking politics in the circumpolar north*. London: I.B. Tauris.

Strandsbjerg, J. (2014): Making sense of contemporary Greenland: indigeneity, resources and sovereignty. In: Dodds, K. and Powell, R. (eds), *Polar geopolitics? Knowledges, resources and legal regimes*. Aldershot: Edward Elgar.

Struck, B., Ferris, K., and Revel, J. (2011): Introduction: space and scale in transnational history. *The International History Review* 33:4:573–584.

Swyngedouw, E. (2007): Impossible 'sustainability' and the postpolitical condition. In: Krueger, R. and Gibbs, D. (eds), *The sustainable development paradox*, New York: The Guilford Press, pp. 13–40.

Tennberg, M. (2000): The politics of sustainability in the European Arctic. In: Hedegaard, L. and Lindström, B. (eds), *The NEBI yearbook 2000: North European and Baltic Sea integration*. Berlin: Springer, pp. 117–126.

Tennberg, M. and Keskitalo, C. (2002): Global change in the Arctic and institutional responses: discourse analytic approaches. In: Käyhkö, J. and Talve, L. (eds), *Understanding the global system: the Finnish perspective*. Turku: FIGARE, pp. 225–228.

Tennberg, M., Vola, J., Espiritu, A.A., Fors, B.S., Ejdemo, T., Riabova, L.; Korchak, E., Tonkova, E., and Nosova, T. (2014): Neoliberal governance, sustainable development and local communities in the Barents Region. *Barents Studies: Peoples, Economics and Politics* 1(1):41–72.

Warde, P. (2011): The invention of sustainability. *Modern Intellectual History* 8(1):153–170.

WCED (1987): *Our Common Future, Report from the 'Brundtland' World Commission on Environment and Development*. Oxford: Oxford University Press.

Wenzel, G. (2009): Canadian Inuit subsistence and ecological instability: if the climate changes, must the Inuit? *Polar Research* 28:89–99.

Williams, L. (2011): Canada, the Arctic, and post-national identity in the circumpolar world. *The Northern Review* 33.

Wilson, E. (2007): Arctic unity, Arctic difference: mapping the reach of northern discourses. *Polar Record* 43(2):125–133.

2 The sustainability of what?

Stocks, communities, the public purse?

Rikke Becker Jacobsen

Introduction

Having formed the backbone of Arctic economies and communities (AHDR II 2004) since colonial times and during postcoloniality, fisheries provide a unique window for observation of the ways in which competing aspirations for development, change, and preservation are enacted in Arctic societies in change. Political exchanges on the development of the Greenlandic fishery sector often involve debates about whether to 'change', 'preserve', 'cancel' or even 're-insert' largely technical entities such as 'yearly quotas', 'quota shares', 'the landing obligation', etc. The starting point of this chapter is that all of these technical entities reflect politically negotiated prioritizations of specific societal developments in which the sustainability of a range of 'referent objects' for sustainability are weighed against each other: fish stocks, communities and the public purse.

As referent objects for sustainability, these regulations all link to a range of legitimate, institutionalized development priorities in Greenlandic society. Meanwhile, and perhaps more often than not, referent objects tend to clash with each other at the management level. For example, competing rationalities, interests and discourses seek to shape policy-making and management in Greenlandic fisheries (Danielsen *et al.* 1998; Becker Jacobsen and Raakjær 2014; Rasmussen 2003). Danielsen *et al.* (1998) and Rasmussen (2003) have described the competition between different rationalities in the coastal and offshore fleet and problematized the unequal power relationship of the two segments in their capability to influence the decisions of Greenland's Home Rule Government. They argue that the large-scale and frequently publicly owned seafood companies have been more influential. On the basis of an analysis of actors, discourses and networks in the recommendations of the 2009 fishery commission and in the subsequent fishery reform of 2011–2012, Becker Jacobsen and Raakjær (2014) identified a 'grand reform network' which pushed for greater profitability in fisheries in order to sustain the Greenlandic economy. Concretely, the 'grand reform network' called for a comprehensive fishery law reform, which was only partly implemented from 2011 to 2012. Nonetheless, the changes were still significant: (1) individual transferable quotas (ITQs) were imposed on coastal Greenland halibut fishery; (2) the maximum allowable quota ownership was raised from 10 per cent to 15 per cent; and (3) the 'rubber boot' paragraph was removed from the fishery law, thus permitting ownership in the Greenlandic fishery by others than fishers to boost investment in coastal fisheries (Becker Jacobsen and Raakjær 2014).

The characteristics of this discourse in general and of its technical remedies in particular are not confined to Greenlandic fishery governance. ITQ regulations have achieved the status of a universal panacea for problems of biological and economic sustainability (Degnbol *et al.* 2006). For example, Alaska (Carothers 2010), Denmark (Høst 2012) and Iceland (Eythórsson 1997) have all implemented ITQ regulation in their fisheries, achieving similar levels of high consolidation within the sector in terms of quota ownership, company size and profit. In Greenland in 2009–2013, however, the grand reform discourse also interacted with two other Greenlandic debates regarding topics mentioned above, namely the economic importance of fisheries in sustaining Greenlandic society at large and the economic need for extractive industries to sustain Greenlandic economic independence. For the Fishery Commission of 2009, the context of visions of booming extractive industries was precisely the notion that the fisheries could dedicate themselves entirely to goals of fleet profitability. As the new extractive industries were expected to provide employment alternatives, it was hoped that this could free fishery of its obligation to sustain coastal employment. The Fishery Commission therefore recommended the removal from the fishery law of the 'landing obligation', which had obliged the fishery to land some or all its catches to coastal fish factories.

Political discussions about the societal role of fisheries are not unique to Greenlandic and Arctic space: Governance of the fishery sector has been generically described as an inherently wicked problem because balance cannot be reached once and for all by means of technical solutions (Jentoft and Chuenpagdee 2008; Koimann 2005). Quite the contrary, the same type of dilemmas have a strong tendency to re-emerge again and again. In spite of its technical-sounding language, fishery governance is not a technical problem – it is a political and moral problem which rightly involves political and normative considerations (Koimann 2005). In developing countries in particular, fisheries as development projects are tasked with competing goals which need to be specified (Bailey and Jentoft 1990). In a similar vein, others have characterized the fishery laws of fishery-dependent Northern countries as the materialization of a 'social contract' (Holm *et al.* 2015) between the fishing industry provided with access to the national fishery commons and the welfare societies that govern these resources on behalf of the owners of the fishery commons – the populace.

Approaching sustainability as a political concept (Gad *et al.*, this volume), this chapter revisits previously analysed discourses surrounding Greenlandic fishery reforms to communicate a clear account of selected referent objects at stake when it comes to Greenlandic society and Greenlandic identities. The aim is to create a more simplified account of specific 'reference objects' that

Greenlandic fisheries policy is striving to 'sustain', while at the same time providing an account of the complexities and seeming contradictions involved as actors bring these different reference objects into dialogue with each other. The chapter explicitly addresses three prevalent reference objects of sustainability in fisheries: fish stocks, the public purse and communities.

The empirical material used in the analysis consists of previously published studies, supplemented with desktop research of reports, laws and public debate related to Greenlandic fisheries. Most of the studies cited and the personal observations stem from the author's long-term ethnographic fieldwork in fishery policy-making events taking place in and around Greenland Self-Government 2010–2015 (see also Becker Jacobsen 2013a). Furthermore, results are included from a national survey conducted by the author in cooperation with the national survey agency 'HS Analyse' in January 2016 (Becker Jacobsen 2017). The desktop research also consists of the daily (2015–2017) observation of developments and debates in relation to fishery policy as covered by the Greenlandic media KNR.gl and Sermitsiaq.gl.

The chapter will unfold each of the three reference objects (and to some extent their respective protagonists) one after the other, each within the historical and political context of competing concerns and interest. Subsequent analysis will focus on public support for the three reference objects. The chapter eventually argues that even though reference objects for sustainability are internally contradictory and in competition with each other, these contradictions tend to be internal to many individual actors. The argument is that the dynamic of competitive sustainability discussions serves to uphold ambiguous relations between different reference objects and actors rather than simply sustaining each individual reference object in isolation.

Stocks: biologically sustainable catch quotas in Greenlandic fisheries

As a referent object, biologically sustainable fish stocks are manifest in the stated purpose of the fishery law as detailed in its second paragraph: 'The purpose of the law is to secure an appropriate and biologically responsible use of the fish stocks' (Greenland 1996). The targets for the biological responsible use of the fish stocks are first and foremost defined as total allowable catch quotas (TACs), the size of which are recommended by the Greenland Institute of Natural Resources in cooperation with international fishery organizations. The government's pursuit of biologically informed catch quotas are economically manifest in the yearly budget expenditure to the Greenland Institute of Natural Resources, which has steadily increased from 43.3 million DKR in 2011 to 59.4 million DKR in 2017 (Greenland Self-Government 2017). All in all, biological sustainability is a highly institutionalized referent object in Greenlandic fishery governance and appears to be only increasing in importance.

Still, conflicts over biological sustainable catch quotas have managed to break government coalitions in Greenland and to profile political parties. For example, fisheries laid the foundation for the so-called 'avalanche' election in IA's favour in 2009, resulting in the exclusion of Siumut from government for the first (and only) time since the commencement of Greenland home rule in 1979. Disagreement about whether or not to exempt parts of the prawn catch from the trawlers' quota shares had led to a conflict over the prioritization of biological sustainability in the prawn fishery, which in 2007 broke the government coalition between Greenland's two largest parties – Siumut and Inuit Ataqatigiit (IA) (Becker Jacobsen 2013a). IA opted for a stronger prioritization of biologically sustainable stocks, and when IA later gained government power, party political subject positions were increasingly constructed in relation to the prioritization of biological advice in the implementation of the TAC policy (Becker Jacobsen and Raakjær 2012).

The policy focus on biological sustainability is largely defined as adherence to scientific advice in the TAC policy. Meanwhile, strict TACs as referent objects in fisheries were contested by Siumut politicians who portrayed the IA/Demokraatit government as indifferent to the social and economic impact of a strict TAC policy on coastal fishers. IA/Demokraatit was criticized for prioritizing international concerns about sustainability over the more immediate concerns of Greenlandic fishers and communities (Becker Jacobsen and Raakjær 2012). With the global emergence of the Marine Stewardship Certification (MSC) eco-labelling scheme, the prioritization of sustainability was increasingly advocated by both the IA/Demokraatit government and the offshore fleet: if the Greenlandic fisheries were to secure and expand their international market shares, they would need to be profiled as biologically sustainable within the emergent eco-labelling paradigm of MSC, and this would require management plans for the certified fisheries. It was in this context that the fishing industry established the 'Sustainable Fisheries Greenland' organization, thereby explicitly employing the sustainability concept to communicate to an international audience of consumers. From this new platform, the fishing industry has managed to obtain MSC certification for the prawn fisheries, the (offshore) Greenland halibut fisheries and the lumpsucker fisheries, whereby it has committed itself to a management agenda that prioritizes stock sustainability to secure market shares with European consumers. The already highly institutionalized reference object of sustainable fish stocks is thus increasingly being situated within the context of a specific scientific regime and an emergent consumer paradigm emphasizing not just sustainable fish stocks, but also sustainable fisheries in terms of other ecosystem impacts. The strict TAC paradigm, however, is sometimes contested by the everyday knowledge and concerns of fishers and those who represent them as a constituency.

In contrast to issues related to some of the sea mammal and bird species in Greenland, there are no inherent conservationist concerns related to the commercially exploited fish stocks, with the exception of Atlantic salmon. Biologists at the Greenlandic Institute of Natural Resources have consistently presented the argument to politicians that stock sustainability is first and foremost an economic concern for Greenlandic fisheries and Greenlandic society (see for example KNR 25 February 2016). Sustainable fish stocks are considered essential to a stable and ongoing fishery, allowing the realization of a range of economic goals in the long term, in contrast to cycles of resource boom and bust. In pursuing this other referent object for sustainability, the Greenlandic economy, the chapter will now look further into a related reference object in Greenlandic fishery governance, namely the public purse. Healthy fish stocks and sustainable TACs are temporarily left aside: this debate centres on the ability of the industry to sustain Greenlandic society in economic terms.

Public purse: economic sustainability in Greenlandic fisheries

Greenland is a typical Arctic case, in the sense that its economy is based on the large-scale exploitation of natural resources for export. Greenlandic fisheries emerged in the context of trade-based colonialism in the Arctic. The Danish colonialization of Greenland was trade colonialism based on the extraction of marine resources by the Danish 'Royal Greenlandic Trade Company'. In the earliest stages of colonialization, the Royal Danish Trade Company purchased mammal products from the Inuit in Greenland. From the 1920s onwards, however, fish gradually became an increasingly important commodity, and by the 1950s and 1960s, cod had been selected as the primary trade commodity around which Greenlandic industrialization would evolve (Dahl 1986).

Once established under colonial rule, the fisheries came to play a leading role in the movement for independence and the establishment of Greenland Home Rule in 1979. Greenland voted for devolution, but it certainly also voted for an exit from the European Union's common fishery policy, to which it was connected through Denmark's EU membership. Independence from Denmark thereby became a central means to achieve sovereignty over Greenlandic fish resources that had been overfished by foreign fleets (Rebhan 2016). In this respect, Greenland has succeeded in circumventing the Arctic trend for resource exploitation to be conducted and controlled by the South. Like other states in the North Atlantic, Greenland Home Rule – and now Greenland Self-Government – has managed its own fisheries alone or in cooperation with other states within the relevant international fisheries governance organizations (Becker Jacobsen and Raakjær 2012). However, traces of colonial history are still visible – for example, in the current organization of production, which is still dependent on monopoly structures and export routes with Denmark as the first stop for fish en route to the world market.

Today, in the context of modern self-government and postcolonial aspirations for full independence, the single most influential discourse about the significance of fisheries for Greenlandic society centres on their share of the national export: over 80 per cent of the country exports has come from fish and shellfish products within the last eight years. This fact is commonly used to explain how the fishery is intrinsically connected to Greenlandic society on a national economic scale.

This discourse about the significance of fisheries in sustaining the Greenlandic economy is nuanced by the block grant of 3,766.4 billion DKR that Greenland receives directly from the Danish state each year to sustain Greenlandic society. This economic dependence on southern governments is a typical feature of Arctic economies (AHDR II 2014), and, in a postcolonial context, the block grant acts as a constant reminder that Greenland is economically dependent on so much more than fish and shellfish. In spite of the impressive export share, Greenlandic fishery has not managed to single-handedly build a self-sustaining economy. 'Economic diversification' is therefore a recurrent goal in Greenland, where four pillars for economic development have been pursued: fishery, tourism, extractive industries and 'other land based industries' (Danielsen *et al.* 1998). While the goal of establishing extractive industries is far from recent, the necessity of diversification became particularly urgent with the Self-Government arrangement in 2009 and the renewed aspirations for full-fledged independence from Denmark (see also Bjørst, this volume; Sejersen, this volume). In that political context, the special position of the fishery sector in Greenlandic society, economy and politics is contested. The economic significance of fisheries for Greenlandic society is a discourse that exists side by side with a discourse of their shortcomings in achieving economic independence and their prioritization as *the* most important industry in Greenland (Becker Jacobsen 2013a).

It is perhaps in the context of renewed competition that the fishing industry explicitly repositioned itself on the political and societal agenda in a 2013 report prepared by the Danish consultancy company Copenhagen Economics, *The Economic Footprint of the Fishery in Greenland*. The report was commissioned by the Greenlandic Business Association, an interest organization which represents the large-scale segment of the fishery. Emphasizing export figures, it described the fisheries in the following manner:

> So far the extractive industries remain an unrealized potential. Tourism … exports amount to 340 million DKR, corresponding to 8% of total exports. Fishery exports were 2,4 billion DKR in 2011, corresponding to 57% of total exports[1] amounting to 4,2 billion DKR, which makes the industry the main exporting sector in Greenland. Income from exports makes imports possible and is a precondition for growth and self-support. Therefore, the fishery is the economic standard bearer. Both raw materials and tourism must exploit their full potential for export, but the fishery is already a strong industry in terms of export.
>
> (Copenhagen Economics 2013: 8; author's translation)

As the offshore fish companies which commissioned the report worked to position the fisheries centre-stage in aspirations for a self-sustaining Greenlandic economy, society and politics, it can be argued that they were simultaneously engaged in another political discussion about sustainable development that was internal to the fishery sector itself. The fishery law divides the TAC quota for all major fisheries between two socio-administrative categories: the *coastal* fleet segment and the *offshore* fleet segment. Both categories result from administrative categorization but they also correspond to two very different types of vessels that operate in different spatial zones under different administrative conditions and even, to some extent, in different sociocultural contexts. The relationship between these two fleet segments is complex. The offshore segment is dominated by two seafood groups, Polar Seafood and Royal Greenland (Copenhagen Economics 2013). While these companies have the access rights to the offshore share of the overall TACs, they also own most of the landing facilities and processing factories that buy catches from the coastal fisheries in Greenland. Furthermore, they have invested in the coastal fisheries that deliver fish to their factories. In reality, this knits the large-scale companies in the offshore fleet together with the smaller-scale fisheries to such an extent that small-scale fishers do not necessarily perceive a conflict between coastal fishery and offshore fishery (Snyder *et al.* 2017). At the same time, however, the two segments can also be observed competing for quota access. When this competition unfolds, the arguments in favour of the seafood groups are couched in a discourse of economic sustainability which links the national purse to the corporate bottom line obtained by means of large-scale production, efficiency and concentration of fishing rights (Becker Jacobsen 2013a; Becker Jacobsen and Raakjær 2014).

This unfolding argument is central to the report by Copenhagen Economics (2013). Having established the direct and indirect contribution of the fisheries to the Greenlandic economy due to their high export levels, the report subsequently analyses the relative productivity of the offshore and coastal fleet. It shows that the vast majority of the offshore TAC is fished by just two companies, Royal Greenland and Polar Seafood, which together with their affiliated companies accounted for the majority of production in the fisheries sector (77 per cent) in 2011. While rendering visible the economic footprint of the fisheries in general, the analysis also explicitly presents an argument about the societal values of the two large-scale seafood groups in Greenland. This value is described in terms of higher productivity, profitability, wage levels and contributions to GNP and the public purse. The report notes that:

> In Greenland as in other countries, a high level of prosperity is created through high productivity, meaning that each employee must generate the highest possible value. Here we find that the Greenlandic fishery actually consists of two completely different parts: a quite productive part, which among other things operates in the offshore fishery, and the remaining fishers, who primarily practice coastal fishery. The productivity of the Royal Greenland and Polar Seafood groups is almost three times as high as that of the other group of fishers, measured in value creation per person employed.
> (Copenhagen Economics 2013)

The relatively 'superior' societal value of the offshore fleet vis-à-vis the coastal fleet has been used more or less directly to argue for a political prioritization of the large-scale fish companies in the distribution of the TAC between the coastal fleet and the offshore fleet, most recently in connection with the industry's resistance to the most recent proposal by the Greenland Government to re-allocate prawn quota from the offshore to the coastal fishery (Government of Greenland 14 August 2017). The bill was 'intended as an initiative aimed at fighting the increased inequality in Greenlandic society, to secure employment and to increase income opportunities in the fisheries'. Therefore, the bill proposes proposed 'a range of adaptation in the ownership of the fish resources with the goal of giving more fishers a greater share in the profit that is created from the publicly owned resources' (Government of Greenland 14 August 2017). This proposal has been strongly criticized, not only by the Greenlandic Business Association and Polar Seafood (e.g. Sørensen 5 May 2017), but also by economic experts and the media (*Sermitsiaq* 18 August 2017), on the grounds that this restructuring will be detrimental to the national economy because it will reduce the efficiency and profitability of the fisheries.

That the offshore fleet has to stay on its toes to expand or simply secure its present quota share makes sense in a Self-Government context where the coastal fishery, though often criticized by the economic elite in Nuuk, expresses competing concerns that continue to find political resonance. The next section will seek to unfold the argument that coastal communities constitute the third recurrent reference object in Greenlandic fisheries.

Communities: employment and culture in Greenlandic fisheries

In the 1950s and 1960s, the development of fisheries was intended to build a self-sustaining economy and improve the living conditions of Greenland, which had by then been formally decolonized and integrated into Denmark as a county. But with growing awareness of a Greenlandic ethno-national identity and the establishment of Greenland Home Rule in 1979, the fisheries were also tasked with sustaining yet another object of sustainability, namely a 'Greenlandic way of life'. The Greenlandic way of life was considered to be closely connected to life in settlements where, in contrast to urban centres, fishing and hunting were perceived as playing an important economic, social and cultural role. The smaller-scale coastal fisheries and the coastal fleet were viewed as particularly important in sustaining this way of life (Rasmussen 1998). As well as promoting sustainable resources and the relationship between resources and fleet capacity, administration of the fishery law is to take into consideration 'the recreational needs of the population' and 'economic and employment related considerations within the fishing industry, the processing industry and auxiliary businesses' (Greenland Self-Government 1996). In its present form, the fishery law already takes employment in coastal communities into consideration in a number of ways. One way is the division of the TAC into a coastal segment and an offshore segment. It is the distribution of the TAC for prawns specifically that the government, at the time of writing, has suggested to tip further in favour of the coastal segment by levelling out the distribution key from 43/57 to 50/50 per cent.

The aforementioned landing obligation is another central regulation upholding coastal employment and livelihoods as it obliges fishers to land their catches to the land-based fish factories on the coast of Greenland. The coastal fisheries must land 100 per cent and the offshore fishery 25 per cent. This difference in landing obligations reflects the submission of coastal fishery to regional/community development concerns while offshore fishery has been allowed to pursue more profit-oriented goals by processing the catch onboard the vessels and exporting it directly without further processing in Greenlandic towns.

In addition, Becker Jacobsen and Delaney (2014) have argued that the community perspective is articulated from within social scientific studies of the Arctic. They argue that social scientific studies have positioned hunting and fishing as social and cultural phenomena to be understood and appreciated in sustainability-related discussion. On the basis of a literature review, Becker Jacobsen and Delaney (2014) have argued that social scientific explorations of social sustainability and related themes in Greenland and the Arctic have developed definitions and indicators that centre around self-control, the connection to land, the continuity of livelihood and culture and a close connection to nature and the land. Social scientific studies have illustrated or argued that as well as the sharing and consumption of catches, hunting and fishing involve important social and cultural activities. These activities are understood as social and cultural in the sense that they reproduce, reconfirm and sustain meaningful social relations in the coastal communities (Dahl 2000; Nuttall 2000; Olsen 2000).

The community appears to be a relatively resilient reference object, which, in spite of considerable contestation around the time of the establishment of the Greenland Self-Government and the Fishery Commission report (2009), continues to reappear in different forms and from a variety of actors. This is especially so within the changing political scene of Greenland Self-Government. For example, although employment considerations had been assigned lower priority with the ITQ reform in 2012 and previously in the underlying discourse of economic sustainability initiated by the Fisheries Commission in 2009, the employment perspective re-emerged as a relevant reference object in the 2015 business strategy of another government coalition (Naalakkersuisut 2015). In this strategy, fishery on the coast is repositioned as a source of increased employment through active landing facilities, increased self-sufficiency and fish refinement being relocated in Greenland.

In the aforementioned report from Copenhagen Economics published in 2013, the seafood groups Royal Greenland and Polar Seafood had even started to lower their previous articulation of the productivity-over-employment perspective. They too began to engage in discussion of employment in Greenland as something to be sustained and developed further by the seafood groups themselves. In light of the previous emphasis on productivity over employment in the context of the 2011–2012 fishery reforms, this report may indicate a slight shift in discourse. Now, the seafood companies have also begun to profile themselves in terms of numbers employed (e.g. Sørensen 5 May 2017; Suluk April 2017). For example, Polar Seafood recently announced that

> We have created many workplaces in Greenland within the last 10 years. We are very conscious of our societal responsibility in the groups management, and, as the largest private operator in Greenland, we will continue to focus on strengthening our position.
>
> (Suluk April 2017:64)

A related shift in discourse is the way that seafood groups have even begun to profile themselves as sustaining the coastal fishery and coastal communities. The Copenhagen Economics report (2013) explicitly argues that the seafood groups help the smaller fishers by investing in larger and more efficient vessels as well as in product and fishing innovation by establishing export channels and giving loans to smaller fishers that the banks are unwilling to grant. The seafood companies are also noted for securing the *existence of some of the small communities* because they can keep factories running even if they do not produce a profit for a number of years (Copenhagen Economics 2013:6–7). Here, a new prioritization is visible; small communities are engaged with as relevant reference objects of a sustainability narrative which positions the large corporations as the subjects doing the sustaining. A new type of hierarchy emerges in which the seafood groups offer themselves as a means to sustain not just the national economy but also the coastal fleet and communities. One might argue that by engaging in the community discourse, the offshore fishery manages to position itself at the centre of political prioritization. The argument seems to be that if politicians want to sustain local communities, they need to sustain the seafood groups and their large-scale operations first; the seafood companies have thereby become indispensable mediators of coastal development.

Thus the offshore fleet has quite successfully found new ways to manoeuver multiple reference objects by positioning itself as the 'maintainer' of complex and intertwined relationships in Greenlandic fisheries. The two large seafood groups help to sustain the coastal fisheries by means of investments in the coastal fleet, for example, and by doing so they simultaneously partake in the discourse that coastal fisheries and coastal communities are relevant reference objects to sustain. As the large-scale fishing companies have positioned themselves, however, the coastal fisheries and economic sustainability as such remain de-linked, glued together only by the interference of the seafood groups. In the economic view that emphasizes large-scale efficiency, the coastal fisheries are worth sustaining but are still not considered sustain*able* in themselves and are certainly not seen as contributing to the sustainability of the national economy. One might argue that the large-scale corporations represent themselves as lifting a burden off the shoulders of self-government politicians and the public purse by sustaining communities on their behalf. However, this argument downplays the aforementioned competition that also characterizes the relationship between the coastal and offshore segment in terms of access rights to the fish resources, which the financially stronger seafood companies have tended to dominate.

On one hand, the reference objects of the national purse and coastal employment are in competition for political prioritization and there has been a tendency for different actors to emphasize one over the other. On the other hand, the same actors are increasingly recognizing the importance of both reference objects in their self-representation. The competition of reference objects is not really confined to a competition between different actors. Instead, the different reference objects increasingly integrate into the competitive sustainability discourse of each individual actor.

Public opinion and the co-existence of competing referent objects

Approaching fisheries governance as a means to achieve certain societal goals, a recent survey study (Becker Jacobsen 2017) set out to explore public opinion on the societal role of fisheries in the light of the many 'wicked problems' (Jentoft and Chuenpagdee 2008) that so characteristically define recurrent battles over economic and social sustainability in Greenlandic fisheries. The study first condensed various competing discourses into positive normative statements as to the importance of different goals in fisheries and then asked a representative sample of Greenlandic citizens over 18 years old ($N = 689$) to evaluate the extent to which they agreed with the statements (on a Likert scale of 1–10, 5 being neutral).

The survey included several questions related to the three reference objects examined here – stocks, national purse and communities – and is therefore able to shed light on the Greenlandic public's evaluation of their relevance. The sustainability of fish stocks is first and foremost understood as the adherence to biological advice on TACs within a biological research paradigm. In the meantime, biological advice within the coastal fisheries is not always observed. Decisions to increase TAC quota are often taken by government with reference to fisher knowledge that contest the conclusions (and methods) of the biologist (Becker Jacobsen and Raakjær 2012). When asked about the prioritization of biological advice vs fisher knowledge and experience, few respondents (7 per cent) objected to greater inclusion of fishers' knowledge, whereas the proposed need to listen more to the biologists was met with more divergent attitudes, with 30 per cent disagreeing. This does not imply that 30 per cent of the public is critical of the relevance of 'sustainable stocks' as a reference object as such. It only implies that the definition of 'sustainable stocks' and its epistemological basis remains highly contested and that TACs defined solely by biologists are not always strongly prioritized.

Public opinion is less divided on the sustainability of the public purse as 89 per cent agreed that fisheries should support the national economy and Greenlandic independence, while only 4 per cent disagreed. However, there was remarkably strong resistance to further quota concentration as a means to increase the profitability of fisheries: 53 per cent of the respondents disagree to some extent with this type of prioritization. It thus appears that public opinion operates with the public purse as a relevant reference object but does not agree with the method selected by the Greenland Self-Government for securing revenue from fisheries. The maximum profitability of relatively few fishers is a highly contested reference object.

When it comes to public support for the community perspective, 80 per cent of the Greenlandic public was supportive of the notion that Greenlandic fisheries should sustain employment; 79 per cent also agreed that citizens should always have access to fisheries for their own consumption, while only 5 per cent disagreed. This indicates that coastal livelihood options remain relevant reference objects in both economic and cultural terms.

The survey results are interesting in that they show that public opinion refuses to prioritize between the two reference objects that so consistently characterize recurrent conflict and debate in Greenlandic fishery policy: the community perspective and the public purse. They are, however, slightly more willing to 'take side' with the discourse that argues for increased fisher influence in the assessment of stock sustainability.

Conclusion

For the sake of argument, sustainability concerns in Greenlandic fisheries can be grouped under three 'reference objects': *fish stocks*, *the public purse* and *community*. The main advantage of operating with these three categories has been to illustrate the unique types of concerns and discourses defining each of them individually. At the same time, however, the analysis shows that these three reference objects are most vividly articulated when confronted with each other. The three reference objects seem to force themselves into dialogue with each other when in competition for political prioritization.

The Greenlandic fishery law has tended to operate with all three reference objects at once. The recent ITQ reforms of the fisheries towards larger entities and greater profitability can be seen as a process whereby the relevant reference objects for sustainable development have been gradually reduced to economic profitability in the fisheries, national economy and Greenlandic independence alone.

This chapter and the survey results in particular draw attention to the fact that even though we recognize the political nature of sustainability and the way it is easily applied by actors to position themselves in a policy context as they compete for fish quota, profit or votes, it may still be worthwhile to remember the original focus of the concept of sustainability on achieving ecological, economic and social goals (WCED 1987). Analytically, we certainly have the option of focusing on the way sustainability is monopolized by competing discourses. But it may also be fruitful to actively seek and explore events and incidents in which actors refuse to confine sustainability to a single reference object or two.

The central message of this chapter, using Greenlandic fisheries as a case, is that conflicts of sustainability may even be internal to individual actors who are able to recognize and value competing referent objects at the same time. Complex relationships between reference objects and actors have evolved where, for example, biological sustainability is promoted in terms of economic sustainability and offshore fisheries in terms of coastal fisheries. Still the question remains as to whether and how exactly to hierarchize, prioritize or balance the different reference objects against each other in the concrete cases of fisheries management. The concept of sustainability may not provide a final answer to this question. But it may provide the ambiguity necessary to facilitate the continued co-existence of multiple reference objects.

Note

1 Copenhagen Economics operate with the total export of not just goods, but also services, and their number is therefore lower than the 86–87 per cent presented above.

Bibliography

AHDR II (2014). *Arctic Human Development Report: Regional Processes and Global Linkages*. Larsen, J.N. and Fondahl, G. (eds). Copenhagen, Nordic Council of Ministers.

Bailey, C. and Jentoft, S. (1990). Hard choices in fisheries development. *Marine Policy* 14 (4), pp. 333–344.

Becker Jacobsen, R.B. (2013a). Power and participation in Greenlandic fisheries governance: the becoming of problems, selves and others in the everyday politics of meaning. PhD thesis, Aalborg University, Faculty of Engineering and Science.

Becker Jacobsen, R.B. (2013b). Small-scale fisheries in Greenlandic planning: the becoming of a governance problem. *Maritime Studies* 12 (2). DOI: 10.1186/2212-9790-12-2.

Becker Jacobsen, R.B. (2017). Befolkningens holdninger til fiskeriets rolle i det grønlandskc samfund: Aalborg University. Available from: http://vbn.aau.dk/files/256740043/Gr_nlandske_holdninger_til_fiskeriets_rolle_i_det_gr_nlandske_samfund.pdf (accessed 15 December 2017).

Becker Jacobsen, R.B. and Delaney, A. (2014). When social sustainability becomes politics: perspectives from Greenlandic fisheries governance. *Maritime Studies* 13 (6).

Becker Jacobsen, R.B and Raakjær, J. (2012). A case of Greenlandic fisheries co-politics: power and participation on Total Allowable Catch policy-making. *Human Ecology* 40 (2), pp. 175–184.

Becker Jacobsen, R.B. and Raakjær, J. (2014). Who defines the need for fishery reform? Participants, discourses and networks in the reform of the Greenlandic fishery. *Polar Record* 50 (255), pp. 391–402.

Carothers, C. (2010): Tragedy of Commodification: Displacements in Alutiiq Communities in the Gulf of Alaska. *Maritime Studies* 9 (2), pp. 95–120.

Copenhagen Economics (2013). *Fiskeriets økonomiske fodaftryk i Grønland*. Grønlands Arbejdsgiverforening, 22 October 2013. Available from: www.copenhageneconomics.com/dyn/resources/Publication/publicationPDF/8/248/0/Fiskeriets%20%C3%B8konomiske%20fodaftryk%20i%20Gr%C3%B8nland.pdf (accessed 15 December 2017).

Dahl, J. (1986). *Arktisk Selvstyre. Historien bag og rammerne for det grønlandske hjemmestyre*. København, Akademisk Forlag.

Dahl, J. (2000). *Saqqaq: An Inuit Hunting Community in the Modern world*. Toronto, University of Toronto press.

Danielsen, M.T., Knudsen, T. and Nielsen, O. (1998). *Mål og strategier i den grønlandske erhvervsudvikling*. Sulisa A/S.

Degnbol, P., Gislason, H., Hanna, S., Jentoft, S., Nielsen, J.R., Sverdrup-Jensen, S. and Wilson, D.C. (2006). Painting the floor with a hammer: technical fixes in fisheries management. *Marine Policy* 30 (5), pp. 534–543.

Delaney, A., Jacobsen, R.B. and Hendriksen, K. (2012). *Greenland Halibut in Upernavik: A Preliminary Study of the Importance of the Stock for the Fishing Populace*. Department of Development and Planning, Alborg University. Available from: http://vbn.aau.dk/files/72423753/FINAL.Upernavik_GreenlandHalibut_SIA.final_1.pdf (accessed 15 December 2017).

Eythórsson, E. (1997): Coastal communities and ITQ management: the case of Icelandic fisheries. In: Pálsson, G. and Pétursdóttir, G. (eds), *Social Implications of Quota Systems in Fisheries*. Copenhagen, Nordic Council of Ministers, pp. 107–120.

Government of Greenland (14 August 2017). Bemærkninger til lovforslaget. Consultation material. Available from: http://naalakkersuisut.gl/~/media/Nanoq/Files/Hearings/2017/Fiskerilov/Documents/Bemaerkninger%20til%20Forslag%20til%20Inatsisartutlov%20om%20fiskeri%20DaF22318910272308809 71.pdf (accessed 15 December 2017).

Greenland Self-Government (1996). *Landstingslov nr. 18 af 31. oktober 1996 om fiskeri.* Available from: http://lovgivning.gl/lov?rid={633775EA-C4B9-401C-99D6-892817ED86B1} (accessed 4 December 2017).

Greenland Self-Government (2009). *Fiskerikommissionens betænkning*. Grønlands Hjemmestyre. Available from: http://naalakkersuisut.gl/~/media/Nanoq/Files/Publications/Fangst%20og%20fiskeri/DK/Udgivelser_FA_Fiskerikommissionens%20bet%C3%A6nkning_Februar%202009_DK.pdf (accessed 15 December 2017).

Greenland Self-Government (2017). *Finanslov for 2017*. Available from: http://naalakkersuisut.gl/~/media/Nanoq/Files/Attached%20Files/Finans/DK/Finanslov/2017/FL2017%20DK.pdf (accessed 4 December 2017).

Holm, P. and Nielsen, K.N. (2004). The TAC machine, Appendix B, Working document I. In: ICES, *Report of the Working group for Fisheries Systems (WGFS)*. Annual Report. Copenhagen, Denmark.

Holm, P., Raakjær, J., Jacobsen, R.B. and Henriksen, E. (2015). Contesting the social contracts underpinning fisheries: lessons from Norway, Iceland and Greenland. *Marine policy* 55, pp. 64–72.

Høst, J.E. (2012). Captains of finance: an inquiry into market-based fisheries management. PhD thesis, University of Copenhagen, Faculty of Humanities.

Jentoft, S. and Chuenpagdee, R. (2008). Fisheries and coastal governance as a wicked problem. *Marine Policy* 33 (4), pp. 553–560.

KNR (25 February 2016). Qularpaa. Fiskeriet 25.02.2016. Available from: http://knr.gl/da/tv/qulaarpaa-debate-programme/qulaarpaa-fiskeriet-25022016 (accessed 15 December 2017).

KNR (16 August 2017). GE: Forslag til fiskerilov er farlig for Grønland. Available from: http://knr.gl/da/nyheder/forslag-til-fiskerilov-er-farlig-gr%C3%B8nland (accessed 15 December 2017).

KNR (11 September 2017). Økonomisk råd er stærkt kritisk overfor fiskerilov. Available from: http://sermitsiaq.ag/oekonomisk-raad-staerkt-kritisk-fiskerilov (accessed 15 December 2017).

Koimann, J. (ed.). (2005). *Fish for Life: Interactive Governance Theory*. Amsterdam, Amsterdam University Press.

Naalakkersuisut (2015). *Et trygt arbejdsmarked. Beskæftigelsesstrategi 2015*. Departement for Erhverv, Arbejdsmarked og Handel. Available from: http://naalakkersuisut.gl/~/media/Nanoq/Files/Publications/Arbejdsmarked/DK/Beskaeftigelsesstrategi_DK_01102015.pdf (accessed 15 December 2017).

Nuttall, M. (2000). Choosing kin: sharing and subsistence in a Greenlandic hunting community. In: Schweitzer, P. (ed.), *Dividends of Kinship: Meaning and Uses of Social Relatedness*. London, Routledge.

Olsen, N.L. (2000). Kalaalimernit: en immateriel kulturarv. *Tidsskriftet Grønland* 3 (58), pp. 218–227.

Rasmussen, R.O. (1998). Managing resources in the Arctic: problems in the development of fisheries. Working paper no. 135 Roskilde, Roskilde Universitet.

Rasmussen, R.O. (2003). Havfiskeri/kystfiskeri: magt og afmagt i Grønlands hovederhverv. In: Winther, G. (ed.), *Demokrati og magt i Grønland*. Denmark, Århus Universitetsforlag, pp. 133–161.

Rebhan, C. (2016). *North Atlantic Euroscepticism*. Tórshavn: Faroe University Press.

Sermitsiaq (18 August 2017). Sermitisiaq mener. 'Træk lovforslaget tilbage'.

Snyder, H., Jacobsen, R.B. and Delaney, A.E. (2017). Pernicious harmony: Greenland and the small-scale fisheries guidelines. In: Jentoft, S., Chuenpagdee, R., Barragán-Paladines, M.J. and Franz, N. (eds), *The Small-Scale Fisheries Guidelines*. New York: Springer, pp. 95–114.

Sørensen, B.H. (5 May 2017). Grønlandske kvotekonger dropper investeringer efter politiske trusler. *Berlingske*. section 2, pp. 10–11.

Statistics Greenland (2017). Greenland in figures. Available from: www.stat.gl/publ/da/GF/2017/pdf/Greenland%20in%20Figures%202017.pdf (accessed 15 December 2017).

Suluk (2017). Uge 4. *Tæt på 1000 jobs*.

WCED (United Nations World Commission on Environment and Development) (1987). *Our Common Future*. Oxford: Oxford University Press.

3 Sustainability understandings of Arctic shipping[1]

Kathrin Keil

Introduction

Arctic sea ice is on a downward trend; not linearly, with lower ice extents from year to year, but in the more long-term perspective ice extent and volume are diminishing.[2] The receding Arctic sea ice has sparked great expectations as to the possibility to utilize new shipping routes between Asia and Europe via the Arctic Ocean and its adjacent seas (e.g. Bekkers *et al.* 2015). This concerns especially the Northwest Passage, a series of channels between the Atlantic and Pacific Oceans running between the islands of Canada's Arctic Archipelago, and the Northeast Passage (NEP), a set of routes that run from Northwest Europe around the North Cape and along the coast of Eurasia and Siberia through the Bering Strait to the Pacific. The Northern Sea Route (NSR) is defined in Russian law as a set of marine routes from Kara Gate, south of Novaya Zemlaya in the west to the Bering Strait in the east, with some of the routes running along the coast, and others running north of the islands of the Russian Arctic (Arctic Council 2009:20, 23, 34; Tymchenko 2001:269–271). The NEP and the NSR are often used interchangeably; while they do overlap, the NEP extends farther west than the NSR.[3]

The ultimate 'boom' in commercial Arctic transit shipping has not materialized thus far. Instead, we have seen highly volatile transit numbers and a very low overall traffic density in comparison to other shipping straits. Despite official Russian expectations that traffic on the NSR would increase considerably, the transit numbers recorded over the last few years seem to have been driven by shipping companies utilizing short-term opportunities, and do not reflect long-term interest or strategies for large-scale usage of the route (Moe 2015:25). Arctic routes will not in all likelihood become viable for international container (i.e. trade) shipping because of the limited reliability, consistency, and schedule integrity of Arctic routes, and due to geographical circumstances that are more obstructive on these routes than on major routes and hubs further south (Humpert 2013; Keil 2015). However, regional increases in shipping activity can be expected, especially on the NSR. Such increases may relate to the transportation of bulk cargo such as iron ore, coal, and liquefied natural gas (LNG) from places in the Arctic to markets in Asia, and for the delivery of supplies into the Arctic to support its increasing economic activity, such as the oil and gas projects on the Yamal Peninsula (Huang *et al.* 2015:66; Humpert 2013:4; Keil 2015:29; Moe 2015:25f.). Given that for climatic, economic, and political reasons (Keil 2015:26f.) the (potential for) shipping activity is higher in Russian and Norwegian Arctic waters than elsewhere in the Arctic, this chapter focuses specifically on the NSR and the NEP.

The sustainability debate in Arctic shipping

> given that the term sustainable development can mean anything to anyone, it is important to ask serious questions about the concept of sustainable development itself, and how it can be related to the notion of development in the Arctic.
>
> (Hattingh 2009:263)

As all ongoing and anticipated activities in the warming Arctic, shipping, too, is listed as an activity that must be conducted *sustainably* or must be part of the *sustainable development of the Arctic*, given its environmental impacts (Andersson *et al.* 2016:16–25) and the possible economic opportunities it provides. However, an explicit discussion of what sustainable Arctic shipping is in concrete terms, or how Arctic shipping could be rendered sustainable, is largely absent in current debates on Arctic shipping.[4] An interesting finding from the analysis below is that all actors appear to find that Arctic shipping can be sustainable, in the sense that no actor argues that ships should not go through Arctic waters at all. Some actors issue reservations about Arctic shipping, calling for prohibiting shipping in particularly sensitive marine areas (see below), but generally the statement holds that shipping can be done sustainably.

While the literature on sustainability and sustainable development, and how to achieve these, is vast, scholars and decision-makers still quarrel over the fuzziness of these concepts due to their excessive usage in virtually all policy and societal debates. This also applies to sustainable Arctic shipping, because sustainability as a concept is seldom explicitly defined and exemplified in actors' shipping policies. In order to make sustainability a tangible concept for analysis, it needs to be defined how the concept is used and operationalized. While such a definition may be made in different ways, in line with the analytical strategy of this book (see Gad *et al.*, this volume), this chapter understands sustainability as a relational concept and aims to identify that which should be sustainable in relation to what. This approach requires a thorough investigation of the concrete *referent objects of sustainability*. Such an analysis is in high demand, especially since the objects of what is intended to be sustained have varied greatly over time, ranging from terrestrial and marine resources for economic usage, through the global climate, regional ecosystems, humans and their livelihoods, regional identities and culture, national economies to community viability and future generations.

Given the centrality of sustainability in the discourse about the Arctic's future, we can safely assume that sustainability is implicitly invoked in the debate on Arctic shipping, even where the word is not explicitly used (Bartelson 2001:10). In other words, despite the absence of an explicit definition and debate on sustainable Arctic shipping, this chapter asks how different actor groups involved in Arctic shipping conceptualize sustainability in their arguments and storylines, thus adopting a semasiological analytical approach (see Gad *et al.*, this volume). In addition to identifying the words conveying the meaning of sustainability and then utilizing them to identify what should be sustained or developed to be sustained[5] (i.e. the referent object of sustainability), the analysis must clarify the purpose for which something needs to be sustained (e.g. Hattingh 2009:264). Moreover, the analysis should pay attention to the scale on which sustainability should be achieved, as sustainability can take on different characteristics and focus on various spatial scales (see Gad *et al.*, this volume). In sum, the chapter aims to unravel sustainability understandings of Arctic shipping by showing, first, what it is in concrete terms that should be *sustained* or *developed to be sustained* in Arctic shipping on different scales and, second, for what purpose.

To accomplish this, the chapter analyses how major Arctic shipping actors frame their policies on sustainable Arctic shipping, which necessitates a focus on various scales on which sustainability is articulated, including the international, regional, and national levels as well as perspectives of local populations and industries. Included in the analysis were actors from different groups on various scales (states, NGOs, local communities, industry) as well as actors who have an explicit strategy or policy towards Arctic shipping generally and northeastern routes specifically. The analysis thus includes Arctic strategies of states located along the NEP (Norway and Russia) and non-Arctic countries (China, Japan, Germany, the Netherlands, the United Kingdom) interested in shipping along the route. Among NGOs, the analysis covers the WWF and the Clean Arctic Alliance.[6] All these states and NGOs are engaged in the International Maritime Organization (IMO), either as members or with a consultative status, and are thus expected to show a significant interest in shipping issues. Locally, strategies that represent inhabitants and indigenous peoples living in the Arctic are included as well as business actors operating on northeastern routes. Finally, three shipping companies that are currently using the NSR, have been using it in the past, and/or are planning to send ships through the route in the future have been analysed. Sovcomflot is Russia's largest shipping company; it has long experience in Arctic shipping, specializes in the transportation of crude oil, petroleum products, and LNG, and provides services to offshore oil and gas installations and equipment (Sovcomflot 2017b). Murmansk Shipping Company (MSCO) operates a major part of shipments under the Russian flag in the Russian Arctic sector, focusing on oil transportation and transhipment as well as exploration.[7] Nordic Bulk Carriers is a Danish company specializing in dry bulk shipping with special expertise in ice class bulk carriers.[8]

The chapter analyses primary sources, including governments' Arctic policy documents, the International Maritime Organization's International Code for Ships Operating in Polar Waters (for short, the Polar Code), business strategies on Arctic shipping, statements by NGOs, and the Arctic Marine Shipping Assessment report (AMSA) written under the auspices of the Arctic Council, as well as secondary literature on China's and Japan's Arctic policies. In line with the semasiological approach, the shipping-relevant parts of the texts are analysed; this includes not only passages in which sustainability or sustainable development are explicitly mentioned, but also passages conveying such content, using different words. To detect such passages, the analysis looks for referent objects that are intended to be *sustained* or *developed to be sustained* for a specific purpose.

What is to be sustained in Arctic shipping?

The driving force for establishing the major international legal document for Arctic shipping, the International Code for Ships Operating in Polar Waters, or the Polar Code (IMO 2017), was a concern for sustainability: the underlying rationale of the Polar Code is the understanding that Arctic shipping can only be sustained if it takes into account human safety and environmental protection. Hence, humans on board the ships and the polar marine environment are important referent objects for sustainability.

Inclusion of a broader range of empirical material in the analysis, however, leads to the detection of many more sustainability understandings in Arctic shipping strategies, ranging across living standards and social well-being, health and cultural integrity to economic benefit, political stability/cooperation/peace, national security, and political influence. Accordingly, actors also relate multiple referent objects to sustainable Arctic shipping. These referent objects have been grouped into four categories, Environment and Climate, Economy, People, and Politics (see Table 3.1), which will be analysed in more detail in the following. These four categories are not always clear-cut, since referent objects that are in separate categories are often combined in actors' strategy documents. These linkages are highlighted in the analysis below.

Sustaining environment and climate

Referent objects in relation to the environment and climate are mentioned frequently by almost all actors,[9] mostly referring to marine and coastal environments but also to Arctic and global climates.

Norway, for example, is striving to improve the safety and efficiency of its maritime activities to protect the marine and coastal environments (Norwegian Ministry of Foreign Affairs 2011) through improved map coverage and satellite data usage (Ayhan 2014). Germany emphasizes internationally preventive and effective multilateral action to protect the Arctic against oil pollution from ships and maritime accidents (Federal Foreign Office Germany 2013). The

Table 3.1 Overview of referent objects in shipping actors' Arctic strategy documents

	Environment and climate			*Economy*				*People*				*Politics*		
	Marine (and coastal) environments	*Arctic climate*	*Global climate*	*Local economy*	*National economy*	*International economy*	*Company's market position*	*Humans on board the ships*	*Local (and coastal) communities*	*Indigenous peoples*	*Global communities*	*Regional and international state relations*	*Strategic position*	*Political reputation, standing*
Norway	x		x	x	x	x		x	x			x	x	x
Russia				x	x	x		x	x	x		x		
China	x			x	x	x					x	x	x	x
Japan	x				x			x						
Germany	x				x			x	x					
Netherlands	x	x	x		x			x	x					
United Kingdom	x	x	x		x			x						
WWF	x	x	x	x					x					
Clean Arctic Alliance	x	x	x						x	x				
Indigenous/local population	x								x	x				
Shipping industry	x		x		x	x	x	x					x	

Dutch and UK Arctic policy documents support the creation of marine protected areas in the North Pole region to safeguard sensitive areas from shipping impacts (Ministry of Foreign Affairs of the Netherlands 2011; UK Government 2013). WWF mentions the more specific aim to establish particularly sensitive sea areas (PSSAs) under the IMO in the Arctic, and to implement measures to minimize the risk of invasive species (WWF Global, n.d.). In line with the Clean Arctic Alliance (CAA), WWF is aiming to secure a legally binding phase-out of the use and carriage of heavy fuel oil (HFO) as ship fuel in Arctic waters by 2020.[10] On the one hand, an HFO phase-out is intended to protect Arctic ecosystems from a possible spill of this highly viscous and toxic substance. On the other hand, the phase-out addresses regional (Arctic) and global climates as referent objects with the intention to reduce harmful emissions from burning HFO as ship fuel (CAA n.d.; WWF Global, n.d.), which is a serious concern in terms of the accelerating ice melting. While not explicitly calling for a phase-out, Norway is urging the IMO to establish international requirements for soot emissions from ships (Government of Norway 2014), while the UK mentions the need to reduce black carbon from shipping activities to protect regional (Arctic) and global climates (UK Government 2013), and Dutch Arctic documents are calling for stricter standards for sulphur emissions and ship fuels to reduce soot emissions (Ministry of Foreign Affairs of the Netherlands 2016).

Environmental protection ambitions are often combined with human safety concerns through demands for adequate ship installations and crew training (Government of Norway 2009, 2014). In line with this, Japan's Arctic policy is striving to develop safe, free, and environmentally sound shipping activities in the North through sustaining a balance between the freedom and safety of navigation and the protection and preservation of the marine environment under the principles of international law (Government of Japan 2015).

Nordic Bulk Carriers explicitly includes environmental protection (in connection with economic benefit) and the global climate (related to the company's market position) in its Arctic shipping philosophy. The commercial usage of the NSR would save time, fuel, and CO_2 emissions and open up new business opportunities for mining and shipping industries (Nordic Bulk Carriers A/S 2010). MSCO promises to identify, assess, and minimize possible risks and impacts from shipping on the environment (MSCO n.d.-a). MSCO and Sovcomflot state the necessity to continuously adapt to the most advanced technologies and maritime safety standards to prevent substandard shipping in the Arctic and the resulting damage to the sensitive Arctic environment (MSCO n.d.-b; Sovcomflot 2017a).

Sustaining economy

Second only to the environment and climate, referent objects related to the economy are most frequently used in almost all actors' shipping strategies, and often in a very prominent fashion especially in states' shipping strategies. The referent object most frequently mentioned is the national economy, followed by local and international economies, and companies' market positions.

Norwegian Arctic strategy documents depict Arctic shipping as relevant for fostering high economic growth rates for the national economy and increasing the profitability of companies engaged in shipping (Government of Norway 2014, 2017). Furthermore, shorter transport distances and lower prices may improve the competitive position of Norwegian actors in Asian markets (Norwegian Ministry of Foreign Affairs 2011). In Russia, shipping along the NSR is seen as part of the overarching goal of using the Russian Arctic as the country's 'strategic resource base', which is supposed to satisfy various Russian resource needs (Government of Russia 2008, 2013). Through the development of the NSR, Russia is hoping to diversify the main supply routes for Russian hydrocarbons to world markets, emphasizing economic benefit for the national (and international) economy (Government of Russia 2013).

Increasing usage of the NSR can help satisfy Chinese demands for resource imports, expand and diversify commodity supply chains from various parts of the world, and support China's export activities; these are all activities on which China is heavily dependent for its economic growth (Chen 2012; Hong 2012). Given the centrality of securing economic growth for China's national security, access to shorter shipping routes thus also has a political connotation (Jakobson and Peng 2012). As one of three 'blue economic corridors', the NSR is part of the Silk Road Economic Belt and the 21st Century Maritime Silk Road (for short: Belt and Road) Initiative, which was launched in 2013 (The State Council 2017). Generally speaking, however, Chinese shipping companies have thus far rather adopted a wait-and-see approach, given the uncertainty about the economic feasibility of Arctic routes and the lack of an overall Chinese Arctic strategy (Huang *et al.* 2015:62; Jakobson and Peng 2012:7). In Japan, Arctic shipping is still largely seen as a matter of the future.[11] However, the relevant Japanese ministries and the national shipping industry are interested in verifying the feasibility of the NSR and in using the route once it has proved profitable (Headquarters for Ocean Policy 2013; Ohnishi 2016:177; Tonami 2016:55). To benefit the national economy, systems to support maritime navigation, e.g. sea ice and weather forecasts, should be established to make the NSR interesting for Japanese shipping companies (Government of Japan 2015).

The 2013 German Arctic Guidelines depict benefits from Arctic shipping in the form of new passageways to East Asian trading centres and the creation of new markets for German maritime technologies and shipbuilding industries (Federal Foreign Office Germany 2013). The national economy is also referenced in the UK Arctic shipping policy, with the government striving to sustain UK ports and the shipping industry's capability to take advantage of any commercial opportunities that the expansion of Arctic shipping may present. The document states the intention to promote the UK as a centre of commercial expertise with direct relevance to many industries that are growing in the Arctic, including the maritime sector (UK Government 2013).

Shipping is also often mentioned as a means to sustain and foster economic growth locally – for example in northern Norway (Government of Norway 2017), and internationally by making use of the High North's possibilities for the world's maritime and offshore industries (Ayhan 2014). Moreover, China is expecting economic benefits for local economies – for instance, in China's northeast region and eastern coastal area (Hong 2012) and in the Tumen river area (Jakobson and Peng 2012).

Economic benefit is often combined with environmental protection and human safety concerns. High standards for safety at sea, search and rescue services, and oil-spill response should be ensured to protect the riches of the sea (Norwegian Ministry of Foreign Affairs 2006). Measures to improve maritime safety, oil-spill response, and search and rescue services should ensure a good transport system, particularly for shipping Norwegian petroleum products to world markets (Ayhan 2014). The Dutch Arctic strategy takes a very cautious and critical stance towards Arctic shipping opportunities, pushing for weighing possible distance savings against higher costs of equipment, insurance, mandatory icebreaker assistance, and expensive licences, as well as environmental and safety issues (Ministry of Foreign Affairs of the Netherlands 2016). Nevertheless, shipping is part of the 'sustainable economic development' theme of the current Netherlands Polar Programme (Ministry of Foreign Affairs of the Netherlands 2016:19). Mention is made of stimulus and innovation programmes and Dutch scientific and technical expertise for safeguarding the sustainability of Arctic shipping and offshore technology (Ministry of Foreign Affairs of the Netherlands 2016).

Shipping companies usually combine the economic benefit for national and international economies and the respective company's market position in their strategy papers. For example, Sovcomflot is planning to sustain and further develop its fleet of high ice class vessels to support the transportation needs of energy projects located in the North and Far East of Russia, but also to serve international energy markets (Arctic Economic Council n.d.; Sovcomflot 2017c). By utilizing the NSR as a trade route, Northern Bulk Carriers is aiming to cement their leading position in the industry and foster the business potential of the route (Nordic Bulk Carriers A/S 2010). By ensuring the highest standards of quality and safety in all MSCO activities, the company intends to keep a competitive position in the shipping market and to gain maximum profit (MSCO n.d.-c).

Sustaining people

Humans on board the ships as well as local and coastal communities are frequently mentioned as referent objects of sustainable Arctic shipping. In contrast, global communities and, maybe surprisingly, indigenous peoples are rather seldom referred to in actors' shipping strategies.[12]

Safety is primarily understood as 'sustaining' the life of the humans on board the ships – for example, through ensuring the adequate training of officers and crew and measures for safe navigation on the NSR (Ayhan 2014; Government of Russia 2008, 2013). Germany and the UK are aiming to increase safety standards for passengers, crew, and local communities through the UNCLOS and the IMO, for example by establishing the necessary technical prerequisites for Arctic shipping through the IMO and the improvement of the bureaucratic, infrastructure-related, and legal framework conditions for Arctic shipping. The UK is further calling for safe Arctic tourism, which is to be achieved through providing travel advice and cooperation with the travel industry (Federal Foreign Office Germany 2013; UK Government 2013).

Human safety is also mentioned in relation to local and coastal communities, which the Norwegian strategy is aiming to sustain by developing expertise on oil-spill preparedness and response. Moreover, shipping is intended to sustain good living standards in northern Norway (Government of Norway 2017), and turnover on the NSR is defined as one of the measurements for social and economic development of the Russian Arctic zone (Government of Russia 2008, 2013). WWF stresses the importance of maximizing the benefits of development for northern people who rely on healthy Arctic ecosystems through equipping ships with the best information, best practices, and the latest technology to avoid or minimize the disruption of wildlife, the introduction of invasive species, and the release of pollutants (WWF Canada, n.d.). If good standards and practices are implemented, WWF believes that Arctic shipping can benefit northern communities while protecting the fragile Arctic environment.

By calling for strict sulphur and soot emission standards, the Dutch, WWF, and CAA strategies also keep local communities in focus, given the detrimental effects of these emissions on air quality and consequently on health levels in Arctic communities (CAA n.d.; Ministry of Foreign Affairs of the Netherlands 2016; WWF Global n.d.). A reduction of these emissions would lower health risks from ship emissions, which have been linked to an increased risk of heart and lung diseases as well as premature death (CAA n.d.).

The chapter 'Human Dimensions' in the 2009 AMSA report provides good insight into the concerns of local and indigenous peoples as regards Arctic shipping (Arctic Council 2009:122–133). The local aspects of human dimensions in the context of Arctic marine shipping are manifold, placing local communities and indigenous peoples as well as their dependence on a healthy environment as the focus attention. For example, in terms of cultural integrity, the continuation of indigenous ways of life and indigenous people's right of self-determination need to be sustained, making sure that the communities of indigenous peoples can still exist in the Arctic. A healthy marine environment must be sustained to preserve the environmental foundations of local societies, e.g. access to marine mammals, fishing grounds, and ice for hunting and travelling. Sustainability notions are also often conveyed through the concepts of economic benefit in conjunction with living standards and social well-being. This concerns the development of Arctic shipping and required infrastructures such as port facilities, emergency response capability, and mechanisms of governance which may provide employment opportunities. Longer shipping seasons for local cargo shipping and the usage of larger boats allow for reduced prices for the transportation of goods and greater access by visitors. At the same time, careful regulation of tourism in the Arctic is necessary, since this may become a source of revenue, but also of disruption – for example, due to competition between tourism and traditional activities and excessive demands for local response capacity in case of an emergency on a large cruise ship.

Few other documents explicitly mention indigenous people in relation to shipping. The Russian strategy explicitly aims to improve the quality of life of the Russian indigenous population in the Russian Arctic through modernization and infrastructure development of the Arctic transport system, including shipping (Government of Russia 2013; cf. Wilson Rowe, this volume). Also, the CAA explicitly refers to indigenous peoples, who must be protected from HFO spills that threaten their nutritional, cultural, and economic needs (CAA n.d.).

Chinese documents provide vague hints of the global community as a referent object of sustainable Arctic shipping in that they combine living standard and social well-being with environmental protection concepts. The document 'Vision for Maritime Cooperation under the Belt and Road Initiative' gives for the first time official hints as to China's interests in Arctic shipping. It reflects China's attempt to build unobstructed, safe, and efficient maritime transport channels for green development, ocean-based prosperity, maritime security, innovative growth, collaborative governance for present and future generations, healthy oceans, and societies without poverty (The State Council 2017).

Sustaining politics

Approaching the texts semasiologically, one finds that among the referent objects to be sustained by Arctic shipping are regional and international state relations, countries' and companies' strategic positions, and political reputation or standing. The Norwegian and Russian Arctic strategies mention the relevance of shipping in their efforts to foster regional and international cooperation and the possibility to preserve the Arctic as a zone of peace through the effective use of the NSR for international shipping[13] (Government of Norway 2017; Government of Russia 2008, 2013). Through increased international usage of the NSR, China is hoping to establish and sustain the recognition of the Arctic as an international or inter-regional issue (Hong 2012). Moreover, China is seeking the recognition of interests of non-Arctic states in the region (Hong 2012) and, more generally, to be respected as a major power and responsible member of the international community (Jakobson and Peng 2012).

Political sustainability notions are frequently combined with economic purposes. For example, through increased cooperation and exchange with China, Japan, South Korea, and Singapore, Norway hopes to develop expertise, infrastructure, and networks in relation to shipping, thus improving its strategic position in economic and political relations with these countries (Norwegian Ministry of Foreign Affairs 2011). Norway also sees opportunities to sustain the country's political reputation and standing as a good Arctic shipping actor. In concrete terms, what is at stake is the development of Norway's capacity to deal with accidents in Arctic waters to avoid tarnishing Norway's reputation and jeopardizing business development in the North (Government of Norway 2009; cf. Steinberg and Kristoffersen this volume). For China, Arctic shipping is an implicit part of establishing 'a constructive and pragmatic Blue Partnership to forge a "blue engine" for sustainable development' (The State Council 2017), which reflects both political and economic ambitions.

The Murmansk Shipping Company highlights the importance of enhancing the strategic position of the company and Russia as a shipping nation. Through participation in regional social programmes and social guarantees for its employees, the company is aiming to improve the image of the shipping industry and to represent Russia adequately in the international shipping industry (MSCO n.d.-d). Simultaneously, this provides the company with political credibility to pursue its economic goals.

Discussion: complexities and conflicts of arctic sustainable shipping

> Will Arctic sustainable development be understood as an environmental doctrine or a developmental doctrine? Do we intend this to mean development or non-development of the Arctic primarily for the benefit of non-Arctic residents, or for Arctic peoples?
>
> (Funston 2009:278)

Many actors, usually states, include a large variety of referent objects in their Arctic strategy documents spanning all four categories and ranging from local, national, and regional to international scales, while others have a narrower focus. Maybe surprisingly, the shipping industry has been found to span all four categories. Marine and coastal environments, local and coastal communities, humans on board the ships, and the national economy are the referent objects most frequently mentioned. The regional (Arctic) climate and the global climate are mentioned relatively often as well. Companies' market positions are explicitly mentioned only by the shipping industry itself. However, given the close relation of companies' market positions to the economy of a country, also other (especially state) actors are assumed to attribute importance to companies' market positions as a referent object. Surprisingly, indigenous peoples are rather seldom explicitly mentioned as referent objects of sustainable Arctic shipping. While regional and local scales are strongly represented through frequent reference to the marine and coastal environment as well as to local communities, the national level takes centre-stage in relation to economic benefit. National and international scales are in focus in the Politics category.

It is a surprising finding that some actors depict shipping as a means to sustain and foster regional and international cooperation, political stability, and ultimately peace among states. These political referent objects are seldom brought into the context of Arctic shipping and are marginal in global sustainability understandings generally. The findings show that, at least by some actors, shipping is seen as relatively important for improving political ties and cooperation.

Economically sustainable shipping?

Despite the range of referent objects indicating a broad understanding of sustainable Arctic shipping, the analysis found a particularly strong focus on economic notions. Although environmental protection and human safety are frequently referred to in Arctic shipping strategies, these concepts are often either used in conjunction with economic benefit notions or even employed to achieve the ultimate goal of economic profits. For example, despite Norway's emphasis on 'environmentally sound growth' and on providing 'funds to promote more environmentally sound maritime transport' (Government of Norway 2009:16), their focus is ultimately on economic growth, with references to the environment and sustainability in the form of adjectives and adverbs related to growth (cf. Steinberg, and Kristoffersen this volume). The predominance of the economic value of Arctic shipping is also exemplified by the Norwegian Government's strategy, since it is aiming to '*reduce* [...] the environmental impact of maritime operations under difficult conditions, for example in the High North' (Government of Norway 2009:16; own emphasis). The verb is 'reduce', not 'eliminate' or 'prevent', so a tacit understanding is implied that pollution from maritime activities such as shipping will occur and that a certain amount of this is acceptable.

At first sight, the economic focus does not seem very surprising, since shipping is predominantly an activity undertaken for economic ends. However, one could argue that the economic focus in the Arctic shipping debate is perpetuating the dominant discourse of valuing the Arctic predominantly from a cost–benefit point of view, which is generally regarded as the underlying cause of *un*sustainable development. In other words, the economic emphasis appears to be 'a continuation, and even an intensification, of the instrumental approach to valuing that currently dominates the world [which] allows for only one kind of value to be accounted for – resource or use value' (Hattingh 2009:261, 263). Such an approach is seen to ignore the large number of threats the region is facing in times of climate change, as well as the very causes of peril in the region. In short, from an environmental ethics perspective, the economically focused discourse on Arctic shipping is rather detrimental to a credible sustainable development of the Arctic.

Sustainable Arctic shipping = sustained Arctic shipping

Ultimately, the underlying assumption behind both national and international decisions regarding Arctic shipping in recent years has been that Arctic shipping should or will be enhanced (or, more precisely, that it will occur where it has not yet taken place at all and will increase where it has only taken place to a limited extent). In other words, sustainable Arctic shipping appears to mean sustained shipping. For example, despite the Polar Code's strong focus on safety and environmental protection, the Code (and thus the actors who determined its content) does not conclude that Arctic shipping is too dangerous or risky (i.e. that it would render other valuable assets 'unsustainable', such as the lives of seafarers and the environment) and should therefore not take place. Given the often raised concerns that shipping in the Arctic is afflicted with serious dangers for people and ecosystems, one might have expected at least some actors to be more critical of Arctic shipping.

Even voices that issue concerns about Arctic shipping usually assume that it will take place to an increasing extent. The approach is rather to influence the legal and technical conditions of Arctic shipping so as to bring it in line with a respective understanding of sustainability; not to try to prevent Arctic shipping because of its possible unsustainable features. For example, environmental NGOs do not call for a ban on Arctic shipping due to sustainability concerns;[14] rather, they call for Arctic shipping to be conducted sustainably, e.g. in the form of not using HFO, or through the adoption of pan-Arctic rules and regulations for preventing environmental harm. Ultimately, calls by environmental NGOs for environmental safeguards, such as a ban on HFO, are calls for Arctic shipping to take place and to continue (if certain conditions are complied with).

The overall message is that Arctic shipping per se is not unsustainable and should thus not be stopped or prevented; in other words, Arctic shipping is seen universally as an activity that can be conducted sustainably under certain circumstances. Arctic shipping is considered to be capable of interacting with the natural environment, Arctic communities, and business interests in a way that enables these assets to co-exist over time without threatening the existence of nature, societies, or businesses; thus, their relationship is regarded as fundamentally sustainable.

Taking into account the many obstacles that large-scale Arctic shipping is still facing today (Humpert 2013; Keil 2015; Moe 2015), this is a rather surprising finding, especially against the background that it is doubtful if large-scale Arctic shipping is sustainable even from an *economic* point of view (not to speak of the doubtful social benefits that could be derived from it and how environmental dangers can be prevented). One reason why stakeholders generally perceive shipping as (possibly) sustainable is *precisely because* there are still so many obstacles to Arctic shipping becoming a large-scale, long-term activity. In this interpretation, the obstacles are a means to render shipping levels in the Arctic sustainable by keeping them generally low in spatial and temporal extent. Also the existing legal shipping regime, especially in the form of the IMO's Polar Code, its predecessors, and possible future developments, may contribute to limiting the volume of Arctic shipping since these regulations impose often costly requirements on shipping companies.

Conflicting sustainability notions

As discussed above, the Arctic shipping discourse reveals a general notion of allowing, enabling, and enhancing shipping in the Arctic, and actors may bring this notion in line with their own understandings of sustainability. While this indicates little potential for contention, conflicts arise when considering the concrete referent objects that are to be sustained or developed, and the conditions under which sustainable Arctic shipping can be achieved.

The large number of referent objects is indicative of the inherent conflict potential of the sustainable shipping discourse. For example, environmental NGOs depict the current use of HFO in the Arctic as unsustainable, given the tremendous risk it poses to the local environment, communities, and climate. Some shipping operators with a strong focus on economic benefit and the market position of their company speak out in favour of using the relatively cheap HFO in order not to render the already cost-intensive Arctic shipping business unsustainable. They argue that the risks incurred by HFO usage are acceptable, seeing HFO as a bridging solution until cheaper fuel becomes available (Nilsen 2014; Ship & Bunker 2014). Business actors often underpin the sustainability of their Arctic shipping activities with the prospect of jobs and other economic and social development opportunities for local communities. This contrasts with concerns expressed by local and indigenous communities regarding little economic and social development opportunities because of the 'flying in' of workers from outside the region to work in new industries coming to the Arctic. The shipping industry is also not likely to offer long-term job opportunities, given the volatile development of ship movements through Arctic waters. For example, investors take a cautious approach because of the changing climatic conditions for shipping from year to year. Adding to local concerns are possible environmental damage and social impacts in the form of interruptions to traditional ways of life through conflicts between shipping and culturally important activities such as fishing and hunting. Further social challenges result from immigration to small Arctic communities and misguided investments of new income.

Sustainability as legitimization

The breadth of sustainability concepts and referent objects is a further sign of the legitimating function of sustainability employed by many actors to justify their policy preferences. Generally, as mentioned above, the sustainability concept allows all actors to agree on the general desirability (or at least possibility) of Arctic shipping or, in other words, on a possible future of the Arctic that will entail shipping. Thus, the inherent sustainability notion of Arctic shipping is to keep shipping running. This, in turn, allows a number of actors, particularly those with a strong interest in the expansion of Arctic shipping, to legitimize their actions, since shipping per se is deemed sustainable by all actors. Moreover, the concept of sustainability in relation to shipping is also employed by non-Arctic actors with an interest and ambitions in the region. Thus, pledging to support 'sustainable shipping' is part of the legitimization of their engagement in Arctic affairs; in other words, including sustainability ambitions in one's Arctic strategy can be seen as a 'ticket to the Arctic'.

Notes

1 The author would like to thank the participants of the POSUSA Workshop in Nuuk, Greenland, 22–24 August 2017, the participants at the panel 'Politics of Sustainability' at the 9th International Congress on Arctic Social Sciences (ICASS IX) in Umeå, Sweden, on 10 June 2017, the editors of this volume, and Manuel Rivera from the Institute for Advanced Sustainability Studies (IASS) for helpful comments on earlier versions of this chapter.

2 See Arctic Sea Ice News & Analysis of the National Snow & Ice Data Center, University of Colorado Boulder, http://nsidc.org/arcticseaicenews, last accessed 21 November 2017.

3 See map on p. 17 of the Arctic Marine Shipping Assessment Report (Arctic Council 2009).

4 For a rare discussion of the sustainability of ships, see Cabezas-Basurko *et al.* (2008).

5 For a similar differentiation see Kates *et al.* (2005:11).

6 The Clean Arctic Alliance consists of 17 NGOs, of which some have a consultative status at the IMO (WWF, Friends of the Earth, Pacific Environment, and Greenpeace). See the homepage of the Alliance at www.hfofreearctic.org/en/about, last accessed 24 July 2017.

7 See the homepage of Murmansk Shipping Company 'Main activities' at www.msco.ru/en/company/general/the-main-activities, last accessed 25 July 2017.

8 See the homepage of Nordic Bulk Carriers at www.nordicbulkcarriers.com, last accessed 25 July 2017.

9 Russia is the only actor analysed that does not mention referent objects in relation to environment and climate.

10 Given the dependence of some Arctic communities on HFO for household use, the CAA currently does not address the issue of HFO carriage as cargo (CAA n.d.).

11 According to Japan's Arctic policy document, 'the Arctic Sea Route is not ready yet for safe and reliable use' (Government of Japan 2015).

12 One might assume that indigenous peoples could be implicitly included in the 'local and coastal communities' category. However, given the relatively strong organization of Arctic Indigenous Peoples Organizations, for example in the Arctic Council, and the special rights they have according to international (and often national) law, one would expect an explicit reference to indigenous peoples in case they were meant to be addressed.

13 Although the Russian strategy emphasizes that this has to be under the jurisdiction of the Russian Federation.

14 Only WWF calls for a ban on shipping in – still to be defined and implemented – PSSAs.

Bibliography

Andersson, K., Brynolf, S., Lindgren, F.J., and Wilewska-Bien, M. (2016). Shipping and the environment. In: Brynolf, S., Andersson, K., Lindgren, F.J. and Wilewska-Bien, M. (eds). *Shipping and the Environment: Improving Environmental Performance in Marine Transportation*. Berlin, Springer, pp. 3–27.

Arctic Council (2009). *Arctic Marine Shipping Assessment 2009 Report*. Tromsø, Arctic Portal.

Arctic Economic Council (n.d.). Pao Sovcomflot. Available from: https://arcticeconomiccouncil.com/members/business-profiles/pao (accessed 13 December 2017).

Ayhan, D. (2014). *The Norwegian Government Perspective on Arctic Shipping*. Oslo, Government of Norway.

Bartelson, J. (2001). *The Critique of the State*. Cambridge, Cambridge University Press.
Bekkers, E., Francois, J.F., and Rojas-Romagosa, H. (2015). Melting ice caps and the economic impact of opening the Northern Sea Route. CPB Discussion Paper 307. CPB Netherlands Bureau for Economic Policy.
CAA (n.d.). Clean Arctic Alliance position statement. Available from: www.hfofreearctic.org/wp-content/uploads/2016/10/Final-Clean-Arctic-Alliance-Position-Statement-1.pdf (accessed 13 December 2017).
Cabezas-Basurko, O., Mesbahi, E., and Moloney, S.R. (2008). Methodology for sustainability analysis of ships. *Ships and Offshore Structures* 3 (1), pp. 1–11.
Chen, G. (2012). China's emerging Arctic strategy. *The Polar Journal* 2 (2), pp. 358–371.
Federal Foreign Office Germany (2013). *Guidelines of the Germany Arctic Policy: Assume Responsibility, Seize Opportunities*. Berlin, Government of Germany.
Funston, B. (2009). Sustainable development of the Arctic: the challenges of reconciling homeland, laboratory, frontier and wilderness. In: *Climate Change and Arctic Sustainable Development: Scientific, Social, Cultural and Educational Challenges*. Paris, UNESCO Publishing, pp. 278–283.
Government of Japan (2015). *Japan's Arctic Policy*. Tokyo, Government of Japan.
Government of Norway (2009). *New Building Blocks in the North: The Next Step in the Government's High North Strategy*. Oslo, Government of Norway.
Government of Norway (2014). *Norway's Arctic Policy*. Oslo, Government of Norway.
Government of Norway (2017). *A Strategy to Promote Peaceful, Innovative and Sustainable Development in the Arctic*. Oslo, Government of Norway.
Government of Russia (2008). *The Foundations of the Russian Federation's State Policy in the Arctic Until 2020 and Beyond*. Moscow, Government of Russia.
Government of Russia (2013). *Strategy for the Development of the Arctic Zone of the Russian Federation and National Security up to 2020*. Moscow, Government of Russia.
Hattingh, J. (2009). Sustainable development in the Arctic: a view from environmental ethics. In: *Climate Change and Arctic Sustainable Development: Scientific, Social, Cultural and Educational Challenges*. Paris, UNESCO Publishing, pp. 258–266.
Headquarters for Ocean Policy (2013). *Basic Plan on Ocean Policy*. Tokyo: Headquarters for Ocean Policy.
Hong, N. (2012). The melting Arctic and its impact on China's maritime transport. *Research in Transportation Economics* 35 (1), pp. 50–57.
Huang, L., Lasserre, F., and Alexeeva, O. (2015). Is China's interest for the Arctic driven by Arctic shipping potential?. *Asian Geographer* 32 (1), pp. 59–71.
Humpert, M. (2013). The future of Arctic shipping? A new Silk Road for China? Available from: www.thearcticinstitute.org/wp-content/uploads/2013/11/The-Future-of-Arctic-Shipping-A-New-Silk-Road-for-China.pdf?x62767 (accessed 13 December 2017).
IMO (2017). *International Code for Ships Operating in Polar Waters (Polar Code)*. London, International Maritime Organization.
Jakobson, L. and Peng, J. (2012). China's Arctic aspirations. Stockholm: Stockholm International Peace Research Institute (SIPRI).
Kates, R.W., Parris, T.M. and Leiserowitz, A.A. (2005). What is sustainable development? Goals, indicators, values, and practice. *Environment: Science and Policy* 47 (3), pp. 8–21.
Keil, K. (2015). Economic potential. In: Jokela, J. (ed.), *Arctic Security Matters*. Brussels, European Union Institute for Security Studies (EUISS), pp. 21–31.
Ministry of Foreign Affairs of the Netherlands (2011). *Policy Framework: The Netherlands and the Polar Regions, 2011–2015*. The Hague, Government of the Netherlands.

Ministry of Foreign Affairs of the Netherlands (2016). *Detailed Summary of the Dutch Arctic Strategy 2016–2020: Working Together on Sustainability*. The Hague, Government of the Netherlands.

Moe, A. (2015). The Northern Sea Route: smooth sailing ahead? In: Sinha, U.K. and Bekkevold, J.I. (eds), *Arctic: Commerce, Governance and Policy*. Abingdon, Routledge, pp. 18–36.

MSCO (n.d.-a). Environment protection management system. Available from: www.msco.ru/en/company/manpower-and-social-policy/environmental-protection (accessed 13 February 2017).

MSCO (n.d.-b). Occupational health care management system. Available from: www.msco.ru/en/company/manpower-and-social-policy/safety-and-health (accessed 13 February 2017).

MSCO (n.d.-c). Safety and quality system. Available from: http://msco.ru/en/company/manpower-and-social-policy/safety-and-quality (accessed 13 February 2017).

MSCO (n.d.-d). Social policy. Available from: www.msco.ru/en/company/manpower-and-social-policy/social-policy (accessed 13 February 2017).

Nilsen, T. (2014). Debate on Arctic shipping heats up. *BarentsObserver*. Available from: http://barentsobserver.com/en/arctic/2014/01/debate-arctic-shipping-heats-20-01 (accessed 13 December 2017).

Nordic Bulk Carriers A/S (2010). NSR Project 2010. Newsletter. August, 1–4. Available from: www.nordicbulkcarriers.com/images/Media/Diverse/NSR_newsletter.pdf (accessed 13 December 2017).

Norwegian Ministry of Foreign Affairs (2006). *The Norwegian Government's High North Strategy*. Oslo, Government of Norway.

Norwegian Ministry of Foreign Affairs (2011). *The High North: Visions and Strategies*. Oslo, Government of Norway.

Ohnishi, F. (2016). Japan's Arctic policy development: from engagement to a strategy. In: Lunde, L., Jian, Y., and Stensdal, I. (eds), *Asian Countries and the Arctic Future*. Singapore, World Scientific Publishing, pp. 171–182.

Ship & Bunker (2014). Environmental group calls for Arctic HFO ban. Available from: https://shipandbunker.com/news/world/950375-environmental-group-calls-for-arctic-hfo-ban (accessed 13 December 2017).

Sovcomflot (2017a). Arctic Economic Council holds meeting in Russia for the first time, supported by Sovcomflot. Available from: www.sovcomflot.ru/en/press_office/press_releases/item84362.html (accessed 13 February 2017).

Sovcomflot (2017b). Sovcomflot calls for enhanced safety measures on Northern Sea Route. Available from: www.scf-group.ru/en/press_office/press_releases/item89702.html (accessed 13 February 2017).

Sovcomflot (2017c). Sovcomflot participates in international forum – 'Arctic: territory of dialogue'. Available from: www.sovcomflot.ru/en/press_office/press_releases/item86373.html (accessed 13 February 2017).

State Council, The (2017). Vision for maritime cooperation under the Belt and Road Initiative. Available from: http://english.gov.cn/archive/publications/2017/06/20/content_281475691873460.htm (accessed 13 February 2017).

Tonami, A. (2016). *Asian Foreign Policy in a Changing Arctic*. London, Palgrave Macmillan.

Tymchenko, L. (2001). The Northern Sea Route: Russian management and jurisdiction over navigation in Arctic Sea. In: Elferink, A.G.O. and Rothwell, D.R. (eds), *The Law of the Sea and Polar Maritime Delimitation and Jurisdiction*. The Hague, Martinus Nijhoff Publishers, pp. 269–291.

UK Government (2013). *Adapting to Change: UK Policy Towards the Arctic*. London, Government of the United Kingdom.
WWF Canada (n.d.). Safe and sustainable Arctic shipping. Available from: http://awsassets.wwf.ca/downloads/wwf_arctic_shipping_factsheet.pdf?_ga=2.43058145.59646729.1500898171-1208431312.1491289744 (accessed 13 December 2017).
WWF Global (n.d.). Shipping in the Arctic. Available from: http://wwf.panda.org/what_we_do/where_we_work/arctic/what_we_do/shipping (accessed 13 December 2017).

4 Digging sustainability

Scaling and sectoring of sovereignty in Greenland and Nunavut mining discourses

Marc Jacobsen

To mine, or not to mine

The question of whether or not to mine is often placed centre-stage in discussions about rights and resources in the Arctic. Some primarily perceive mining as the key to a better economy, while others put greater emphasis on the possible threat to nature. Both sides do, however, insist that the concept of sustainability is inevitable to their argument which, on the one hand, unites the conflicting parts, while on the other hand drains the concept of meaning. In contrast with harvesting, hunting, fishing, and the exploitation of other renewable resources (see Thisted, this volume), minerals are non-renewable and, thus, not possible to sustain if simultaneously being extracted. Hence, 'sustainable mining' may be perceived as an oxymoron used by those in favour to co-opt and neutralize critique from environmentalists (Kirsch 2010). Through other lenses, it may instead be a sign of environmental, economic, and social responsibility in a capitalist world where development is imperative. The three responsibilities are, however, seldom weighed equally and it is in these nuances that the political becomes observable. It is the aim of this chapter to render visible how the various actors in Nunavut and Greenland mining discourses prioritize when they articulate the sustainability concept, and to show what possible consequences this may have.

The acquisition of rights to natural resources has been a cardinal point in the evolution of Greenland's and Nunavut's respective self-determination. In Greenland, the right to the subsoil was already central in the negotiations with Denmark prior to establishment of Home Rule in 1979, which, however, resulted in a vague formulation that 'the resident population in Greenland has fundamental rights to Greenland's natural resources' (Sørensen 1983:238; cf. Schriver 2013:51). Thirty years later, the control of all offshore and subsoil resources was transferred to the Government of Greenland on 1 January 2010, following the introduction of Self-Government on 21 June 2009.

In Nunavut,[1] the outcome has so far been more moderate: As part of the negotiations of the Nunavut Land Claims Agreement (1993), nearly 18 per cent of the land ownership and about 2 per cent of the mineral rights were transferred to Nunavut Tunngavik Incorporated (NTI),[2] representing the Inuit of

Nunavut. Though the transfer of ownership was limited, the areas were carefully selected, ensuring that Nunavut gained control of the most commercially prospective mineral deposits. Today, negotiations to transfer more sovereignty of the subsoil is at the core of the ongoing negotiations about devolution, whose goal – like Yukon's and the Northwest Territories' similar agreements – is to evolve as a subsidiary government *within* the Canadian Federation (Speca 2012:6). While Nunavut today is home to three active mines – in Meadowbank, Mary River, and, most recently, Hope Bay – Greenland has, despite a long and proud mining history, only one small active mine opened in Qeqertarsuatsiaat in the spring of 2017. Both neighbours do, however, have well-documented vast mineral deposits that may be mined if the global market prices are right, if the necessary investments are in place, and if the political will is present.

Opportunities and possible threats related to mining are often the centre of public attention and debate about what the perfect future should look like for Greenlanders and Nunavummiut. Some highlight the importance of preservation (sustaining), while others emphasize the possibilities that change (development) may bring. This activates different and occasionally conflicting perceptions of what the collective identity entails, how mining may affect local standards of living, and, sometimes, how it relates to regional and global concerns and movements. By comparing the Greenland and the Nunavut mining discourses horizontally – across national borders – while analysing articulations about sustainability vertically – from the national to the local scale – the analysis will show what kind of meaning is ascribed to the concept of sustainability, how priorities are made, and how responsibility is distributed. This is particularly interesting within the Arctic context, where sustainable development has become an omnipresent buzzword (see Gad *et al.*, this volume; Humrich 2016; Kristoffersen and Langhelle 2017) – particularly prominent in Arctic mining debates (Skorstad *et al.* 2017:14) – and where Nunavut and Greenland as partners in tradition and in transition (Jacobsen and Gad 2017:15) continue to alter the political landscape by gaining more independent voices.

Scaling and sectoring of sustainability-speak

What sustainability means depends on whether and how relevant actors perceive and articulate certain projects and policies in relation to specific geographical scales, constructed by actors through 'scaling, spacing and contextualizing each other' (Latour 2007:183–184). Thirty years ago, when the concept of sustainable development was prominently put on the international agenda, it was scaled as a global concern from the outset: '[w]e came to see that a new development path was required, one that sustained human progress not just in a few pieces for a few years, but for the entire planet into the distant future' (WCED 1987:4). Since then, the concept has been taken up by various types and sizes of entities promoting particular political projects with different geographical foci. One example is the Arctic Council, which adopted 'sustainable development' as one of its main pillars in 1996, broadening the scope from the prior Arctic Environmental Protection Strategy's narrow focus on negative climate change aspects. Today, the double perspective of profiting from the emerging opportunities made accessible by climate change, while trying not to accelerate environmental degradation any further, is visible in many plans and agreements on how to realize new projects within the Arctic region. These documents are products of social practices (Li 2007) that create subject positions and fix certain forms of hegemonic knowledge, privileging some while excluding others (Sejersen 2015:165, 171). In this way, scaling is a contingent hierarchical form of social organization, produced and reproduced by certain 'scale-making projects' that ascribe different degrees of authority to different scales, encouraging imaginations of one scale as being more pertinent than another (Tsing 2000:119–120).

Power and interconnections between different scales are highlighted in the analysis, hence making visible how actors at one scale may be held responsible for sustaining a referent object at a different scale in relation to an environment at a third scale. Hence, in this chapter, individual scales will not be predefined, but rather empirically deduced. The analytical lens of 'scale' means asking, of each discourse, at which scale is something sustained? And who carries the responsibility for making it happen? As the analysis will show, the concept of sovereignty is central in the scaling of the two mining discourses, as the respective relations to Ottawa and Copenhagen play noticeably different roles in the exploitation of Nunavut's and Greenland's subsoil. Whereas 'national' and 'country' are frequently used as labels for Greenland, and Denmark is seldom mentioned, 'territory' and 'Inuit identity' are often articulated in the Nunavut discourse, where respect and the subordination of Canada's sovereignty is explicitly acknowledged in key documents and hearings. These details are important to keep in mind when analysing who determines what should be sustained with what possible concomitant consequences.

In addition to the vertical division of scales, this chapter will use sectors as a heuristic concept to horizontally distinguish which aspect of sustainable development is given highest priority. As both Nunavut's and Greenland's mining strategies refer explicitly to the World Commission on Environment and Development's (WCED) definition, the analytical starting point will use the same hegemonic triangular division, asking whether the economic, environmental, or social sector is deemed most urgent to sustain by the involved actors. Combining attention to scales and sectors allow the observation of a matrix of horizontal and vertical articulations: Is it the national economy, the global climate, or the local social cohesion that is given priority in Greenland and Nunavut mining discussions? Or perhaps another combination? Though here presented as rather distinct, the analyses will show how the prioritization of sectors sometimes overlap and how connections are narrated.

The empirical[3] starting points are Nunavut's and Greenland's official mining strategies for how to achieve development in a sustainable way. The next body of texts constituting the data archive are public hearings and Impact Benefit Agreements (IBAs), which are formal contracts between the involved mining company, authorities, and the local community which outline the impacts, commitments, responsibilities, and benefits of a particular mining project. In this chapter, hearings and agreements about the present Mary River project on Baffin Island and the proposed Citronen Fiord mine in the Northeast Greenland National Park are given special attention as cases where opinions from different scales are voiced. These two cases are chosen because they activate different scaling questions through their respective remote locations and related means of transportation, and because they have not yet been scrutinized in the academic literature on Arctic mining and politics of sustainability. By including the communication about these two projects, the chapter analyses how sustainability is articulated in discussions about how concrete mining projects may change the status quo, and at which scale and sector change is expected to occur.

Greenland: mining for a new nation, stretching 'the local'

In *Greenland's Oil and Mineral Strategy 2014–2018*, 'sustainable development' abounds: Across 102 pages, the sustainability concept is mentioned 37 times, and two chapters (2.5 and 7) are dedicated to explaining how the broad sense of 'sustainable development' – as defined by the WCED – is the guiding principle for developing the vast mineral resources with careful environmental, economic, and social considerations. Much importance is attached to the imagined future mineral exploitation, which is not merely framed as a single significant business opportunity, but as a general development of the whole country that will 'involve an adjustment of our entire way of organising society' (Naalakkersuisut 2014a:14). The Government of Greenland 'believes that it is important that all of us contribute to a sustainable development of the area of mineral resources activities' (Naalakkersuisut 2014a:7), and in its aim to actively involve the broader public, a popularized version of the strategy has subsequently been distributed (Naalakkersuisut 2014b). The responsibility for making this development sustainable is, thus, not limited to the politicians, the mining companies, or the individual citizens, but instead everyone is expected to partake as stakeholders in this national endeavour. As such, the strategy discursively places all potential mining projects in Greenland on the national scale, while making a diffuse and encompassing 'all of us' responsible for fulfilling its unredeemed potential. While this is the case in the national strategy – which per definition is, exactly, national in its scope – it is necessary to dig deeper into the consultation processes of individual projects to get a more nuanced understanding of who is made responsible for what, and on which scale.

One of the strategy's six prioritized projects is the proposed zinc and lead mine in Citronen Fiord (Naalakkersuisut 2014a:58), which the Australian company Ironbark has the licence to exploit. The project is located in the national park, where no one but a couple of handfuls of Danish servicemen and a few polar scientists live. Hence, there is no obvious local place to conduct public hearings, which is why Ironbark, the relevant ministers, and civil servants visited seven towns in Greenland's four municipalities to engage in a dialogue with the public.[4] The selected towns were (attendees/population): Nuuk (30/17,600), Kangerlussuaq (8/500), Ilulissat (8/4,555), Sisimiut (30/5,414), Qaqortoq (?[5]/3,084), Tasiilaq (100/2,010), and Qaanaaq (200/623) (see Vahl and Kleemann 2017:9), of which the latter two were given special attention because they are nearest Citronen Fiord. Though their 'proximity' is limited to approximately 2,000 km to the south and 1,000 km to the west, respectively, the two towns were discursively constructed as the local scale, even if the immediate economic impact will be abroad in Longyearbyen, Svalbard, from where most air transportation for the proposed mine will arrive,[6] and Akureyri, Iceland, which is expected to be the port of departure for most shipping (Ironbark 2015:6). If environmental concerns had been the decisive optic for determining the local scale, Svalbard could also have been a relevant place to conduct public hearings, as it is located less than 1,000 km east of Citronen Fiord.

Flexibility paves the way to independence

During the hearings, environmental concerns were almost entirely articulated in relation to the local and national scales, where the possible risk to marine mammals and fishes constituted the greatest worry,[7] primarily because they are the main ingredients in traditional Greenlandic food, at the core of Greenlandic culture (see Gad 2005; Olsen 2011). This reflects the idea, promoted by the current coalition government, that '[t]he sustainability principle is a Greenlandic invention where the first provisions on sustainable exploitation of animals were formulated in the Thule Laws, and these principles must be upheld and honoured' (Siumut, Inuit Ataqatigiit, Partii Naleraq 2016:14). While the sustainability concept articulated in the Thule Laws points to the importance of environmental and social preservation, the communication regarding Citronen Fiord generally puts greater emphasis on development and economic gain, echoing the decision not to sign the Paris Agreement (see Bjørst, this volume), as the Citronen Fiord mine alone is expected to cause a 20 per cent increase in Greenland's current total CO_2 emissions.[8] Via the Environmental Impact Assessment – made in the process prior to the IBA – the mining company and respective authorities scrutinize how to mitigate the potential environmental impacts, but the responsibility for ultimately doing so is placed on the shoulders of the mining companies – a decision characterized as 'pioneering' in the strategy (Naalakkersuisut 2014a:66).

The prioritization is clear: mining must be realized in order to sustain the national economy (Naalakkersuisut 2014a:45) even though it may have some negative environmental impacts. As put by then Minister for Finance, Energy, and Foreign Affairs, Vittus Qujaukitsoq from the social democratic party Siumut, when interviewed about why mining is now allowed in Citronen Fiord despite the past ban on any economic activities in the area: 'The national park is the most pristine part of our country, and it is important for all animal and plant life. But we have to be flexible if new business opportunities arise' (Breum 2016:42–43). Such flexibility is legitimized by the overall goal of future independence, which is dependent on new significant sources of income. This was repeatedly emphasized by Qujaukitsoq, who stated: 'We will never be independent if we leave our resources untouched',[9] and by then Minister for Business, Labour, and Trade, Randi Vestergaard Evaldsen from the social liberal party Demokraatit, who argued that 'by giving companies the opportunity to establish mining projects we are paving the way towards independence'.[10] One displeased citizen protested that 'I feel Greenland is just being sold',[11] and another asked, 'Could we postpone such projects until Greenland has been declared independent? We hear about many projects which are never being realised.'[12] However, the dominant storyline throughout the hearings was that the Citronen Fiord mine and other similar proposed projects must be established for the benefit of fulfilling the Greenlandic ambition of independence.[13] Hence, the national scale and the economic sector are the ones prioritized.

The panda in the room: possible challenges to social sustainability

Part of sustaining the economy via mining is to ensure that local workers are hired.[14] This is expected to have a significant positive effect on Greenland's social sustainability.

As stated in the IBA, at least 21 per cent of the mine's total workforce must be Greenlandic during the first four years, increasing to 50 per cent during the subsequent two years, and peaking at 90 per cent in year seven out of the expected 15 years of mining activity in Citronen Fiord. This means 63 of the 300 workers in the first few years, and 423 of 470 when productivity peaks (Ironbark Zinc A/S *et al.* 2016:50–52).[15] Of these, as many as 60 per cent are expected to be unskilled workers on quasi-specialist courses (Naalakkersuisut 2014a:82). If there are, however, not a sufficient number of Greenlanders with the required skills, further steps must be taken to upgrade the qualifications of the Greenlandic workforce so the number of local employees stays 'as high as possible' (Mineral Resources Act, §18). Ironbark will establish an educational fund and a business development fund that will help to finance these initiatives (Ironbark *et al.* 2016:72).[16] Furthermore, a social and cultural fund will be established with the purpose to 'promote and support Greenlandic culture' (2016:73).

The relevance of the social and cultural fund is not further explained in the IBA, nor in the strategy, but in line with past years' mining debates in the Greenlandic parliament[17] it may be related to the awareness of the possible need for a large number of foreign workers in order to realize Greenland's mining plans. This need is also briefly articulated in the mining strategy, where it says that 'foreign workers will supposedly be recruited from abroad to some extent, e.g. from Asia' (Naalakkersuisut 2014a:12). In the parliamentary debates, it is widely acknowledged that such imports of, for instance, Chinese workers may encompass a possible risk to the sustainability of Greenlandic culture. However, this is generally accepted by all parties as a necessary means to serve the overarching goal of independence (Jacobsen 2015:112). In cases where the Danish Government co-authors Greenland's imagined future – such as the 2012 debates about the so-called large-scale law and the *Kingdom of Denmark Strategy for the Arctic 2011–2020* – the risk connected to this expected development is more clearly emphasized. The latter highlights that '[i]t will also be a significant challenge for Greenland to develop policies which, apart from the goal of social and societal-related sustainability, deal with the prospect of significant foreign labour migration' (Espersen *et al.* 2011:23). This reflects a perception of how a country like Greenland – with only 56,000 citizens – may be more susceptible to outside cultural influences by even a few thousand immigrants than countries with higher populations. At the same time, it may also be related to a development in international politics where Chinese investments are sometimes seen as possible threats to Western domination, and with Greenland geographically located on the North American continent, increased Chinese presence here is possibly observed with more attention by the US and its ally Denmark than it would be elsewhere.

Danish and Greenlandic sovereignty negotiations

The attention towards China pertains both to the possible physical presence of the many workers, and to the ownership of the mines. As explained at the public hearing in Tasiilaq by the CEO of Ironbark, John Charles Downes, the ownership of Ironbark is shared between Nyrstar (23 per cent) from Belgium, Glencore (10 percent) from Switzerland, and 'various individual persons or companies' (Tasiilaq 2016). When digging deeper into the ownership structure, it appears that one of these 'various companies' is the China Non-Ferrous Metal Mining Group Co. (CNMC), which is expected to obtain 20 per cent of the shares in return for financing 70 per cent of the establishment of the Citronen Fiord mine (Krebs 2014). As CNMC does business 'with the support and kind care of Chinese government and industrial association' (www.CNMC.com), it is clear that the company's involvement is closely linked to the country's policies and strategies for the Arctic and beyond (see Sørensen 2017). Together with similar Chinese engagement in the development of the proposed iron mine in Isua (western Greenland) and the much-debated uranium and rare earth elements project in Kuannersuit (southern Greenland), China's participation in Citronen Fiord has fuelled Danish geopolitical concerns. This is, among others, reflected in the Danish Defence Intelligence Service's risk assessment, where Chinese investments are mentioned as something with significant possible influence on Greenlandic society, both via the mere physical presence and the increased risk for political interference and pressure (Forsvarets Efterretningstjeneste 2017:45).

Denmark's concern with China's geopolitical presence has subsequently led to Greenlandic dissatisfaction with its former colonizer, re-activating questions about sovereignty. Following the news about the Chinese engagement, Denmark's Ministry of Foreign Affairs emailed an unsolicited approval of the proposed Citronen Fiord project to their Greenlandic colleagues, stating that, in its present form, it does not conflict with the Kingdom's collective security, defence, and foreign affairs interests. The Greenland Government perceived the email as an unnecessary provocation and furiously responded that it conflicts with the delegation of competences and it, thus, is without meaningful content (Sørensen 2016). This incident should be interpreted in light of two previous cases fuelling the dissatisfaction: first, the debate about whether uranium export is solely a matter of mineral rights or also a matter of security – and, thus, also a Danish area of responsibility – had challenged official Danish–Greenlandic relations since the fall of 2013. Second – and just one week prior to the email – a journalist revealed how Denmark's Prime Minister, Lars Løkke Rasmussen, had rejected an offer by a Chinese mining company to buy an abandoned military base in Kangillinguit (southern Greenland) (Brøndum 2016).

Altogether, these cases touched upon the limits of Greenland's current level of self-determination and indicated how Denmark still upholds sovereignty in Greenland. On the other hand, Denmark's explicit acceptance of Greenland's decision to allow mining in the national park simultaneously shows that the Self-Government Act pertains to all of Greenland – something that has previously been questioned by a Danish MP from the right-wing Danish People's Party, Søren Espersen (tv2.dk 2008).[18] Named after a Danish freedom fighter of the Second World War (Laursen 1955:265), and part of the geographical foundation of the Danish Realm's extensive territorial claim to the Lomonosov Ridge (Jacobsen and Strandsbjerg 2017), Citronen Fiord is in the hands of Greenland, which has the right to establish a mine in the pursuit of a more sustainable national economy.

The national economy is the prioritized scale and sector in Greenland's mining strategy and in the hegemonic storyline throughout the Citronen Fiord public hearings. The reason for this is that, in the paramount aim for independence, the national economy needs to be sustained in order to render superfluous the block grant from Denmark. Possible threats to social sustainability from, for example, large numbers of foreign workers with a different culture and threats to hunting traditions via possible negative environmental impacts are widely accepted as necessary sacrifices serving this higher purpose. Greenland's independence aspirations are closely linked to a Westphalian sovereignty perception, so when Denmark reminds Greenland of its formal sovereignty by sending unsolicited approval of the mining plans in the national park, it triggers postcolonial concerns in the Greenland administration. As such, to sustain the national economy via mining is closely linked to enhancing and expanding sovereignty – a cardinal concern reflected in the very first lines of the 2016 coalition government agreement, where it says: 'Greenland is irrevocably on its way to independence and this process requires not only political stability but also national unity' (Siumut, Inuit Ataqatigiit, Partii Naleraq 2016:2). As stated in Greenland's mining strategy, everyone should partake in the sustainable development of Greenland's mineral resources to the benefit of the country. Though there are, indeed, opposing voices sceptical towards mining's possible influence on Greenland, these are marginalized in the national strategy and in the public hearings regarding the Citronen Fiord project. Sustainable national unity involves both territorial integrity and societal accord.

Nunavut: social licence to drill towards devolution

In Nunavut's mining strategy, *Parnautit: A Foundation for the Future*, the sustainability concept is mentioned 20 times across 64 pages, often in central parts of the strategy. The first page states that the strategy serves as

> the plan of the Government of Nunavut to create opportunities for the future self-reliance of Nunavut and Nunavummiut through the sustainable development of our mineral resources. It is intended to guide that development in the period leading up to the devolution of management responsibilities for lands and resources from the federal government.
>
> (Government of Nunavut 2009:5)

As in Greenland, mining is ascribed much importance as a key element to increased self-determination, but while the words used in the neighbouring mining discourse when describing Greenland's present and future status are 'country' and 'independence', the terminology used in Nunavut is instead 'territory' and 'devolution'. This reflects the different directions of their respective postcolonial developments,[19] while simultaneously indicating that both governments perceive mining as the most likely economic contributor to increased autonomy.

Sustaining the territorial economy is at the core of Nunavut's official mining discourse, but unlike the 'flexibility' that Minister Qujaukitsoq pleaded for when explaining the necessity of allowing mining in Greenland's national park, environmental concerns are more explicitly addressed in Nunavut. After stating that 'Nunavut's future economic viability, and the improvement of the quality of life of Nunavummiut, will depend on the development of these known and yet to be discovered resources' (Government of Nunavut 2009:44), the strategy emphasizes that 'Nunavut's arctic environment is fragile, however, and Nunavummiut will not tolerate development that has unacceptable environmental impacts' (Government of Nunavut 2009:45). The line between acceptable and unacceptable is not further explained, but mining companies are advised to pay special attention to public opinion, as the social licence is inevitable for a successful operation. The social licence is obtained through involving the local community by, for example, hiring local workers which can benefit social sustainability, and through minimizing the environmental impact by 'treating their project as a "temporary use of the land", and fully rehabilitating the site for other uses after mining or exploration is complete' (Government of Nunavut 2009:45). Thus, the environmental and social sectors are also given high priority in the assessment of how a potential mine may contribute to the development of all of Nunavut and the individual local communities.

Inevitable conflicts and proactive resolutions

It may be a delicate balancing act to sustain both the environment, the economy, and the social sector when mining. This is acknowledged by the Government of Nunavut, which clearly expresses that '[i]nevitably, the push towards development of mineral resources and the need for environmental protection will create conflicts' (Government of Nunavut 2009:45). In this light, a Planning Commission is currently working on a comprehensive Land Use Plan[20] in order to 'proactively resolve potential conflicts between mineral exploration parties requiring access to land, and wildlife and community uses' (Nunavut Planning Commission 2016:42). While this initiative presents a broader scope, including sectors other than mining, the individual projects aim at proactively settling disagreements via public hearings in the areas deemed most affected by the proposed mining activity. In the case of the Mary River iron mine, the five communities of Igloolik (155 km to the mine), Pond Inlet (160 km), Hall Beach (192 km), Arctic Bay (280 km), and Clyde River (415 km) were pointed out as the places in the 'immediate vicinity' with 'long term social, economic and environmental ties to the proposed ERP [Early Revenue Phase] area' (NIRB 2014:31).

In July 2012, public hearings took place in Iqaluit (1,000 km), Igloolik, and Pond Inlet, while in 2014 – when considerable changes to the plan required new hearings – five days of hearings were held in Pond Inlet. The reason why they were only held in one location was to limit the high transportation costs and because Pond Inlet's environment and 1,500 citizens were deemed potentially most affected by the changes brought by the mine, including accompanying infrastructure[21] and the arrival of 1,700–2,700 people in the construction phase and about 950 in the operation phase (George 2012). The other communities identified as being potentially affected were represented by five invited representatives, each from different demographic groups, including hunters and trappers organizations, hamlet councils, youth groups, women's groups, and elders' societies (NIRB 2014:54). A total of 38 community representatives participated, along with the Pond Inlet representatives in the two days of technical presentations and the subsequent three days of community roundtable sessions (NIRB 2014:54). In 2012, a total of 41 community representatives from Arctic Bay, Cape Dorset, Clyde River, Coral Harbour, Grise Fiord, Hall Beach, Resolute Bay, and Kimmirut participated, together with three Iqaluit representatives in the final hearing in Iqaluit (NIRB 2012:57).[22]

When a place is labelled as being in the 'immediate vicinity', it also means that its citizens have better chances of being employed in or in relation to the mine. This was often highlighted by the local representatives as the main reason why they are ultimately in favour of the project despite accompanying potential risks – such as negative environmental impacts especially connected to the shipping[23] of the iron, abusive use of alcohol and drugs (NIRB 2014:137),[24] and other related consequences known from similar fly-in fly-out camp environments (NIRB 2014:138).[25] The representative of Hall Beach, Abraham Qammaniq, for example, said in his closing remark that '[w]e can support the Mary River project, because we have observed that it's already been beneficial to our community' (Pond Inlet 2014: vol. 5, 1065). In a similar manner, but with greater attention directed towards the younger and future generations, Igloolik representative Josiah Kadlusiak emphasized that 'a lot of youth don't have employment … we want the project at Mary to proceed so people can have employment and generate revenue for many' (Pond Inlet 2014: vol. 5, 1068–1069).[26] In this way, the local representatives do, indeed, represent the local interests in their speeches in which the national, regional, and global scales are almost completely left out of the communication in favour of a unilateral focus on social sustainability through meeting the basic needs of locals by offering new job opportunities.

IQ: *'sustainable development is not a fixed understanding'*

In the North Baffin region where Mary River is located, 94 per cent of the population are Inuit, and Inuktitut is the prevalent language with some monolingual speakers, ranging from 6 per cent in Hall Beach to 24 per cent in Igloolik (NIRB 2014:31–32).[27] The ethnic majority and respect for Inuit traditions are central in the Nunavut mining discourse, mirrored in the name of the IBA, which has an extra *I* indicating that it is, indeed, an *Inuit* Impact Benefit Agreement (IIBA) (see Nunavut Land Claims Agreement article 26). Though this makes it clear who ultimately *should* benefit from the mine, a discrepancy between the modern industry's way of thinking and traditional Inuit knowledge is where most of the dissonances were articulated during the hearings. As Gamailie Kilukishak, an Elder from Pond Inlet, said: 'I think we have two different views. I'm 80 years old. … I have been out hunting as far as 2005. I have lived two different lifestyles, the old and the new. I have never done any research, but I have just had my traditional knowledge, and that's my research' (Iqaluit 2012:232). In response, Mike Setterington from Baffinland Iron Mines Corporation, ensured that Inuit Qaujimajatuqangit – meaning 'Inuit knowledge' and often abbreviated 'IQ' – played a crucial role in developing the mining plans in a sustainable manner: 'All that information that I shared in the baseline report was basically passed on and shared with me from the hunters in the communities, so I would say that baseline report is probably 80 percent local and traditional knowledge' (Iqaluit 2012:230). This shows how respect for Inuit traditions is an important component in the social contract necessary before drilling can commence. At the same time, it is, however, not clear from the hearings or the background documents exactly what the ideal balance between IQ and science is. Instead, the assessment seems to be in the hands of the local representatives at the public hearings.

In Nunavut's official mining strategy and in the Land Use Plan, the dominating Western WCED understanding of sustainable development is particularly visible, but when IQ is included, the hegemonic perception is contested: '[s]ustainable development is not a fixed understanding. As communities change, their relationship with the land and with each other will continue to develop and evolve' (Nunavut Planning Commission 2016:41). Despite this contestation, the Inuktitut translation of 'sustainable' is quite clear. *Ikupik*, as it is called, means 'to conserve and not take all at once; what is brought in from a hunt. Everyone takes a piece for their family, ensuring there is enough to go around' (Nunavut Planning Commission 2016:42). As in the Thule Laws, the sustainability principle is closely linked to food security and social sustainability through covering the basic needs of the local community.[28] This definition makes sense for hunting and harvesting of renewable resources, but in the case of non-renewable minerals this traditional understanding is open to interpretation – and even more so when adding the perception that the concept is in flux.

During the hearings, neither the traditional Inuit definition nor the contestation of the hegemonic WCED understanding were articulated. Instead, utterances about sustainability usually referred to one particular sector, where local representatives emphasize the importance of sustaining the local communities, and the official authorities put more weight on the sector closest to their respective area of responsibility. In this way, the executive director of the Nunavut Planning Commission, Sharon Ehaloak, encouraged *economic* sustainable development (Pond Inlet 2014: vol. 1, 323), and the environmental specialist of Parks Canada, Allison Stoddart, only talked about *environmental* sustainability (Pond Inlet 2014: vol. 4, 842, 863, 872), while Karen Costello – the resource director of the Nunavut Office of Aboriginal Affairs and Northern Development – said that her mandate is to 'improve social and economic well-being and develop healthier, more sustainable communities' (Pond Inlet 2014: vol. 2, 407). While they all agree that if the mine should be established it needs to be in a sustainable manner, the nuances of the different prioritized sectors demonstrate how they may represent different political agendas. Consequently, a compromise needs to be found to balance these different considerations in the development of Nunavut – a development which is predominantly human-centred, as all parts in favour of the mine agree that economic gain should benefit some size of a social entity; the individual, the community, the company, the territory, and the nation.

Explicit acceptance of Canadian sovereignty

While the local scale is the one most often emphasized in the communication about Mary River, the Canadian Federation's sovereignty is also present in the communication regarding this and other similar projects in Nunavut. Only once was the issue directly mentioned during the public hearings, when Caleb Sangoya from the Pond Inlet Mary River Project Committee said: 'Baffinland has already stated that Mary River will benefit Canada, and it will help with Canadian sovereignty in the north … the federal government is going to make a lot of money of this' (Pond Inlet 2014: vol. 4, 876). In a similar manner, the Nunavut Land Use Plan briefly, but clearly, states: '[t]he Commission's Objective is to respect and provide for Canada's sovereignty over Canadian Arctic Waters' (2016:39), leaving no doubt that the Government of Nunavut explicitly accepts that Nunavut is a territory under the Canadian Federation, who at the end of the day has the final say.

The waters referred to are a part of the Northwest Passage (NWP), where the shipping of iron from the Mary River mine takes place. The NWP is claimed by Canada as 'internal waters' via the 1970 Arctic Waters Pollution Prevention Act – which made Canada responsible for the environmental regulation of Arctic seas up to 100 miles from the extended Canadian coastline – and contested by, among others, the US, which perceives it as an international strait (Byers 2009).[29] The increased commercial use caused by the Mary River mine contributes to Canada's occupancy of the area in a more peaceful manner than, for example, the Distant Early Warning Line and other military-related installations do. In this way, the project does not only sustain the social and economic sectors on the local and territorial scales in Nunavut, but also contributes to sustaining Canada's sovereignty in the Arctic.

Though Ottawa's indirect role in the Mary River project is officially accepted by Nunavummiut, other cases of direct involvement have caused widespread dissatisfaction with the subordination if Nunavut has not been welcomed in the decision-making process. A recent example is Barack Obama and Justin Trudeau's 2016 moratorium on oil and gas drilling in North American Arctic waters, which then Premiers of Nunavut and Northwest Territories – Peter Taptuna and Bob McLeod[30] – called a step backwards in the devolution progress as they were only given two hours' notice before the official announcement (Dusen 2016). In response, the two Premiers issued a joint statement arguing that

> [t]he economies of the two territories are small and depend heavily on resource development as the major contributor to GDP and source of jobs and income for their residents at the present time … All Canadians deserve to share in the opportunities and benefits of living in a sustainable and prosperous Canada.
>
> (Taptuna and McLeod 2016)

In this way, they protested against the Trudeau administration's overruling of the two northern territories' interests, while simultaneously acknowledging that they were, indeed, also Canadians. This serves as a good example of the noticeable difference between the Nunavummiut and the Greenlandic discourses about sustainable development.

Nunavut's mining strategy and draft Land Use Plan both clearly articulate the inevitable conflicts that the question of mining cause in a society, where the preservation of hunting grounds traditionally equals the survival of the people inhabiting the territory. As such, the Nunavut Government makes room for the exchange of different opinions, while trying to proactively resolve some of the expected disputes. The cultural specific considerations are underlined by the fact that an IIBA must be conducted, while respect for Inuit knowledge is essential in the process towards concluding a social licence, necessary before any mining activity may take place. The sustainability concept is frequently mentioned in both the key documents and the public hearings in connection with the Mary River mine. While the WCED definition is contested, and described as a concept in flux, the communication generally adopts the hegemonic definition in its aim to balance the considerations of the environmental, economic, and social sectors in the development of the individual, the local community, the company, Nunavut, and Canada – a development which essentially is human-centred as these different scales of social entities are the ones expected to benefit from the mining activities. The Canadian Federation is only mentioned a few times in the communication, but when it is, Nunavut's subordination under Canada's continuous sovereignty is noticeably underlined. Following this, Nunavut is discursively placed on a territorial and – unlike Greenland – not on the national scale, which is reserved for Canada. Instead, territorial devolution from Ottawa and Indigenous cultural considerations are the main concerns, while economic gain for the benefit of the local community, single family, and individual is more often articulated as the main reasons by those in favour of the Mary River mine.

Conclusion

Nunavut's and Greenland's postcolonial developments have been closely related to the acquisition of the right to exploit their own mineral resources. In Greenland, this parallel process seems to continue, while mining in Nunavut is subject to undisguised conflicts between primarily economic and environmental priorities, proactively addressed by the authorities in their aim to direct the territory towards a future where all Nunavummiut feel at home. In both cases, the social sustainability is given high priority. While hunting traditions are emphasized as central to each respective collective identity – which in Nunavut's case unquestionably equals an Inuit identity[31] – the striving for independence is the overarching goal legitimizing mining in Greenland's national park, despite a past ban on any economic activities in the area. In Nunavut, more attention is ascribed to the local community that shall benefit economically and socially, while 'the national' is a label exclusively reserved for the Canadian Federation. Instead, Nunavut is described as a territory that loyally respects Canada's sovereignty, even though decisions such as the 2016 moratorium on hydrocarbon exploitation in the North American Arctic waters was signed without prior consultation with the territorial governments.

The analyses show how sovereignty is closely connected to the question of who gets to decide what to sustain, which is particularly visible in the cases of Nunavut and Greenland, where exploitation of their natural resources is perceived as a potential core contributor to a more viable economy and increased self-determination. The concept of sustainability is, to some degree, contested by Indigenous definitions, but in both discourses the WCED's 1987 description of sustainable development is generally the main point of reference, hence reproducing the hegemonic perception. The defining nuances rendering visible how priorities are made, responsibilities are distributed, and what consequences these decisions may have in the future are, however, found in the weighing of different scales and sectors, as one often takes precedence over the others. In Greenland, this weight is primarily put on the national economy, while the Nunavut mining discourse gives more precedence to local social sustainability. These prioritizations do, however, sometimes overlap, causing the spillover of impacts and benefits across sectors and scales.

Notes

1 I owe special thanks to Letia Obed for her generous hospitality and wise assistance in connection with my field trip to Iqaluit, Nunavut, during November 2017. Thanks to her and the many interviewees from the Government of Nunavut's Departments of Executive and Intergovernmental Affairs, Economic Development and Transportation and the Devolution Secretariat. Thank you also to Stephanie Meakin from the Inuit Circumpolar Council, June Shappa and Bruce Uviluq from Nunavut Tunngavik Incorporated, Kenn Harper, and – not least – Michael Byers, who helped me in the right direction and provided some decisive supervision during my research stay at University of British Columbia in the spring of 2017.

2 NTI also has the right to half of the first $2 million of any resource revenues collected from public lands and 5 per cent of any additional resource revenues, which they consider as Inuit patrimonial property (Nunavut Tunngavik Incorporated 2011:7.4).

3 References to the Greenland mining documents have been translated from Danish to English by the author of this chapter.

4 The public hearings have, however, been widely criticized for being a monologue in which the public's possibility for participation has been quite limited, rather than a dialogue, (Ackrén 2016:3–4, 12; Nuttall 2015:105).

5 The number of attendees has not been made public.

6 As asked by Kim Petersen during the public hearing in Qaanaaq.

7 Articulated by Marius Didriksen in Qaanaaq, Henrik Lyberth in Kangerlussuaq, Gunnar Frederiksen and Johanne Høegh in Qaqortoq, and Niels Amiinnaq in Tasiilaq.

8 Stated by Minister Vittus Qujaukitsoq at the public hearing in Sisimiut.

9 Vittus Qujaukitsoq, at the public hearing in Ilulissat. He made a similar statement at the public hearing in Nuuk.

10 Randi Vestergaard Evaldsen, at the public hearing in Qaqortoq.

11 Hans Josvassen, in Tasiilaq.

12 Niels Nielsen, in Iulissat. A similar utterance was made by Sofie Kielsen at the public hearing in Qaqortoq.

13 Also pointed out by Bjørst (2015:39), Nuttall (2012:116; 2013:380; 2015:110), Kristensen and Rahbek-Clemmensen (2017:48), and Rasmussen and Merkelsen (2017:84).

14 Emphasized by Johan Uitsatikitseq in Tasiilaq, Astrid Bro in Kangerlussuaq, and Jens Peter Kielsen and Nikolaj Joelsen in Qaqortoq.

15 Ironbark has also agreed to have at least eight Greenlandic apprentices at any given time.

16 The three funds are expected to be 600,000 DKK in the first year, 1.2 million DKK in the subsequent three years, and 1.5 million DKK thereafter, based on an expected export of 360,000 tonnes of ore concentrate (Ironbark Zinc A/S *et al.* 2016:72).

17 For example, the FM2014/68 debate in the Greenlandic parliament, Inatsisartut, about the future need for foreign labour if large-scale mines are established.

18 In the interview, Espersen said '[n]obody has ever lived there, and it has never been Greenlandic'.

19 Also pointed out by, among others, Loukacheva (2007:16).

20 The final Nunavut Land Use Plan will be published in 2022 (Frizzell 2018). The one referred to is the latest draft from 2016.

21 The 2012 plans to construct a 110km railway were cancelled in 2014 due to a fall in ore prices. Instead, the iron is transported on 200-tonne dump trucks via a tote road to a port in Milne Inlet, where it is then shipped. Following new expansion plans for Phase 2 of the project, the suggested railway is again on the table, along with five locomotives, 176 rail cars, and winter sealifts of freight through ice, altogether making possible an increase in production from 4.2 to 12 million tonnes per year (Mtpa). In the fall of 2017, interested parties and persons must submit their comments, upon which the planning commission will determine whether new consultations are necessary (Bell 2017).

22 The number of civilians present at the hearings is only registered on handwritten documents that can be accessed online with assistance from nirb.ca. (One must expect to wait several months before receiving an answer from NIRB.) Some names are not possible to identify because of unclear handwriting and/or because they were written in Inuktitut. Furthermore, some civilians also wrote their occupation even though they only represented themselves at the meeting. Altogether, this makes it difficult to determine the exact number of civilians present at the meeting. From the transcripts, it is, however, clear that it was usually only the selected representatives who spoke on behalf of the communities, which is different from the public hearings in Greenland.

23 Articulated by Rhoda Katsak on behalf of the Government of Nunavut (Pond Inlet 2014, vol. 2, 378), Caleb Sangoya on behalf of the Pond Inlet Mary River Project Committee (Pond Inlet 2014, vol. 4, 758–759), Tim Anaviapik-Soucie representing the youth of Pond Inlet (Pond Inlet 2014, vol. 5, 1053–1054), Jaypetee Akeeagok representing Grise Fiord (Pond Inlet 2014, vol. 5, 1061), and Simon Idlout on behalf of Resolute Bay (Pond Inlet 2014, vol. 5, 1062). A master's thesis concludes that the negative impact on caribou in the area was not thoroughly assessed in the Environmental Impact Assessment which, according to the author, privileged the perspective of the mining industry (Williams 2005).

24 Also articulated by Tim Anaviapik-Soucie representing the youth of Pond Inlet (Pond Inlet 2014, vol. 3, 683).

25 Bowes-Lyon *et al.*'s study of the past Polaris and Nanisivik mines in Nunavut also point to alcohol as one of the reasons why the Inuit did not become more involved in the mines (2009:384). They furthermore write that education was the only positive long-term social impact of the mines (Bowes-Lyon *et al.* 2009:385), but that the signing of the Nunavut Land Claims Agreement is more aligned with the sustainable development principles, which they see as 'fortunate' (Bowes-Lyon *et al.* 2009:392).

26 A similar statement was also made by the representative of Grise Fiord, Jaypetee Akeeagok (Pond Inlet 2014, vol. 5, 1062).

27 In response to a question by Clyde River representative Jacobie Iqalukjuak, Baffinland promised that there will also be jobs available for monolingual Inuktitut speakers (Pond Inlet 2014, vol. 5, 1074–1075).

28 Also emphasized by Arctic Bay representative Olayuk Naqitarvika (Pond Inlet 2014, vol. 5, 1071–1072) and in the Nunavut Land Use Plan, where the second goal summarizes: 'the environment, including wildlife and wildlife habitat, is of critical importance to the sustainability of Nunavut's communities, Inuit culture and the continuation of a viable long term economy' (Nunavut Planning Commission 2016:18).

29 According to Steinberg (2014), a possible solution to the dispute could be to designate the NWP as Canada's territorial sea, which would satisfy both political objectives.
30 At the Arctic Circle conference in Reykjavik in 2017, McLeod went a rhetorical step further in his speech, stating that the moratorium was an example of colonialism still present in Canada (McLeod 2016).
31 As newly appointed Premier of Nunavut, Paul Quassa stated that independence is more likely to be a pan-Arctic Inuit project led by ICC than something Nunavut should strive for (Jacobsen 2017).

Bibliography

Ackrén, M. (2016). Public consultation processes in greenland regarding the mining industry. *Arctic Review on Law and Politics*, 7 (1), pp. 3–19.

Bell, J. (2017). Nunavut Planning Commission gets started on Mary River expansion. *Nunatsiaq Online*, News, 6 September 2017. Available from: www.nunatsiaqonline.ca/stories/article/65674nunavut_planning_commission_gets_started_on_mary_river_expansion (accessed 28 December 2017).

Bjørst, L.R. (2015). Saving or destroying the local community? Conflicting spatial storylines in the Greenlandic debate on uranium. *The Extractive Industries and Society*, 3, pp. 34–40.

Bowes-Lyon, L.-M., Richards, J.P. and McGee, T.M. (2009). Socio-economic impacts of the Nanisivik and Polaris mines, Nunavut, Canada. In: Richards, J.P. (ed.), *Mining, Society, and a Sustainable World*, New York: Springer, pp. 371–396.

Breum, M. (2016). Verdens største nationalpark. *National Geographic*, 6, pp. 20–45.

Brøndum, C. (2016). Danmark forhindrer kinesisk opkøb af marinebase i Grønland. *Defence Watch*, 14 December 2016. Available from: www.defencewatch.dk/danmark-forhindrer-kinesisk-opkoeb-marinebase-groenland (accessed 20 December 2017).

Byers, M. (2009). *Who Owns the Arctic? Understanding Sovereignty and International Law in the North*. Berkeley, CA: Douglas & McIntyre.

Dusen, J.V. (2016). Nunavut, N.W.T. premiers slam Arctic drilling moratorium. *CBC News North*, 22 December 2016. Available from: www.cbc.ca/news/canada/north/nunavut-premier-slams-arctic-drilling-moratorium-1.3908037 (accessed 29 December 2017).

Espersen, L., Johannesen, K.L.H. and Kleist, K. (2011). *Kingdom of Denmark Strategy for the Arctic*, Denmark's Mininstry of Foreign Affairs, Greenland's Department of Foreign Affairs, Faroe Island' Ministry of Foreign Affairs. Available from: http://um.dk/en/foreign-policy/the-arctic (accessed 28 July 2017).

Frizzell, S. (2018). Nunavut isn't getting a territory-wide land use plan any time soon, mining symposium hears. *CBC News*, 13 April 2018. Available from: www.cbc.ca/news/canada/north/nunavut-land-use-plan-1.4606832 (accessed 16 April 2018).

Gad, U.P. (2005). *Dansksprogede grønlænderes plads i et Grønland under grønlandisering og modernisering*. Copenhagen: Eskimologis Skrifter.

George, J. (2012). Nunavut's Baffinland iron mine project moves into final hearings. *Nunatsiaq Online*, 1 March 2012. Available from: www.nunatsiaqonline.ca/stories/article/65674nunavuts_baffinland_iron_mine_project_moves_into_final_hearings (accessed 28 December 2017).

Humrich, C. (2016). Sustainable development in Arctic international environmental cooperation and the governance of hydrocarbon related activities. In: Pelaudeix, C.

and Basse, E.M. (eds), *The Governance of Arctic Offshore Oil and Gas*. Farnham: Gower Publishing.

Jacobsen, M. (2015). The power of collective identity narration: Greenland's way to a more autonomous foreign policy. In: Heininen, L., Exner-Pirot, H., and Plouffe, J. (eds), *Arctic Yearbook 2015: Arctic Governance and Governing*. Akureyri, Iceland: Northern Research Forum, pp. 102–118.

Jacobsen, M. (2017). Nunavut's new premier is Inuk with capital I. *High North Dialogue*, 20 November 2017. Available from: www.highnorthnews.com/nunavuts-new-premier-is-inuk-with-capital-i (accessed 17 April 2018).

Jacobsen, M. and Gad, U.P. (2017). Setting the scene in Nuuk: introducing the cast of characters in Greenlandic foreign policy narratives. In: Kristensen, K.S. and Rahbek-Clemmensen, J. (eds), *Greenland and the International Politics of a Changing Arctic: Postcolonial Paradiplomacy between High and Low Politics*. London: Routledge, pp. 11–27.

Jacobsen, M. and Strandsbjerg, J. (2017). Desecuritization as displacement of controversy: geopolitics, law and sovereign rights in the Arctic. *Politik*, 20 (3), pp. 15–30.

Kirsch, S. (2010). Sustainable mining. *Dialectical Anthropology*, 34 (1), pp. 87–93.

Krebs, M.L. (2014). Citronen indgår aftale med kinesisk mineselskab. *Kalaallit Nunaata Radioa*, 16 April 2014. Available from: http://knr.gl/da/nyheder/citronen-indgår-aftale-med-kinesisk-mineselskab (accessed 28 July 2017).

Kristensen, K.S. and Rahbek-Clemmensen, J. (2017). Greenlandic sovereignty in practice: uranium, independence and foreign relations in Greenland between three logics of security. In: Kristensen, K.S. and Rahbek-Clemmensen, J. (eds), *Greenland and the International Politics of a Changing Arctic: Postcolonial Paradiplomacy between High and Low Politics*. London: Routledge, pp. 38–53.

Kristoffersen, B. and Langhelle, L. (2017). Sustainable development as a global-Arctic matter: imaginaries and controversies. In: Keil, K. and Knecht, S. (eds), *Governing Arctic Change: Global Perspectives*. Basingstoke: Palgrave Macmillan, pp. 21–41.

Latour, B. (2007). *Reassembling the Social: An Introduction to Actor–Network-Theory*. Oxford: Oxford University Press.

Laursen, D. (1955). Træk af Nordgrønlands opdagelseshistorie III. *Tidsskriftet Grønland*, 7, pp. 258–265.

Li, T.M. (2007). *The Will to Improve: Governmentality, Development and the Practice of Politics*. Durham, NC: Duke University Press.

Loukacheva, N. (2007). *The Arctic Promise: Legal and Political Autonomy of Greenland and Nunavut*. Toronto: University of Toronto Press.

Nuttall, M. (2012). Imagining and governing the Greenlandic resource frontier. *The Polar Journal*, 2 (1), pp. 113–124.

Nuttall, M. (2013). Zero-tolerance, uranium and Greenland's mining future. *The Polar Journal*, 3 (2), pp. 368–383.

Nuttall, M. (2015). Subsurface politics: Greenlandic discourses on extractive industries. In: Jensen, L.C. and Hønneland, G. (eds), *Handbook of the Politics of the Arctic*. Aldershot: Edward Elgar, pp. 105–127.

tv2.dk (2008). DF sår tvivl om Grønlands grænser. *Nyheder*, 11 March 2008. Available from: http://nyheder.tv2.dk/nytomtv/article.php/id-10774642%3Adf-s%C3%83%C2%A5r-tvivl-om-gr%C3%83%C2%B8nlands-gr%C3%83%C2%A6nser.html (accessed 28 July 2017).

Olsen, N.L. (2011). Uden grønlandsk mad er jeg intet. In: Høiris, O. and Marquardt, O. (eds), *Fra vild til verdensborger*. Aarhus, Aarhus Universitetsforlag.

Powell, R.C. and Dodds, K. (eds) (2014). *Polar Geopolitics? Knowledges, Resources and Legal Regimes*. Aldershot: Edward Elgar, pp. 259–576.

Rasmussen, R.K. and Merkelsen, H. (2017). Post-colonial governance through securitization? A narratological analysis of a securitization controversy in contemporary Danish and Greenlandic uranium policy. *Politik*, 20 (3), pp. 83–103.

Schriver, N. (2013). *Grønlands råstofdebat: an komparativ diskursanalyse fra 1975–2012*. Kandidatspeciale, Institut for Statskundskab, Københavns Universitet. Available from: http://ir.polsci.ku.dk/research_projects/arctic-politics/bilag/opg.3.pdf (accessed 16 January 2018).

Sejersen, F. (2015). *Rethinking Greenland and the Arctic in the Era of Climate Change*. London: Routledge.

Skorstad, B., Dale, B., and Bay-Larsen, I. (2017). Governing complexity: theories, perspectives and methodology for the study of sustainable development and mining in the Arctic. In: Dale, B., Bay-Larsen, I., and Skorstad, B. (eds), *The Will to Drill: Mining in Arctic Communities*. New York: Springer, pp. 13–32.

Sørensen, A.K. (1983). *Danmark – Grønland i det 20. århundrede – en historisk oversigt*. Viborg: Nyt Nordisk Forlag Arnold Busck.

Sørensen, C.T.N. (2017). Chinese investments in Greenland: promises and risks as seen from Nuuk, Copenhagen and Beijing. In: Kristensen, K.S. and Rahbek-Clemmensen, J. (eds), *Greenland and the International Politics of a Changing Arctic: Postcolonial Paradiplomacy between High and Low Politics*. London: Routledge, pp. 83–97.

Sørensen, H.N. (2016). Grønland og Danmark uenige om regler på råstofområdet. *Kalaallit Nunaata Radioa*, 23 December 2016. Available from: http://knr.gl/da/nyheder/grønland-og-danmark-uenige-om-regler-på-råstofområdet (accessed 28 July 2017).

Speca, A. (2012). Nunavut, Greenland and the politics of resource revenues', *Policy Options*, 1 May 2012. Available from: http://policyoptions.irpp.org/magazines/budget-2012/nunavut-greenland-and-the-politics-of-resource-revenues (accessed 30 July 2017).

Steinberg, P. (2014). Steering between Scylla and Charybdis: the Northwest Passage as territorial sea. *Ocean Development & International Law*, 45 (1), pp. 84–106.

Tsing, A.L. (2000). Inside the economy of appearances. *Public Culture* 12 (1), pp. 115–144.

Vahl, B. and Kleemann, N. (2017). *Greenland in Figures*. Nuuk: Statistics Greenland.

World Commission on Environment and Development (WCED) (1987). *Our Common Future, Report from the 'Brundtland' World Commission on Environment and Development*. Oxford: Oxford University Press.

Williams, A. (2005). Governmentality and mining: analyzing the environmental impact assessment for the Mary River mine, Nunavut, Canada. Thesis submitted to the Faculty of Graduate and Postdoctoral Affairs in partial fulfillment of the requirements for the degree of Master of Arts in Geography, Carleton University.

Official documents (Greenland)

CNMC.com 'From the President', China Nonferrous Metal Mining (Group) Co., Ltd. Available from: www.cnmc.com.cn/outlineen.jsp?column_no=12 (accessed 15 January 2018).

FM2014/68 *Forslag til Inatsisartutbeslutning om at Naalakkersuisut til EM14 pålægges at udarbejde en redegørelse, der skal belyse og analysere det fremtidige behov for udefrakommende arbejdskraft i forbindelse med mulige råstofprojekter og storskalaprojekter*. *Redegørelsen bør ligeså beskrive de mulige sociale, kulturelle og praktiske konsekvenser ved en stor tilgang af udefrakommende arbejdskraft. Redegørelsen bedes afsluttet med en perspektivering, der sammenligner arbejdskraftsbehovet med den grønlandske arbejdskrafts kvalifikationer, arbejdsparathed og mobilitet*, Medlemmer af Inatsisartut, Knud Kristiansen og Gerhardt Petersen, Atassut.

Forsvarets Efterretningstjeneste (FE) (2017). *Efterretningsmæssig Risikovurdering 2017*. Available from: https://fe-ddis.dk/SiteCollectionDocuments/FE/Efterretningsmaessige Risikovurderinger/Risikovurdering2017.pdf (accessed 28 December 2017).

Ilulissat (29 November 2015). Available from: http://naalakkersuisut.gl/~/media/Nanoq/Files/Hearings/2015/Ironbark_SIA_EIA_NSI/Referater%20og%20praesentationer%20fra%20moeder/Referat%20Ironbark%20borgerm%C3%B8de%20Ilulissat%20DK.pdf (accessed 25 July 2017).

Ironbark (2015). *Citronen Base Metal Project: Environmental Impact Assessment (Volume 1)*, January 2015 (Rev 6). Available from: http://naalakkersuisut.gl/~/media/Nanoq/Files/Hearings/2015/Ironbark_SIA_EIA_NSI/Documents/4%20Citronen%20EIA%20Ikke-teknisk%20resume_ENG.pdf (accessed 24 July 2017).

Ironbark Zinc A/S, Qaasuitsup Kommunia, Qeqqata Kommunia, Kommunearfik Sermersooq, Kommune Kujalleq, and the Government of Greenland (2016). *Impact and Benefit Agreement (IBA), IBA no. 2016/01, concerning the Citronen Fjord Zinc/Lead project*. Available from: http://naalakkersuisut.gl/~/media/nanoq/files/publications/erhverv/ironbark%20iba/iba%20ironbark%20citronen%20fjordengsep%202016%20signed%20version.pdf (accessed 31 July 2017).

Kangerlussuaq (27 November 2015). Available from: http://naalakkersuisut.gl/~/media/Nanoq/Files/Hearings/2015/Ironbark_SIA_EIA_NSI/Referater%20og%20praesenta tioner%20fra%20moeder/Referat%20Ironbark%20borgerm%C3%B8de%20Kangerlus suaq%20DK.pdf (accessed 25 July 2017).

Naalakkersuisut (2009). Greenland Parliament Act of 7 December 2009 on Mineral Resources and Mineral Resource Activities (the Mineral Resources Act). Unofficial translation. Available from: www.govmin.gl/images/stories/faelles/mineral_resources_act_unofficial_translation.pdf (accessed 31 July 2017).

Naalakkersuisut (2014a). Greenland's Oil and Mineral Strategy 2014–2018. Available from: http://naalakkersuisut.gl/~/media/Nanoq/Files/Publications/Raastof/ENG/Green land%20oil%20and%20mineral%20strategy%202014-2018_ENG.pdf (accessed 25 July 2017).

Naalakkersuisut (2014b). Vores råstoffer skal skabe velstand. Grønlands Grønlands olie- og mineralstrategi 2014–2018. Sammenfatning. Available from: http://naalakkersuisut.gl/~/media/Nanoq/Files/Publications/Raastof/DK/Olie%20og%20Mineral strategi%20DA.pdf (accessed 25 July 2017).

Nuuk (23 November 2015). Available from: http://naalakkersuisut.gl/~/media/Nanoq/Files/Hearings/2015/Ironbark_SIA_EIA_NSI/Referater%20og%20praesentationer%20fra%20moeder/Referat%20Ironbark%20borgerm%C3%B8de%20Nuuk%20DK.pdf (accessed 25 July 2017).

Qaanaaq (20 January 2016). Available from: http://naalakkersuisut.gl/~/media/Nanoq/Files/Hearings/2015/Ironbark_SIA_EIA_NSI/Referater%20og%20praesentationer%20fra%20moeder/Referat%20Ironbark%20borgerm%C3%B8de%20Qaanaaq%20DK.pdf (accessed 25 July 2017).

Qaqortoq (4 February 2016). Available from: http://naalakkersuisut.gl/~/media/Nanoq/Files/Hearings/2015/Ironbark_SIA_EIA_NSI/Referater%20og%20praesentationer%20fra%20moeder/Referat%20Ironbark%20Qaqortoq%20DK.pdf (accessed 25 July 2017).

Sisimiut (28 November 2015). Available from: http://naalakkersuisut.gl/~/media/Nanoq/Files/Hearings/2015/Ironbark_SIA_EIA_NSI/Referater%20og%20praesentationer%20fra%20moeder/Referat%20Ironbark%20borgerm%C3%B8de%20Sisimiut%20DK.pdf (accessed 25 July 2017).

Siumut, Inuit Ataqatigiit, Partii Naleraq (2016). Koalitionsaftale 2016–2018: Lighed, Tryghed, Udvikling. Available from: http://naalakkersuisut.gl/~/media/Nanoq/Files/Attached%20Files/Naalakkersuisut/DK/Koalitionsaftaler/Koalitionsaftale_S_IA_PN_2016_2018.pdf (accessed 25 July 2017).

Tasiilaq (28 January 2016). Available from: http://naalakkersuisut.gl/~/media/Nanoq/Files/Hearings/2015/Ironbark_SIA_EIA_NSI/Referater%20og%20praesentationer%20fra%20moeder/Referat%20borgerm%C3%B8de%20i%20Tasiilaq%20DK.pdf (accessed 25 July 2017).

Official documents (Nunavut)

Government of Nunavut (2009). *Parnautit: A Foundation for the Future – Mineral Exploration and Mining Strategy*. Iqaluit: Department of Economic Development & Transportation. Available from: http://gov.nu.ca/sites/default/files/Parnautit_Mineral_Exploration_and_Mining_Strategy.pdf (accessed 29 July 2017).

Iqaluit (16 July 2012). Nunavut Impact Review Board's final hearing regarding Baffinland's Mary River project proposal, file: 08MN053, Volume 1, Dicta Court Reporting Inc. 403-531-0590.

McLeod, B. (2016). Bob McLeod: 2017 Arctic Circle Assembly, 31 October 2017, Government of Northwest Territories. Available from: www.gov.nt.ca/newsroom/news/bob-mcleod-2017-arctic-circle-assembly (accessed 29 December 2017).

Nunavut Impact Review Board (NIRB) (2012). Final hearing report 2012, Mary River project. Baffinland Iron Mines Corporation. NIRB file no.: 08MN 053.

Nunavut Impact Review Board (NIRB) (2014). Public hearing report Mary River project: early revenue phase proposal. Baffinland Iron Mines Corporation. NIRB file no. 08MN053. Available from: www.nunavut.ca/files/amendments/140317-08MN053-NIRB%2012%208%202%20Public%20Hearing%20Report-OEDE.pdf (accessed 29 July 2017).

Nunavut Planning Commission (2016). Nunavut land use plan (draft). Available from: www.nunavut.ca/files/2016DNLUP/2016_Draft_Nunavut_Land_Use_Plan.pdf (accessed 29 July 2017).

Nunavut Tunngavik Incorporated (2011). Resource revenue policy. Available from: www.tunngavik.com/documents/staffdocs/30%20-%20Resource%20Revenue%20Policy.pdf (accessed 31 July 2017).

Nunavut Tunngavik Incorporated (2013). The Minister of Indian Affairs and Northern Development & Federal Interlocutor for Métis and Non-Status Indians 2010, Agreement Between the Inuit of the Nunavut Settlement Area and Her Majesty the Queen in Right of Canada as Amended. Ottawa, 2010. Available from: www.tunngavik.com/documents/publications/LAND_CLAIMS_AGREEMENT_NUNAVUT.pdf (accessed 30 July 2017).

Qikiqtani Inuit Association & Baffinland Iron Mines Corporation (2013). The Mary River project Inuit impact and benefit agreement. Available from: http://qia.ca/wp-content/uploads/2017/02/qia_-_baffinland_-_iiba.pdf (accessed 31 July 2017).

Pond Inlet (27 January 2014). Nunavut Impact Review Board's hearing regarding reconsideration of Baffinland's Mary River project, certificate number 005, file: 08MN053, Volume 1, Dicta Court Reporting Inc. 403-531-0590.

Pond Inlet (28 January 2014). Nunavut Impact Review Board's hearing regarding reconsideration of Baffinland's Mary River project, certificate number 005, file: 08MN053, Volume 2, Dicta Court Reporting Inc. 403-531-0590.

Pond Inlet (29 January 2014). Nunavut Impact Review Board's hearing regarding reconsideration of Baffinland's Mary River project, certificate number 005, file: 08MN053, Volume 3, Dicta Court Reporting Inc. 403-531-0590.

Pond Inlet (30 January 2014). Nunavut Impact Review Board's hearing regarding reconsideration of Baffinland's Mary River project, certificate number 005, file: 08MN053, Volume 4, Dicta Court Reporting Inc. 403-531-0590.

Pond Inlet (31 January 2014). Nunavut Impact Review Board's hearing regarding reconsideration of Baffinland's Mary River project, certificate number 005, file: 08MN053, Volume 5, Dicta Court Reporting Inc. 403-531-0590.

Taptuna, P. and McLeod, B. (22 December 2016). NWT and Nunavut Premiers react to federal announcement of Arctic oil and gas moratorium. Government of Nunavut. Available from: www.gov.nu.ca/eia/news/nwt-and-nunavut-premiers-react-federal-announcement-arctic-oil-and-gas-moratorium (accessed 29 December 2017).

5 'Without seals, there are no Greenlanders'

Colonial and postcolonial narratives of sustainability and Inuit seal hunting

Naja Dyrendom Graugaard

Since the resurgence of the anti-sealing movement resulted in the 2009 European Union import ban on sealskin products, advocates of Inuit sealing have increasingly described Inuit seal hunting as a sustainable practice. Sustainability has thus surfaced as a concept that invites differing narratives on the meaning and definition of 'sustainable sealing'. In response to the accusations of animal welfare organizations that sealing amounts to 'a cruel, unnecessary waste' (IFAW 2016), the sustainability concept serves as a counter-narrative that articulates the relationship between seals, seal hunting, and Inuit livelihoods as 'sustainable' and thereby destabilizes anti-sealing propositions. In 2006, Greenland's president of the Inuit Circumpolar Council at the time, Aqqaluk Lynge, stated that: 'Our traditional Greenland seal hunt is sustainable. The meat plays an important part of the healthy Arctic diet and the trade of our sealskin coats to Europe helps our local village economies enormously' (Lynge 2006). Similarly, the Inuit Sila organization suggested that Greenlandic seal hunting is '100% sustainable' in its counter-campaign, which was primarily created to inform European citizens about Inuit sealing. Since this latest 'sealing dispute', concepts of sustainable Inuit seal hunting have also made their way into the EU Seal Regime to designate 'accepted exceptions' in a regulation that has otherwise banned imports of all seal products since 2009. This 'Inuit Exception' legalizes import of sealskins to the EU as long as they are products of traditional subsistence hunting (European Commission 2016). The exemption is currently criticized by Inuit: Not only does it deny the right of Inuit seal hunters to engage in the global economy on similar terms as others, but it also fails to recognize that Inuit seal products have long been part of commercial markets due to the processes of colonization (*Angry Inuk* 2016).

Scientific assessments have contradicted anti-sealing narratives by stressing that the majority of seal populations in the North Atlantic are abundant and thriving (Garde 2013:41; Rosing-Asvid 2011). Meanwhile, conceptualizations of seal hunting as 'economically sustainable' in Arctic economies are challenged by the international sealskin market collapse, which rendered sales a deficit enterprise. These present discourses on sustainability and Inuit sealing may appear as a recent convergence, but they have origins in Arctic colonial histories. Thus, this chapter traces 'the seal' through Greenland's history and suggests that varying *sustainability narratives* concerning Inuit seal hunting have been pivotal in sustaining Danish colonization. In this sense, the emergence of concepts of sustainability in relation to Greenland's sealskin industry has not occurred in a vacuum. It is also conditioned by the particular development of Inuit sealing as a colonial undertaking which has sought to transform an Indigenous practice into a monocultural commercialized occupation. This process has been developed and guarded by a colonial narrative that coupled Greenlandic identity with seal hunting (e.g. Rud 2006; Thomsen 1998). In this aspiration for a narrative genealogy, it is possible to elucidate traces of colonial narratives in postcolonial approaches to Inuit seal hunting and to explicate how these narratives are being challenged and reinvigorated. Such interrogation may differ in major ways from other Arctic histories, yet it provides an example of how contemporary sustainability narratives of Inuit seal hunting are shaped by the particular mechanisms of colonization.

In line with the analytical framework of the volume, I treat sustainability as a *political concept* which 'defines and shapes different discourses about future developments' (Gad *et al.*, this volume), here seen through the lens of Greenland's colonial history. In all epochs, the continuity of colonization becomes the central aspiration in the colonial equation of 'sustainability'. Meanwhile, the strategies for sustaining colonization implicate the narratives on Inuit sealing. While the word 'sustainability' does not appear in relation to seal hunting before the postcolonial era, a semasiological approach facilitates an account of the ideas, concepts, and strategies which the (different) meaning(s) of sustainability have produced 'without speaking the word' (Gad *et al.*, this volume). Here, words such as 'maintenance' ('Vedligeholdelse'; Instrux 1782:24) and 'equilibrium' ('Ligevægt'; Rink 1862:16) used in Danish statements about Greenlandic seal hunting convey (some) meaning of the concept of sustainability in the particular colonial structures. In order to explicate how such narrative elements have dominated, transformed, or been reinstituted, I have analysed five periods with different sustainability narratives and policies concerning Inuit sealing: (1) the early colonial period of commercial exploration, (2) the following century of solidifying and 'protecting' sealing as the primary colonial commercial focus, (3) the Rinkian period of preserving and privileging sealing in the late nineteenth century, and (4) the transition from sealing to fishing as the main colonial enterprise from the early twentieth century to the 1960s. Lastly, I have engaged Greenlandic political (re-)narrations of Inuit seal hunting by the Home Rule through the business enterprises of *Great Greenland* (sealskin) and *Puisi A/S* (seal sausage). Consulting primary sources as well as relevant scholarly works, this research is mainly confined to the narrative positioning undertaken by the Danish colonial administration and subsequently by Greenland's Home Rule.

Overall, my analysis primarily interrogates the *dominant* sustainability narratives at a certain level of deconstruction. While this analysis undoubtedly suffers from an unsettling silence of those who lived with the consequences of Danish colonial strategies for 'sustainability', I have attempted to account for some of the ways in which Inuit hunting practices counter colonial sustainability narratives as well as disrupt some historicisms related to sealing. In my forthcoming work (n.d.), I engage how local knowledges on Inuit sealing counter dominant narratives of 'sustainability'. While this present chapter will likely not change the terms of the conversation, I hope that the silences that are left may 'build arguments to confront' (Mignolo 2009:4) neo-coloniality in knowledge production about Inuit seal hunting.

Sustaining Danish colonization

Throughout the colonial history of Greenland, the Danish administration sought ways to exploit Greenlandic resources in order to maximize profits or, at the very least, to 'break even' (Gad 1976, II: 299). From the early colonial period starting in 1721 to the period after the establishment of the Royal Greenlandic Trade Company (KGH[1]) in 1776, Denmark explored different commercial opportunities in her Greenlandic colony (Gad 1976, II). In KGH's quest for profitable commercial exploitation, Danish colonial administrators and traders were commanded 'at all times and in all instances, to seek the [...] advancement of the Royal Trade' (own translation, Instrux 1782:3), as is spelled out in the company's rules and regulations: *Instruction of 1782*. This implied encouraging Greenlanders to generate various products of hunting for the Royal Trade: seals, whales and other marine mammals, foxes and bears, fish, and eiderdown (Instrux 1782). Most efforts in the early colonial period were devoted to establishing a whaling industry with a particular focus on 'procur[ing] a profitable share of the Greenlander's whaling catch' (Gad 1976, II: 282) – but this attempt failed due to poor equipment, a lack of colonial expertise and capital, and difficulties in recruiting skilled crew members. The Danish whaling model was dependent on local engagement in the hunt, but the promise of a little pay and a share of the catch did not seem to attract help from Inuit hunters. Furthermore, 'the work interfered with their seal hunt by which they were to sustain their lives' (Gad 1976, III: 53; own translation).

Indeed, seal hunting proved to be the most significant industry among all the trade stations along the coast. It 'was and became the most important component of the economy – the source of supply, one always fell back on' (Gad 1976, II: 186; own translation). Seal hunting already performed an essential role in Inuit livelihoods, providing food as well as materials for heat, light, boats (*qajaq* and *umiaq*), clothes, boots, and tents. The seal also embodied central social and spiritual meaning for Inuit communities. For example, a boy's first-caught seal was (and still is) celebrated as a rite of passage to manhood (Marquardt 1999a: 8; Peter *et al.* 2002). For such reasons, sealing was already a major part of Inuit hunting cycles (Petersen 1991:19–20, 64–68), and, as byproducts, blubber and sealskins became the most stable supply for KGH.

Blubber was profitable in European markets as a source of train oil, and sealskins also fared well until the late nineteenth century, when European market prices fell. The price gap for these products between Greenland and Copenhagen ranged from 1:4 to 1:7, generating substantial gross profits for KGH. These profits covered the wages of a large number of KGH employees, the shipping expenses, the administration and construction costs, and an annual subsidy to the Danish mission, yet they still returned substantial net profits for much of the nineteenth century (Marquardt 1999a: 10–11). In addition to producing trade items for KGH, seal hunting sustained the livelihoods of Inuit in considerable ways, relieving Denmark of the need to consider Greenlandic food security (Marquardt 1999a; Thorleifsen 1999). Arguably, until European market prices fell in the late nineteenth century, the Danish colonial administration approached seal hunting as the most 'sustainable' investment. In this perspective, it seems that in the first two centuries of Danish colonization, seal hunting constituted the 'sustaining backbone' of KGH, while other potential commercial opportunities were being trialled.

'Protecting' Inuit seal hunting: a desirable sustainability narrative

In this specific colonial matrix, ensuring the continuity of seal hunting was a central focus of Danish colonial policies from the beginning of colonization. According to Marquardt, 'KGH settled for a policy of protecting, as well as commercially exploiting, the Greenlanders' seal hunt' (Marquardt 1999a: 9). As a consequence, the idea of 'protecting' Inuit hunters became central to Danish colonial discourse. As articulated in the *Instruction of 1782*, it was the colonist's task to

> guard that the country's children (besides preventing that they, from young, get used to a European diet, foreign drinks, and too warm clothing) are raised as Greenlanders, on the foundation of the inhabitants' old hardiness and natural way of life – and are not developing desires which will result in the ruin of the country and the trade.
>
> (Instrux 1782:16; own translation)

The *Instruction* here seems to suggest a colonial concept of 'sustainability' which relates the wealth of the Royal Trade to the protection of Greenlanders' way of life. In the framework proposed by Gad *et al.* (this volume), KGH sought what may be termed a 'sustainable relation' with its colonial subjects. Characterized as an interdependence to be sustained over time (Gad *et al.*, this volume), this relationship would ensure *and* legitimize continuous exploitation. Noticeably, this Danish colonial strategy to prevent a potential shortage of hunting resources (primarily seals) draws a parallel to the eighteenth-century German concept of *nachhaltigkeit* in forestry. In the face of scarcity, political-economists at the time called for sustained yields of timber in order to secure continuous exploitation of this key resource in the European economies (Grober 2007).

The strategy of sustaining exploitative Danish–Greenlandic relations became instituted in KGH's policies and regulations on seal hunting. In the *Instruction*,

KGH's board of managers ordered their employees only to buy those hunting products that were not needed in the daily lives of Inuit. The *Instruction* urged a 'clever and cautious trade' which did not 'strip them of their hunting drive and the true necessities in their household maintenance, but fairly barters the surplus of their own needs' (1782:24; own translation). A century later, in 1873, additions to the original *Instruction* specified that the trade of sealskins prepared as boat skins should not exceed the estimated levels of local consumption, and that skins used to waterproof the *qajaq* and *umiaq* were not to be viewed as commodities (Instrux 1873:4). This addition to the *Instruction* also encouraged hunting from *qajaq* as well as the teaching of kayaking skills to sons of mixed marriages (Instrux 1873: 8; Rud 2010). Arguably, the *Instruction(s)* of the eighteenth and nineteenth centuries established the colonial proposition to strike a golden balance, ensuring a future in which Inuit lives were continuously rooted in hunting and thus securing the trade.

This colonial sustainability narrative assisted a conviction that Danish colonial policies were not just intended to sustain Denmark's present and future undertakings in Greenland, but also to sustain the Indigenous population's ways of life – which, in the eyes of the colonist, constituted a desirable Rousseauesqe 'natural state' (Graugaard 2008:10; Petterson 2012; 2014:124). The reinforcement of a racial distinction which placed Greenlanders in a 'sphere of nature' and Danes in a 'sphere of civilization' granted colonialists the power to define the identity of those 'natural Greenlanders' (Petterson 2012:32–33) – and consequently of what should be 'protected'. Stressing the protectionist character of Danish colonial policies regulating sealing (e.g. Marquardt 1999a: 11; Sørensen 2007:12; Thomsen 1998:23), contemporary Danish scholarly works often imply that the strategy of protection amounted to protection of Inuit ways of life. Such historical analysis fails to account for the consequences of the colonial system of control and neglects to recognize that safeguarding seal hunting through colonization was a protection of *Danish notions* of Inuit hunting practices, and not necessarily of Inuit hunting practices themselves. As Petterson demonstrates, *the Instruction* (1782) urged colonists to instruct Greenlanders in how to optimize their hunt and the best ways to store food, thereby subsuming the existing Greenlandic practices (e.g. Petterson 2014:124).

Furthermore, the process of commercialization which transformed hunting into a 'national occupation' (Instrux 1873; own translation) entailed a disruption of Indigenous practices, requiring Inuit families to produce a surplus in order to access imported goods. Arguably, the colonial requirement of excess production contradicted Inuit conceptions of 'sustainability', according to which, specific hunting practices such as redistribution of the catch (Peter *et al.* 2002:168) upheld the balance of living and ensured the return of the animals. In pre-colonial Greenland, hunting was solely governed by Indigenous knowledge systems, in which animals and humans partook in shared cycles of reincarnation (Fienup-Riordan 1994; Gulløv and Toft 2017:28–29; Peter *et al.* 2002). Commercialization, however, enforced a colonial system of mercantilist capitalist logics, thus pervading the existing relations between hunter and prey. In the colonial system, hunted animals were described as 'Greenlandic products' and 'objects for the Monopoly' (Instrux 1873:4; own translation) to be measured, numbered, rated, and priced under the auspices of Danish colonists. The seal, which performed a central social and spiritual role in the daily lives of Inuit, was reduced to 'skin' and 'blubber' for the Royal Trade (Instrux 1783; 1873). In effect, the colonial administration dissected the seal between its commercial uses, which came to belong to KGH, and its subsistence uses, which remained in Inuit households.

Without seals, there are no Greenlanders: a Rinkian sustainability narrative

The protectionist narrative, which held seal hunting as constitutive to sustaining colonial relations, gained a stronghold in the second half of the nineteenth century. This period is often termed *the Rinkian Era* because H.J. Rink occupied a series of influential positions in KGH which enabled him to frame the basis of Denmark's Greenland policy at the time (Marquardt 1999b: 11). Rink was a prominent driving force behind ideas of cultural preservation and a proponent of political reforms presented as the means to ameliorate what was described as[2] a nation-wide socioeconomic crisis in Greenland from the 1850s to the 1880s (Marquardt 1999b: 11; Rud 2006:507–8). Rink reported poverty, confusion, and misery in Inuit communities, and explained the situation as the consequence of Westernization (Rink 1862). According to Rink, 'through the trade and the encounter with Europeans, the indigenous condition is brought out of balance' (Rink 1856, quoted in Rud 2010:112; own translation). Rink's descriptions reinforced the racialized Danish discourse, defining 'Greenlanders' nature' as noble but childishly careless, and 'indisposed for resisting the temptations of Western luxuries' (Marquardt 1999b: 11) – particularly sugar and coffee. In Rinkian discourse, this led to self-destructive behaviours of Greenlanders: trading excessively in skins and blubber while neglecting to store food for the winter or repair hunting equipment (Marquardt 1999b:31; Marquardt *et al.* 2017:216).

According to Rink, seal hunting was crucial to the *original balance* in Inuit society. Frequently associating 'the national occupation' with seal hunting, Rink described it as 'a sort of conscription which is necessary for the society's continued existence' (Rink 1862:22; own translation). Rink and his associates proposed policies to preserve and restore the prevalence of sealing as a strategy to overcome the national 'decline' (Rink 1862). Increasingly, colonial representations cemented Greenlandic identity as being intrinsically connected to seal hunting (Rud 2006:490). This is exemplified in the words of Emil Bluhme, member of the Danish Parliament, who had been considered a 'Greenland expert' in Danish politics since wintering involuntarily in Greenland in 1863–1864. To support Rinkian policies, Bluhme stated:

> It is to Denmark's credit that it has been able to maintain the Eskimos. No other state could have done it. For the Danes it was totally impossible to

> live in Greenland without the Eskimos. Seal hunting will be the most important occupation for as long as one can see. Without seals, there are no Greenlanders.
>
> (Quoted in Thorleifsen 1999:72)

Thomsen (1998) proposes that this discourse was intimately connected to colonial interests in assuring and maximizing KGH profits – at a time when the KGH monopoly was under pressure from a wave of free trade supporters. This is exemplified in the workings of the *Boards of Guardians*, which were bodies established in each of the colony's trading stations, as part of the Rinkian strategy. The most proficient seal hunters were given an opportunity to occupy a seat on the boards, along with Danish colonists and a minority of fellow Inuit hunters. The Boards administered council funds which provided both relief to the poor *and* rewards ('*repartition*') to successful hunters (Goldschmidt 1987:214–215; Marquardt 1999b: 12; Petterson 2014:139; Thomsen 1998:25). In practice, this meant that the less the council paid to the poor, the more would be left for the hunters, providing the Greenlandic board members with 'an incentive not to distribute [poor relief] excessively' (Petterson 2014:139). The colonial administration found Inuit redistribution practices problematic since a great part of the hunters' catch was bound to the community instead of becoming products for the trade. The Rinkian policies can thus also be viewed as the culmination of years of efforts on the part of the colonial administration to channel more of the hunters' products into KGH (Thomsen 1998:25).

Evidently, this strategy only preserved seal hunting culture to the extent that it could serve the Royal Trade. In this sense, it reproduced the logics of the existing colonial policy, as is described in the previous section. Yet, whereas the *Instruction(s)* (1782 and 1873) described the Greenlandic national occupation as the hunting of various animals (and often from *qajaq*), Rink ascribed *sealing* a central role, repeatedly referring to the importance of 'the seal hunt' ('sælhundefangsten') and 'the seal hunter' ('sælhundefangeren') (Rink 1862). Seal hunting was thus singled out in the Rinkian narrative of sustainability as essential to the *original balance* that had been destabilized by Westernization. Defined as Greenlandic identity, seal hunting thus came to be articulated as *the* Inuit way of life. I argue that this constituted a narrative shift by which, *the Inuk hunter*, who engaged in a plurality of activities, became *the Greenlandic seal hunter* with a singular purpose of hunting seals. Even though the concept of seal hunter is not usually deployed in Kalaallisut (the West-Greenlandic Inuit language), it is uncritically employed in most contemporary scholarly works (e.g. Langgård 1999; Marquardt 1999a; Rud 2006; Thomsen 1998). In Greenlandic, seal hunting is *a part* of being a hunter, but is not singled out as an identity in itself.[3] A *piniartoq*, which literally means 'someone who wants something' (Nuttall 2016:302) or 'someone who goes with the intention of getting something',[4] engages in a variety of activities, including hunting for reindeer, muskox, whales, seals, birds, and fish, and collecting eggs and berries. The practice of hunting is necessarily conditioned by the cycles of seasons and migratory patterns of the animals (see Petersen 1991:63), including seals. Thus, one of the consequences of the Rinkian sustainability narrative was a monoculturalization of Inuit hunting practices, if not in practice then as a colonial desire. In this light, the Rinkian call to re-traditionalize can be viewed as a colonial strategy to sustain Danish–Greenlandic relations as being *mutually* dependent on *seal* hunting.

Changing the colonial sustenance: from seals to fish

In the twentieth century, the Greenlandic colony and the possibilities of commercial fishing became the subject of frequent debates in the Danish parliament. KGH's monopoly (relying on skins and blubber) had been under pressure since the late 1880s due to falling world market prices, as European and North American dependence on train oil was replaced by the use of mineral oils and paraffin (Marquardt 1999a: 16; Thorleifsen 1999:63). Meanwhile, Inuit hunters were reporting decreasing numbers of sea mammals (Langgård 1999:42). As climatic changes had led to rising temperatures in the North Atlantic waters, fish stocks expanded (Marquardt 1999a: 16). Arguably, these changes challenged the Rinkian narrative in which a sealing-based colony had been the strategy for 'sustainability' of the colonial project.

In the Danish debates, reformists criticized both the seal-based monopoly and KGH's double role as both the Royal Trade *and* the colonial administration. Furthermore, Danish financial and commercial bodies argued for liberalization and private investment in the state's colonial possessions. In general, critics complained that Denmark kept Greenlanders in the traditional industry, thus preventing the development of fishing (Thorleifsen 1999). On returning from an expedition to Greenland, Danish journalist Mylius-Erichsen directed a harsh critique at Denmark's colonial policies and the underdevelopment of her Arctic colonies (Marquardt 1999a: 16). He stated that 'KGH considers the kayak seal hunters as the gem of society and calls fishermen cowards' (quoted in Thorleifsen 1999:70). Up to this point, Greenlandic fishing had primarily been for domestic consumption, though shark and cod liver was also traded for use in KGH's train oil production. Anthropologist William Thalbitzer also questioned KGH's condemnation of fishing, stating that Inuit had fished in pre-colonial Greenland and that it was therefore an old 'national occupation' – just like sealing (Thorleifsen 1999:71).

However, the KGH directorate defended the existing policy on the grounds that opening Greenland to business and foreign fishermen would be detrimental to Greenlanders because it would endanger seal hunting and increase dependency on foreign goods (Sørensen 2007:27; Thorleifsen 1999:73). KGH's defence reproduced the colonial racialized discourse that constructed Greenlanders as 'less developed, children of nature' who were not mature enough to handle the full scale of 'civilization' (Thorleifsen 1999:73). Yet, whereas Rinkian policies were committed to re-stabilizing a lost *original balance* in Inuit hunting lives, the colonial administration now emphasized their investment in 'ripening [Greenlanders] in the long term for the demands and conditions of the modern world' (Thorleifsen 1999:66). In this period, as Thomsen points out, the colonial administration endorsed a new language of 'maturing', later adopted by the Greenlandic political elite (Thomsen 1998:38; own translation). The concept of maturing arguably signified a narrative shift to accommodate fishing as a future to (cautiously) progress towards, while reinscribing colonial control in the transition. I argue that, as a consequence, hunting and fishing became conceptualized as two distinct occupations, where hunting increasingly came to represent the *past* and *tradition*, and fishing came to represent the *future* and *progress*. This, in turn, relieved fishing from KGH's disapproval.

The Danish debates resulted in commercial fishing experiments (Marquardt 1999a:16), and these evolved into a fishing industry that depended on cod. In 1926, the value of the cod trade surpassed the trade in blubber (Rud 2017:261), thus signalling changes in occupational patterns. However, the conceptual divisions between fishing and hunting were (and are) transcended in practice: many hunters fished in the fishing seasons and hunted for the rest of the year (Petersen 1991:78). Nonetheless, the colonial administration insisted that fishing was merely an occupational opportunity for Greenlanders 'who could not sustain themselves by seal hunting' (Rud 2017:261; own translation). Danish colonial policy has probably also been influenced by the costs of transitioning to fishing – for example, providing fishing equipment and losing the food security provided by seal hunting.

With the new grand development programmes of the 1950s and 1960s (known as G-50 and G-60), the notion of *cautious* progress was deducted from the existing strategy for sustainability. Instead, a scheme for intensive modernization was introduced with the primary goal of developing the fisheries. The trade monopoly was lifted to create opportunities for private initiatives and capital. However, as the strategy failed to attract sufficient private capital, Danish state funding and intervention increased (Jensen 2017:325–326; Skydsbjerg 1999:16). This new development programme was heralded as a 'social experiment on large scale' (Danish Chair of Commission, quoted in Caulfield 1997:35), necessary 'for the Greenlanders' growth into a greater cultural and economic maturity' (G-50 programme quoted in Sørensen 2007:102). It was, seemingly, not only 'the evolution' of Greenlanders which was of concern to the colonial administration. The Danish prime minister at the time, Hans Hedtoft, stated that employing a new business perspective was 'a huge national task that has to be done to preserve Greenland for Denmark' (Sørensen 2007: 96). Notably, Greenland's colonial status was formally abolished through annexation as the UN pressured for decolonization; nevertheless, this epoch has been characterized as a neo-colonial period in which, more than ever, Greenland was governed politically, economically, intellectually, and physically by Denmark (Petersen 1995:121).

The consequences for the hunting sector were concentration and relocation programmes, an economy that was strongly dependent on the export of fish (particularly shrimp), and the diminishing economic role of seal hunting in the new fishing towns (Dahl 1986:24–25; Heinrich 2017:313–314; Jensen 2017:349–350; Petersen 1995:121). In the attempt to assimilate Danish standards (Jensen 2017; Skydsbjerg 1999; Sørensen 2007), fishing replaced sealing as the new 'sustaining backbone' of a modern Greenland (Heinrich 2017:313). Arguably, this development further reinforced the binary of seal hunting as *tradition* vs fishing as *progress* and *modernity*. In place of the old sustainability narrative that had constructed *seal hunting* as essential to securing colonial relations and the continuity of the colonial project, the neo-colonial narrative featured *fishing* as a central necessity in the new equation of sustainability. In this equation, Danish–Greenlandic relations were to be sustained in the implementation of a modernized Greenlandic capitalist-welfare economy similar to Denmark's.

Sealing under Home Rule

When it came into effect in 1979, the Home Rule Act signified a major transition towards enhancing Greenlandic self-determination. Shifting the terms of production from Danish to Greenlandic hands (particularly the fisheries) was a main focus in the new policies that were intended to *Greenlandicize* (Caulfield 1997:42). In the aftermath of G-50 and G-60, encouraging the way of life of the settlements and developing the hunting sector were core ambitions of the new Greenlandic Home Rule government (Sørensen 2007:165–166). As Greenlandic politician Anders Andreasen stated, the settlements were 'bearers of some of the indigenous life qualities of our country' and were regarded as important for 'the cultural development of the whole country' (Jensen and Heinrich 2017:377; own translation). In practical terms, however, the Home Rule government invested heavily in fish-processing-plants in the towns and in building a high-sea fishing fleet (Sørensen 2007:165–167).

Eventually, seal hunting (and those whose lives depended on it) was destabilized in various ways. Hence, in postcolonial Greenland, attempts were made to change the new and pressuring narratives of sealing as being *unsustainable* in relation to either the economy or the environment. However complex, Great Greenland and Puisi A/S are examples of postcolonial undertakings to redirect sealing and engage other narratives describing *sealing as sustainable*. In some ways, they have challenged colonial concepts of 'sustainability', and in other ways they have been conditioned by the very same concepts.

Great Greenland

By the end of the 1970s, the prices of sealskins had plummeted due to anti-sealing campaigns and subsequent American (1972) and European (1983) import bans. This led to a sealskin market collapse in the early 1980s (ICC 1996; Wenzel 1996), and in 1985, Home Rule subsidies made up three-quarters of the price that a hunter received per skin. In response, the Home Rule government launched a 'sealskin programme' whose aim was to increase the value of and demand for Greenlandic sealskins by modernizing the tannery in Qaqortoq,[5] by increasing production, and by inviting international designers to create new sealskin coat designs (Skydsbjerg 1999:91–92). Renamed *Great Greenland* in 1991, the new Home Rule-owned sealskin company played a role in the general strategy of 'development' on Greenlandic premises, seeking to establish a relation between business and cultural revitalization (Boehm 1994:59–61; Skydsbjerg 1999). Great Greenland was established with a combination of business, socioeconomic, and cultural purposes: to buy subsidized sealskins from hunters, to tan skins, to produce sealskin products, and to sell the various products. The company also assumed responsibility for maintaining a number of the trade stations along the coast which allowed hunters to sell their sealskins (Namminersorlutik Oqartussat 2010:16–17).

Despite these efforts, global anti-sealing sentiments have had lasting effects on the sealskin market. In 1994, Danish economist Martin Paldam argued that the Greenlandic hunting sector was 'economically irrational' due to small net returns and the introduction of sealskin subsidies. Paldam deemed it a 'museum industry [representing] a connection to the past that one will not sever' (Paldam quoted in Caulfield 1997:44–45). Paldam's arguments were criticized in Greenland for overlooking the contribution of hunting to contemporary livelihoods, in terms of country foods, household economy, and cultural identity, *and* for ignoring the underlying devastating effects of anti-sealing campaigns on the sealing economy (Caulfield 1997). Yet, Great Greenland has been a deficit enterprise throughout its existence, relying on substantive economic Home Rule support (and now Self-Government support).[6] Taarsted notes how, during a guided tour of its headquarters, that Great Greenland was described as a 'culture-sustaining company' (Taarsted 2010:780–781), while the Government of Greenland underlines Great Greenland's function as a socioeconomic support to hunters and settlements (Namminersorlutik Oqartussat 2010). The Inuit Circumpolar Council (ICC) furthermore argues that Greenlandic sealskin subsidies (channelled through Great Greenland) make economic sense, considering the costs of establishing alternative industries and social services in hunting settlements (ICC 1996:16). Despite their variations, these statements all point to Great Greenland's performative role in sustaining the way of life in the settlements. Facing the challenges of reviving sealing as a profitable endeavour, seal hunting has seemingly been positioned as a peripheral and supplemental occupation to be *supported* and *protected* through Great Greenland. In the light of the central position of the fishing industry in the Home Rule's development plans, the establishment of Great Greenland reflected a conception of 'sustainable sealing' as primarily related with a cultural purpose.

Described as 'the last remains of our cultural heritage' (MP Godmand Jensen, Inatsisartut 2001; own translation), this postcolonial approach ascribed hunting an urgent and requisite role in Greenlandic visions of cultural continuity. It arguably also upheld the *hunting–fishing*, *past–future*, *tradition–progress* binaries of the G-50 and G-60 policies. In this sense, the postcolonial sustainability narratives surrounding Great Greenland draw parallels to colonial narratives and the resulting policies. In the Rinkian era, seal hunting was aligned with the prevention of a so-called cultural 'decline'. In the modernization schemes of the 1950s, hunting was deducted from the plans of Greenlandic progress. Similarly, Great Greenland did not become part of the Home Rule's visions of future economic independence (from Denmark); it was positioned as guarding a last-remaining heritage needed for Greenland's cultural continuity.

The narrative positioning of Great Greenland also operated on an international level, as the task of strengthening the market entailed a challenge to worldwide anti-sealing sentiments. In its information flyer from 1993, Great Greenland portrayed its task as 'securing the survival of the old cultural heritage' (Boehm 1994:60; own translation). Choosing to wear a Great Greenland sealskin was described as enabling 'Greenlanders to continue their life in harmony with nature' (Boehm 1994:60; own translation). Overall, these claims can be considered a narrative to counter the discourse on sealskins as 'unethical', instead emphasizing the role of sealskin in sustaining Inuit culture and thereby the environment at large – with Inuit as its guardians. While this business profile indirectly targeted international anti-sealing discourse, it was arguably also shaped by the Danish colonial logics which relocated seal hunting in *the past* and in *tradition*. In consequence, Great Greenland's marketing image ignored the fact that Inuit seal hunting had been involved in the global market for centuries due to processes of colonization. Alluding to an essentialized image of Inuit as 'natural conservationists', it reproduced the notion of the 'noble savage living on in a changing world' (Caulfield 1997:18). In this international profile of Great Greenland, the relations between contemporary livelihoods, socioeconomic conditions, and hunting were silenced. Arguably, this provided a window for animal welfare organizations to reduce Inuit seal hunting to 'traditional', 'subsistence', and 'non-commercial' activities – in effect solidifying criteria that are impossible to fulfil, as seen in the Inuit Exemption in the EU Seal Regime (European Commission 2016; Graugaard, unpublished). On the other hand, Great Greenland's business profile (under Home Rule) can also be seen as an act to raise international awareness and demand by relaying alternative narratives of the 'sustainable' relations between seal hunting, Greenlandic cultural identity, and the environment. In this light, the establishment of Great Greenland was arguably also a political act intended to re-humanize Inuit hunters who had borne the brunt of Western anti-sealing condemnation, originally aimed at the hunt for seal calves in New Foundland.

Puisi A/S

The case of Puisi A/S differs somewhat from the postcolonial narratives, which mainly related Great Greenland and seal hunting to cultural sustainability.

Puisi A/S was a Greenlandic company established in 1995 to create a lucrative enterprise by producing seal sausages and seal oil capsules for the Chinese market. In the following years, the Home Rule backed the project with approximately 20 million DKK, and leading Greenlandic politicians Jonathan Motzfeldt and Lars Emil Johansen were engaged as board members. Despite high expectations, Puisi

A/S crashed shortly after the first trial sausage production in its factory in Nanortalik. The board was accused of misconduct and the budget was criticized for being highly unrealistic – particularly as the economic statistics regarding the Chinese demand for seal sausages were as yet unknown. Other aspects of the project's credibility have been questioned: the required veterinary, export, and import permissions were not available; and David Stevens, an American who was the company's connection to China, appeared to be unreliable (Holmsgaard 1999:6; Netredaktionen 2010; Sørensen and Ipsen 2003).[7] Greenlandic media criticized Puisi A/S for being an 'unnatural' ('naturstridigt') project because its ambitions were disproportionately large for a new company (Redaktionen 1999).

However, Puisi's promise of a lucrative seal meat export was initially welcomed by hunters, politicians, and the general public in Greenland. It was believed that the project would bring hunters a more favourable outcome from hunting seal by extending the meat's role as subsistence to one that yielded a cash income. At the same time, Puisi A/S was expected to generate substantial income for Greenland's national economy (Lichtenberg 2000). Nanortalik's mayor and citizens had supported and invested in the project. As mayor Ludvigsen stated, 'all this is a clear expression of our belief in Nanortalik in a sustainable seal-project' (*Sermitsiaq* 1999c:11; own translation). The reference to 'sustainability' in relation to Puisi A/S strummed on multiple strings. Sustainability referred to Puisi's role in sustaining the economy of local hunters in Nanortalik in particular and in Greenland in general, but it also supported the political strategy of 'steering development' on Greenlandic premises. Implicitly, the project entailed a strategy to sustain Greenlandic hunting culture and make use of a local resource. Furthermore, the project seemed to tap into a public debate, at the time, on the issue of discarded, wasted seal meat (*Sermitsiaq* 1999a:6–7). In this sense, sustainability also referred to making better use of seals. After a guided tour with the manager of Puisi's seal sausage factory, a journalist reported: 'All of the seal is used. Nothing is wasted. This will result in the best budget, both in economic and ethical terms' (*Sermitsiaq* 1999b:14; own translation). Puisi's seal processing promised to minimize waste, producing sausages from the main meat and dog food from the byproducts. The sealskins were intended for the local sewing workshops, and the excess skins would be sold to Great Greenland. Notably, the concept of a more sustainable use of seals was now part of the framework of a commercial industry. Puisi was enthusiastically compared with the Danish pig industry: 'The efficiency of the Danish pig slaughterhouses is soaring high .... Everything from the pig is used in production. Puisi A/S ... has the same level of ambition' (*Sermitsiaq* 1999b:14; own translation).

In many ways, Puisi A/S diverged from the narrative of preserving an age-old cultural heritage which encapsulated seal hunting in *past tense*. Approaching seal meat as a new lucrative export challenged the accusation that sealing was 'disguised social welfare assistance'. Puisi A/S reflected a new narrative in which seal hunting sustained not only Greenlandic culture but also the national economy, thus contributing to greater economic independence from Danish subsidies. The vision of Puisi A/S promised a new solution to the challenges of the viability of settlements *and* the national economy. Here, the settlement way of life was no longer a political hurdle but a potential contribution to greater national independence. As Mayor Ludvigsen explained, while hoisting a seal sausage wrapped in plastic, 'I hold Greenland's future in my hand' (*Sermitsiaq* 1999c:11; own translation). Envisioned as a central part of Greenland's future development, seals and seal hunting became a panacea imagined to enhance Greenlandic cultural and economic sustainability at one and the same time.

While it was a postcolonial undertaking aimed at a more independent and local-based economy, the conception of Puisi A/S as a 'sustainable seal-project' also carried a colonial resemblance. In some ways, revitalizing seal hunting as a strategy to sustain the national economy mimicked the colonial model. The commercialization of seal hunting had financed the Danish colonial apparatus, and was now set to regain a central role in financing a postcolonial Greenland. Through a more efficient and industrial use of seals, Greenlandic sealing was to reinvent itself as a 'sustainable investment' in national economic terms. This would arguably complicate the sustainability narratives that had emphasized the role of seal hunting in subsistence and as an Inuit cultural practice. In contrast to *Sermitsiaq*'s enthusiastic comparison of Puisi A/S with Danish pig slaughterhouses, Greenlandic politician and author Finn Lynge has in fact criticized the European meat industries in his defence of Greenlandic seal hunting practices (Lynge 1992). This raises the question of whether industrializing seal meat for export may also have compromised Greenlandic hunting culture, while sustaining (some parts of) it. In metaphorical terms, Puisi's attempt at an export seal meat industry involved extracting a taste of Greenland. Puisi A/S had developed a technique to change the taste of the meat and add more desirable flavours. As the factory manager explained, 'in the centrifuge, the characteristic taste of seal meat – which we value so much in Greenland but that other countries frown upon – is slowly washed out' (*Sermitsiaq* 1999b:14; own translation). Void of the taste of seal, the flavour of exported seal meat was 'hot'n'sweet' (ibid.:15; own translation).

Rearview: the trace of seal

Evidently, this interrogation of the sustainability concept in the history of Greenland does not provide a conventional *assessment* of the sustainability of Inuit sealing (then and now). Biologist Eva Garde (2013) concludes that Greenlandic seal hunting *is* sustainable, based on biological levels of seal stocks and socioeconomic parameters. Instead, I have traced the seal through *sustainability narratives* in Greenland's history to convey how specific colonialities condition the ways in which Greenlandic sealing is narrated in terms of sustainability today – and the criteria by which the practice is claimed to be sustainable or unsustainable. In other words, this attempted genealogy may uncover more on how we have come to speak of 'seal hunting' the way we do; how this speaking is anchored in specific historical legacies; and how our speaking sometimes reproduces or counters them. In this light, contemporary Greenlandic articulations and the ways in which Inuit seal hunting is deemed *sustainable* or *unsustainable* are also conditioned by the narratives constituted in particular colonial structures through the course of Greenland's history. This study then begs the question of whether the concept of sustainability can be considered a colonial concept in and of itself, and, if so, how its employment shapes or confines postcolonial visions of the future (of seal hunting).

The title of this chapter, 'Without seals, there are no Greenlanders', should not be read as a simple reproduction of Danish politician Erik Bluhme's century-old statement. Rather, it is intended to exemplify the ways in which the relations between Inuit and seals have been conceived and defined in Danish conceptual realms. Furthermore, Bluhme's statement – which centres seal hunting as a determinant of 'sustainable relations' – elucidates the underlying violence in colonial conceptions of sustainability. Many times I have pondered on a veiled aggression in Bluhme's suggestion: What if Greenlanders stopped hunting seals? Is the message also a warning? And how does this depiction of Greenlanders' relation to seal hunting insinuate an erasure of other options? My title therefore seeks to reflect the ways in which colonial logics of sustainability have co-opted the existing and multiple relations between Inuit, the land, and the sea – and not only in the Rinkian approach (supported by Bluhme) but throughout the history of Danish colonization.

My genealogy of 'sustainable sealing' has traced a complex process by which, in their strategies to sustain the colony, the workings of colonial *sustainability narratives* have produced (or at the least, aspired to produce) monoculturalization in a variety of ways. Initially, a 'golden balance' of sealing was to finance the colonial apparatus, and later on, fishing was to perform the same function. During this transition, sustainable relations shifted. Facing the pressures on the sealing economy at the end of the eighteenth century, Rinkian policies had sought to preserve Inuit culture in a reduced identity of 'the seal hunter', thereby compromising the diversity in Inuit ways of life and hunting. In the transition to a fishing-based economy, hunting and fishing were constituted as two distinct occupations (even though the boundary continued to be perceived as permeable in practice). This produced a binary where the (un)sustainability of sealing was equated with *the past, tradition, and culture* and fishing with *the future, progress, and economy*. It seems apparent that contradictions and complexities are left out of the conceptual terms of 'sustainability', and this may be explained as a reduction from plural sustainabilities to a singular. Consequently, the diversity of relationships to (seal) hunting is silenced in the colonial grammar of sustainability, be it aspects of sewing, cooking, spirituality, local Inuit knowledges, non-occupational hunting, human–nature relations, etc. Revisiting *Great Greenland* and *Puisi A/S* under Greenlandic Home Rule, I suggest that the colonial criteria by which Inuit seal hunting is deemed sustainable or unsustainable may also confine postcolonial sustainability narratives and their visions for the future of seal hunting. Inuit seal hunting has increasingly been referred to as culture-sustaining in postcolonial Greenlandic politics. Being a deficit enterprise, Great Greenland did not become part of the Home Rule's visions of future progress while it was positioned as a last-remaining heritage to be protected for cultural continuity. Puisi A/S attempted to reinvent sealing by industrializing seal meat for export and emulating the efficiency of a Danish slaughterhouse. Had the project succeeded, it might have compromised with major cultural aspects of seal hunting by washing away all Greenlandic.

Returning to my point of departure, the 'sealing dispute' has stimulated differing narratives on the meaning and definition of sustainable seal hunting. As a response to animal welfare organizations who have condemned sealing as *unsustainable*, the concept of 'sustainability' has also served an Inuit counter-narrative presenting Inuit ways of hunting as *sustainable*. For many years, Great Greenland's counter-narrative relied on classifications of Inuit hunting as *traditional*, *old cultural heritage*, and *subsistence*. While defending a Greenlandic hunting way of life and targeting anti-sealing sentiments with some effect, this narrative also recreated some of the essentialisms that grew out of colonial sustainability narratives. The EU Seal Regime, only allowing seal products from non-commercial traditional hunting, reinforced this effect. The #sealfie campaigns (Hawkins and Silver 2016) and the recent film documentary *Angry Inuk* by Alethea Arnaquq-Baril can be seen as contemporary counter-narratives which seek to decolonize colonial binaries such as the traditional–commercial binary. Directing a decolonial critique at anti-sealing organizations, these counter-narratives insist on 'the presence' of redistributing the seal catch in present-day Iqaluit (Nunavut) *and* of producing sealskins for the global market. What is left to consider here is the *untranslatability* of sustainability as a Eurocentric term, and the ways in which local and Indigenous knowledges may provide alternative perspectives on the meaning and use of such concepts.

Notes

1 KGH is the abbreviation of the Royal Greenlandic Trade Company in Danish: 'Kongelige Grønlandske Handel'.
2 The Rinkian interpretation of the socioeconomic crisis has been challenged by Marquardt (1999b), who argues that 'the crisis' was more likely due to a general decrease in seal stocks and climatic changes. Marquardt argues that hunters actually 'considered hunting for subsistence to be more important than hunting for commercial reasons' (Marquardt 1999b:30–31).
3 Thanks to Tina Kûitse for discussions on this topic.
4 Thanks to students in our *History of the Seal* class (Ilisimatusarfik, spring 2017) for discussion and translation.
5 The tannery in Qaqortoq was established in 1977 (Great Greenland 2017).
6 In 2015, the Self-Government's subsidies amounted to 26.7 million DKK (Statistics Greenland 2016:33).
7 Stevens secured himself a favourable deal, an annual pay of 1.5 million DKK plus extra profit shares, before Puisi A/S crashed (Lichtenberg 2000; Sørensen and Ipsen 2003).

References

Angry Inuk (2016). [Documentary]. Iqaluit, Nunavut: Arnaquq-Baril, Alethea. National Film Board of Canada/Unikkaat Studios Inc.

Boehm, C. (1994). Great Greenland. *Jordens folk*, 2, pp. 59–64.

Caulfield, R. (1997). *Greenlanders, Whales and Whaling: Sustainability and Self-determination in the Arctic*. Hanover: University Press of New England.

Dahl, J. (1986). *Arktisk Selvstyre*. Viborg: Akademisk Forlag.

European Commission (2016). The EU seal regime. [online]. The European Commission. Available at: http://ec.europa.eu/environment/biodiversity/animal_welfare/seals/pdf/factsheet/EN.pdf (accessed 7 November 2016).

Fienup-Riordan, A. (1994). *Boundaries and Passages: Rule and Ritual in Yup'ik Eskimo Oral Tradition*. Norman, OK: University of Oklahoma Press.

Gad, F. (1976). *Grønlands Historie*, three vols., København: Nyt Nordisk Forlag Arnold Busck

Gad, U., Jacobsen, M., Graugaard, N. and Strandsbjerg, J. (forthcoming). Politics of sustainability in the Arctic: postcoloniality, nature and development. Framework text for the POSUSA kick-off workshop, Copenhagen 27–28 May 2016.

Garde, E. (2013). *Seals in Greenland: An Important Component of Culture and Economy*. Report. København: WWF Verdensnaturfonden. Available at: http://awsassets.wwfdk.panda.org/downloads/seals_in_greenland__wwf_report__dec_2013.pdf (accessed 5 August 2017).

Graugaard, N.D. (2008). National identity in Greenland in the age of Self-Government Working Paper CSGP 09/5. Centre for the Critical Study of Global Power and Politics. Available at: www.trentu.ca/globalpolitics/documents/Graugaard095.pdf.

Great Greenland (2017). Om Great Greenland furhouse. Available at http://b2b.great-greenland.com/page_info.php?pages_id=13 (accessed 13 November 2017).

Grober, U. (2007). Deep roots: a conceptual history of 'sustainable development' (Nachhaltigkeit). Discussion Papers, Presidential Department P 2007-002. Social Science Research Center Berlin (WZB). Available at: https://bibliothek.wzb.eu/pdf/2007/p07-002.pdf (accessed 10 October 2017).

Goldschmidt, D.B. (1987). Danmark – moder eller koloniherre?, *Fortid og Nutid*, 15, pp. 209–224.

Gulløv, H.C. and Toft, P. (2017). Den førkoloniale tid. In: H.C. Gulløv (ed.), *Grønland – Den arktiske koloni*. Bosnia Herzegovina: Gads Forlag, pp. 16–43.

Hawkins, R. and Silver, J.J. (2016). From selfie to #sealfie: Nature 2.0 and the digital cultural politics of an internationally contested resource. *Geoforum* Available at: http://dx.doi.org/10.1016/geoforum.2016.06.019 (accessed 14 November 2016).

Heinrich, J. (2017). Krig og afkolonisering 1939–53. In: H.C. Gulløv (ed.), *Grønland – Den arktiske koloni*. Bosnia Herzegovina: Gads Forlag, pp. 282–317.

Holmsgaard, E. (1999). Ingen tillid til guldrandet Puisi-budget. *Sermitsiaq*, 20, p. 6.

Inatsisartut (2001). *Punkt 63. Forslag til forespørgelsesdebat vedrørende sikring af at sælskindsprodukter fortsat danner indtjeningsgrundlaget for fangerdistrikterne*. [Minutes from debate in the Greenlandic Parliament] *Inatsisartut*. Available at: https://ina.gl/dvd/cd-rom/samlinger/EM-2001/Dagsordens%20punkt%2063-1.htm (accessed 2 November 2017).

Instrux (1782). *Instrux, hvorefter Kiøbmændene eller de som enten bestyre Handelen eller forestaae Hvalfanger-Anlæggene i Grønland i Særdeleshed, saavelsom og alle de der staae i Handelens Tieneste i Almindelighed sig for Fremtiden have at rette og forholde*. Royal

Greenlandic Trade document. Bibliotheca Regia Hafnienses. Digitized by the Danish Royal Library, Copenhagen.

Instrux (1873). *Instrux udfærdiget af Directoratet for den kongelige grønlandske handel til Iagttagelse samt Veiledning for de i handelens Tjeneste staaende Indvaanere i Grønland*. Royal Greenland Trade document. Bibliotheca Regia Hafnienses. Digitized by the Danish Royal Library, Copenhagen.

International Fund for Animal Welfare (IFAW) (2016). Canada's commercial seal hunt: a cruel, unnecessary waste. Available at: www.ifaw.org/united-states/our-work/seals/canada%E2%80%99s-commercial-seal-hunt-cruel-unnecessary-waste (accessed 4 August 2017).

Inuit Circumpolar Council (ICC) (1996). The Arctic sealing industry: a retrospective analysis of its collapse and options for sustainable development. The Arctic Environmental Protection Strategy Task Force on Sustainable Development and Utilization.

Inuit Sila (2013). Inuit Sila. Available at: http://inuitsila.rglr.dk (accessed 4 August 2017).

Jensen, E.L. (2017). Nyordning og modernisering 1950–1979. In: H.C. Gulløv (ed.), *Grønland – Den arktiske koloni*. Bosnia-Herzegovina: Gads Forlag, pp. 320–371.

Jensen, E.L. and Heinrich, J. (2017). Fra hjemmestyre til selvstyre 1979–2009. In: H.C. Gulløv (ed.), *Grønland – Den arktiske koloni*. Bosnia-Herzegovina: Gads Forlag, pp. 374–421.

Langgård, K. (1999). 'Fishermen are weaklings:' perceptions of fishermen in *Atuagadliutit* before the First World War. In: O. Marquardt, P. Holm, and D. Starkey (eds), *From Sealing to Fishing: Social and Economic Change in Greenland, 1850–1940*. Esbjerg: Fiskeri og Søfartsmuseets Forlag, pp. 40–61.

Lichtenberg, H.H. (23 July 2000). Grønlands økonomi: Sællerten. *Jyllandsposten* Available at: www.ytr.dk/gr%C3%B8nland-s%C3%A6llerten (accessed 10 July 2017).

Lynge, A. (2006). Regarding Paul McCartney's misinformation on CNN: Inuit ask Europeans to support its seal hunt and way of life. Inuit Circumpolar Council Canada. Available at: www.inuitcircumpolar.com/inuit-ask-europeans-to-support-its-seal-hunt-and-way-of-life.html (accessed 23 May 2017).

Lynge, F. (1992). *Arctic Wars, Animal Rights, Endangered Peoples*. Dartmouth: University Press of New England.

Marquardt, O. (1999a). An introduction to colonial Greenland's economic history. In: O. Marquardt, P. Holm, and D. Starkey (eds), *From Sealing to Fishing: Social and Economic Change in Greenland, 1850–1940*. Esbjerg: Fiskeri og Søfartsmuseets Forlag, pp. 7–17.

Marquardt, O. (1999b). A critique of the common interpretation of the great socio-economic crisis in Greenland 1850–1880: the case of Nuuk and Qeqertarsuatsiaat. *Etudes/Inuit/Studies*, 32(1–2), pp. 9–34.

Marquardt, O., Seiding, I., Frandsen, N.H., and Thuesen, S. (2017). Koloniale strategier i en ny samfundsorden 1845–1904. In: H.C. Gulløv, ed., *Grønland – Den arktiske koloni*. Bosnia-Herzegovina: Gads Forlag, pp. 172–235.

Mignolo, W.D. (2009). Epistemic disobedience, independent thought and de-colonial freedom. *Theory, Culture & Society*, 26(7–8), pp. 1–23.

Namminersorlutik Oqartussat (2010). *Redegørelse om Great Greenland A/S' fremtid*. The Department of Fishing, Hunting and Agriculture. Available at: http://naalakkersuisut.gl/~/media/Nanoq/Files/Publications/Fangst%20og%20fiskeri/DK/Udgivelser_FJA_Redeg%C3%B8relse%20om%20Great%20Greenlands%20fremtid_februar%202011_DK.pdf (accessed 23 February 2017).

Netredaktionen (2010). AG mener: Pølsedramaet. *Sermitsiaq*.AG. Available at: http://sermitsiaq.ag/ag-mener-poelsedramaet (accessed 24 April 2017).

Nuttall, M. (2016). Living in a World of movement: human resilience to environmental instability in Greenland. In: M. Nuttall and S. Crate (eds), *Anthropology and Climate Change: From Encounters to Actions*. New York: Routledge, pp. 292–310.

Peter, A., Ishulutak, M., Shaimaiyuk, J., Kisa, N., Kootoo, B., and Enuaraq, D. (2002). The seal: an integral part of our culture. *Etudes/Inuit/Studies*, 1 (26), pp. 167–174.

Petersen, H.C. (1991). Fangererhvervet. In: H.C. Petersen (ed.), *Grønlændernes Historie før 1925*. Nuuk: Atuakkiorfik, pp. 63–79.

Petersen, H.C. (1991). Grønlændernes identitet og en kortfattet historisk oversigt frem til år 1900. In: H.C. Petersen (ed.), *Grønlændernes Historie før 1925*. Nuuk: Atuakkiorfik, pp. 11–41.

Petersen, R. (1995). Colonialism as seen from a former colonized area. *Arctic Anthropology*, 32 (2), pp. 118–126.

Petterson, C. (2012). Colonialism, racism and exceptionalism. In: K. Loftsdottir and L. Jensen (eds), *Whiteness and Postcolonialism in the Nordic Region: Exceptionalism, Migrant Others and National Identities*. New York: Routledge, pp. 29–41.

Petterson, C. (2014). *The Missionary, the Catechist and the Hunter: Foucault, Protestantism and Colonialism*. Leiden: Brill.

Redaktionen (1999). Fænomenet Puisi. *Sermitsiaq*, 43.

Rink, H.J. (1862). *Om Aarsagen til Grønlændernes og lignende af Jagt levende, Nationers Materielle Tilbagegang ved Berøringen med Europæerne*. Universitetsbiblioteket, Lauge Kochs Samling, MCMXXXIX. Digitized by the Danish Royal Library, Copenhagen.

Rosing-Asvid, A. (2011). *Grønlands sæler*. Nuuk: Ilinniusiorfik Undervisningsmiddelforlag/Pinngortitaleriffik.

Rud, S. (2006). Erobringen af Grønland: Opdagelsesrejser, etnologi og forstanderskab i attenhundredetallet. *Historisk Tidsskrift*, 2 (106), pp. 488–520.

Rud, S. (2010). *Subjektiveringsprocesser i metropol og koloni: København og Grønland i 1800-tallet*. PhD dissertation, Københavns Universitet, Det Humanistiske Fakultet.

Rud, S. (2017). Grønland til debat: 1905–1939. In: H.C. Gulløv (ed.), *Grønland – Den arktiske koloni*. Bosnia-Herzegovina: Gads Forlag, pp. 238–279.

Skydsbjerg, H. (1999). *Grønland – 20 år med Hjemmestyre*. Nuuk: Atuagkat.

Sermitsiaq (1999a). Tonsvis af sælkød smides i havet. *Sermitsiaq*, 42, pp. 6–7.

Sermitsiaq (1999b). Kun skriget går til spilde. *Sermitsiaq*, 21, pp. 14–15.

Sermitsiaq (1999c). Jeg holder Grønlands fremtid i min hånd. *Sermitsiaq*, 21, p. 19.

Statistics Greenland (2016). Fiskeri, fangst og landbrug. *2016 statistisk årbog*. Available at: www.stat.gl/publ/da/SA/201607/pdf/2016%20statistisk%20%C3%A5rbog.pdf (accessed 30 October 2017).

Sørensen, B.H. and Ipsen, J.M. (2003). Sælpølser koster landsstyre millioner. *Berlingske*. Available at: www.b.dk/danmark/saelpoelser-koster-landsstyre-milloner (accessed 24 April 2017).

Sørensen, A.K. (2007). *Denmark–Greenland in the Twentieth Century*. Copenhagen: Museum Tusculanum Press.

Taarsted, M.N. (2010). Great Greenland – en arbejdsplads med en lidt anderledes målsætning. *Geografisk Orientering*, 6, pp. 778–783.

Thomsen, H. (1998). Ægte grønlændere og nye grønlændere. *Den jyske historiker*, 81, pp. 21–55.

Thorleifsen, D. (1999). The prelude to Greenland's commercial fishery. In: O. Marquardt, P. Holm, and D. Starkey (eds), *From Sealing to Fishing: Social and Economic Change in Greenland, 1850–1940*. Esbjerg: Fiskeri og Søfartsmuseets Forlag, pp. 62–83.
Wenzel, G.W. (1996). Inuit sealing and subsistence management after the E.U. sealskin ban. *Geographische Zeitschrift*, 84(3–4), pp. 130–142.

6 Scaling sustainability in the Arctic

Frank Sejersen

The concept of sustainability is malleable as it can be scaled in a variety of ways. Just by asking the question, 'Sustainability for what, whom, and for how long?', one may appreciate the concept's physical, social, and temporal dimensions. By scaling the concept in different ways, it may become productive for management purposes, but the scaling also in itself produces particular time-spaces in which the social world emerges and is configured. Consequently, it is important to be analytically sensitive towards how and what kind of understanding of the social world a certain take on sustainability produces and to appreciate that the scaling and application of sustainability are not without political implications.

In this chapter, scales are seen as particular time–space configurations produced and mobilized by communities of scale-makers. This means that the scales themselves are constructed, conjured, and evoked by people and communities with political agendas. The term scale-makers (Sejersen 2015) points to this creative, political, and productive activity of scale-making. Thus, scales can be seen as an inherent aspect of the structuring of discourse. In line with this, Erik Swyngedouw argues that 'the multiplicity of scalar levels and perspectives also suggests that scale is neither an ontological given and a priori definable geographical territory nor a politically neutral discursive strategy in the construction of narratives' (Swyngedouw 1997:140). Not only do scales offer a frame for explanations, a perspective, and the production of contexts for sustainability, but they also carve out possibilities to make causal relations, events, and agencies emerge. Subsequently, scales are productive when addressing responsibilities and when acts of claiming and blaming are pursued. When scale-making is understood as inherent in the construction and mobilization of morality and political identities, scale-making creates the foundation for the construction of power geometries. In practice, this means that social relations of empowerment and disempowerment are closely linked to how sustainability is scaled and how this scaling is politically controlled in order to maintain particular configurations of socio-spatial relations, as argued by political ecology (Biersack and Greenberg 2006). Consequently, scalings of sustainability may indeed frame the way the social world emerges. Understandings of what constitutes the social world in sustainability discourses are also carving out the room for political manoeuvres and the subject positions of people. In some cases, people emerge as 'publics', 'locals', 'communities', 'households', 'populations', 'users', and 'consumers' or 'stakeholders' as the agents that the social world is made up of and the objects of government. This chapter will explore how the scaling of sustainability is intimately linked to the production of social worlds (or social ontologies, more precisely). When sustainability and scale-making are seen together, the evocation of social worlds becomes anything but innocent. Tim Rowse (2009:34) suggests that the configurations of the social world can be seen as 'the ontological politics'. Rowse argues that the tools to produce inquiries into the social world like censuses, hearings, and polls constitute a political technology that in fact shapes political possibilities and political ontologies. The ability to mobilize these people-making tools when talking about sustainability can indeed be seen as a political achievement in itself, because it is just *one* out of many ways of organizing the social world. Thus, power operates through the construction of the social worlds in configurations of sustainability.

This chapter investigates how the social world is creatively constructed and used in sustainability discussions related to the construction of a large-scale industrial plant in Greenland. At the beginning of the millennium, the Greenlandic authorities initiated a comprehensive political strategy to transform their society by inviting in extractive and other large-scale industries. Following this initiative, they set in motion a comprehensive rethinking of Greenlandic society (Sejersen 2015). The social contract between authorities and citizens was re-negotiated, so to speak. In this process, different configurations and categorizations of the social world were used strategically and became centre-stage in public and political debates about how to construct a new sustainable society.

In the Arctic, as elsewhere in the world, it is common to see extractive industries as a solution to regional financial problems and to carve out new community futures. However, Greenland, along with other parts of the Arctic, can be seen as a postcolonial setting in which the future takes up a particular point of orientation for indigenous peoples because they strive for more self-determination. This striving is a constant frame and mobilizer for many discussions. Due to this, the future seems to be crammed into the present. It is within this postcolonial striving and mobilization for a self-determined and sustainable future, combined with the political and economic straightjackets of Arctic societies (AHDR 2004), that particular configurations of the social world appear and are disciplined.

The future social world and the political space for sustainability

Indigenous peoples have striven strongly to create different futures than the ones carved out by the colonial authorities. By invoking an understanding of a new social world located and found in the future they creatively use social anticipation, vision, and hope in order to mobilize and legitimize present action. Sustainability as a concept agglomerates the 'here' and 'now' with the future 'there' and 'then', and infuses action with direction, momentum, and a moral order. Thus, the concept of sustainability links hope with action and promise. The former Greenlandic premier Aleqa Hammond, for example, stated that

> [m]ining is crucial for our financial independence. We will cooperate with external consultants to ensure the best agreements with the mining companies … But independence is about far more than economy. Our culture and our entire society must also be culturally sustainable.
>
> (Cited in Søndergaard 2013, author's translation)

The promises of a better future are used to energize the contemporary discourses and to produce particular 'personhoods' whose life-worlds, problems, potentialities, choices, and knowledge are already confined to be understood within such a frame. The point is that the creative construction of a future social world is used to understand, guide, and mobilize the present. The future casts a productive shadow over how the present is perceived and in some cases even 'forces us to act today' (Ling 2000:253). Contemporary politics are indeed framed and legitimized in the constructions of the future social world. According to Thomas Robinson (2015:63), '[t]he future preoccupies us to such an extent that it has colonized the present as an optic or discourse, and it affects everything from the most banal to the very big decisions' (author's translation). He concludes that the pivotal position of the (imagined and desired) future as a policy driver and the future as the goal of reforms frames and evaluates contemporary values and problems only in relation to how they may contribute to or hinder the proposed and formulated future. In that way, the future turns out to be the primary source for social discipline (Robinson 2015: 62).

In discussions of future sustainabilities, we thus need to have a critical reflection on how the definition, demarcation, and scaling of the social world are pursued in order to make up people (Hacking 1986; Osborne and Rose 1999:368) and to offer certain kinds of political room for agency. Too often, people living in the Arctic are understood to be 'located', 'caught in environmental relations', and 'bound by culture', which generates a perception of their agency as restricted to responses to external forces. Consequently, people emerge as historical 'locals' reacting rather than 'acting' (Ortner 1984:154).

Mega-industries in the Arctic

Living off the land is important in the Arctic, and consequently sustainability discussions are closely linked to wildlife management (see Becker Jacobsen, this volume; and Thisted, this volume). However, the Arctic has had, for decades, the attention of large-scale extractive industries, which have influenced and shaped socio-cultural-political developments, as well as ecosystem dynamics.

In this chapter, the focus is on questions of sustainability in relation to understandings of the social world as they appear from one such large-scale project. In 2007, Greenland initiated a cooperation with the company Alcoa to plan the construction of a huge aluminium smelter powered by hydro-electricity produced partly from water originating from melting glaciers. It is a huge project to be located in the vicinity of the town of Maniitsoq, which has 2,600 inhabitants. It involves the construction of large hydroelectric facilities, a 1.4 km long factory, infrastructure that will double the size of Maniitsoq and the invitation of thousands of foreign workers to do the construction. This chapter explores how the social world is evoked in Greenland when the discussion centres on inviting in, assessing, imagining, and disciplining a project of this magnitude in order to make it sustainable.

The encouragement of large-scale extractive industries can be seen as a huge step taken by the regional government in Greenland. The island's population of roughly 56,000 people (approximately 90 per cent of these are born in Greenland and are thus considered Greenlanders (Inuit)) are anticipating major societal changes in the wake of this political initiative. In 2009, Greenland's right to its subsoil resources was acknowledged by the Danish parliament and, furthermore, it was acknowledged that Greenland could get independence if they so wanted. That was seen as a huge opening for the possibility to pursue a more elaborate path of self-determination, to break the colonial legacy and to define the future course and development of the country. The ambition of the government (and the population) is to become more financially independent of Denmark, and to create new productive arenas that may supplement the fishing industry, which currently is the basis of the Greenlandic economy (apart from the economic support from Denmark). Thus, the extractive industries are considered one of the primary means to establish a better economy, and a large number of extraction licences have been issued. Apart from an aspiration to gain increased independence from Denmark, the Greenland Government is struggling with increasing financial problems (Danmarks Nationalbank 2017). The future discrepancy, ever increasing, between public income and public expenses is represented in Figure 6.1, promoted by economic experts as 'Dødens Gab [the Jaws of Death]' (Økonomisk Råd 2010:19).

According to the government, the graph shows that changes *have* to be made *fast* if *national* economic problems are to be avoided. Indeed, it is a graph indicating a future *national* unsustainability. The social world is constructed in two ways here. First, the social world is evoked as a postcolonial state with an endeavour to sustain its national welfare for 'the inhabitants of Greenland' which emerges as an imagined and manageable national 'us' to be cared for. Second, the graph produces a social world that is closely linked to a perception of the productivity of different social groups in Greenland. The future is immanent in these mobilizations, demarcations and scalings of the social world. In particular, social groups demarcated as dependent upon state support can be seen as hindering a smooth transformation into a more self-determined and economically sustainable future of the nation. In the elaborate report by the Self Rule Commission (Selvstyrekommissionen 2003), this social group is referred to as those who do not in a sufficient degree contribute to the economy of society (chapter 2.5, 'All should contribute to society's economy'). The Commission includes hunters and unemployed people in this group explicitly.

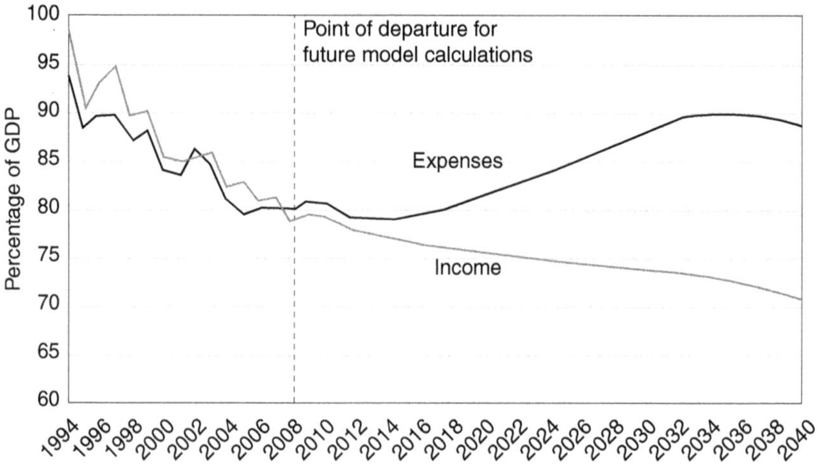

Figure 6.1 The 2010 Greenlandic Economic Council diagram known to the public as 'The Jaws of Death' of public income and expenditures. The lower line is income, the upper line expenses, both in percentages of GDP. Dotted line (2008) marks point of departure for future model calculation.

Source: © Grønlands Økonomiske Råd.

The social world emerging in the graph and the subsequent discussions link individual productivity with national economy, and people are categorized based on their economic contributions to the common welfare. When the social world is configured in this way, particular (political) understandings of problems, urgency, risk, and interventions can be mobilized. Furthermore, the scaled and defined social world carries its own normative project. The *future* social world is creatively made to talk to the *present* in a way that underpins a *contemporary* political discourse emphasizing the emergency of change and reform; 'Status quo is not an option' as it has been stated in the report by the Tax and Welfare Commission in 2011 (Skatte- og Velfærdskommissionen 2011:17). This extrapolated future gives the imaginary 'nation-us' a new point from which a perspective on the *present* may produce and demarcate certain contemporary qualities of the social world. An example could be a hunter, who has an intimate knowledge of his hunting territory and the movement of game. Often, he has been celebrated in the public discourse for his long-term productive contributions to his family and the local community as well as to the underpinning of the national Greenlandic culture (Nuttall 1992). However, in the new perspective where the future casts its shadow on the present he may be reduced to a burden for national development due to his supposed immobility on the national labour market (Nordregio 2010). Framing the future social world primarily as a national project based on mega-industries also translates 'the hunter' into 'a citizen' and 'a worker' by making him a part of 'a national workforce'.

According to Greenlandic politicians, people in Greenland should start – already today – to conform more to the requirements and dynamics of the future labour market and the requirements of the future industries (not yet there!). Seen from this imagined future point of perspective, this implies to be mobile, to be educated, and to work for the national good. In line with this the former premier, Kuupik Kleist, appealed to the population by saying that:

> [i]f these projects are to be realized, it is of the utmost importance that the population as such – the fellow citizens of Greenland – exhibit, to a significant extent, active citizenship in the sense that everyone will be contributing by shouldering the burden.
>
> (Cited in Lynge 2011, author's translation)

By mobilizing an imagined future sustainable Greenland, a transformative productive space is constructed in which new social identities and social contracts can be formulated, but also a space where the understanding of the *contemporary* social world is infused by the anticipated and imagined future. In this national discourse, the citizens are to serve and protect the national interests and a particular *up-scaled* social imperative emerges. This is a scale of the social world which departs dramatically from the discourses in Alaska and Canada where the social world is closely linked to a *down-scaled* understanding of community and identity (Sejersen 2012, 2015). It is within this up-scaled future nation-social that the sustainability of mega-projects in Greenland are to be understood, disciplined, and tamed.

Taming the social

The graph-case indicates that even before the particular mega-project was put on the table, the social world was framed and in overall terms linked to ideas of future national economic sustainability, growth, and independence. Thus, the future is invited into the present, so to speak, by an act of temporal scaling. However, when the initial and more practical planning of the smelter was set in motion, the project as such had to be scrutinized in more detail in order to make it economically, environmentally, and culturally sustainable. Consequences for the environment and society, health, investments, legal framework, economic output, and infrastructure, among other things, had to be assessed. In this sense, it did not depart from any other big construction project. A mega-project can be approached as an act of knowledge assemblage to use David Turnbull's concept (2000), where long lists of focus points are assembled in ways that are considered meaningful (and actually makes it possible for us to think about it as a coherent, manageable 'project'). The contingent nature of this 'meaningfulness' is also reflected in the way the social world is conceptualized. I approach the planning of the mega-project as a narrative that has to be developed and controlled. All the bits and pieces that are linked to and constitute 'the project' have to be tamed, ordered, and disciplined through narrative. There are official ways to discipline a project (stipulated in laws and regulations) in order to address questions of sustainability (understood to include viability, liability, and profitability). Strategic Impact Assessments, Social Impact Assessments, and Environmental Impact Assessments are important instruments for facilitating sustainability discussions and choices. In relation to, among others, these assessments, one of the major official ways to discipline a project and to create a coherent narrative is to use 'phases' as the organizing device. However, the project uses and deals with the social world in ways so differently that we might say that entirely different social worlds are produced when sustainability is narrated in each of the phases.

In relation to large-scale projects, the following phases are often applied, and thus used to configure the project:

- phase of investigation;
- phase of construction;
- phase of production;
- phase of closure.

When we look at these phases, we are already feeling more comfortable and we get a sense of control and overview. For each phase, we are now able to infuse particular kinds of people involvement, consultants, political processes, legal frameworks, economic calculations, and infrastructural requirements, etc. Each of the phases has its own protagonists, dynamics, decisions to be taken, and risks to be calculated. Questions of sustainability are thus made more real and manageable. As the following analysis will show, the phasing of projects is also a way to construct and discipline the social world in particular ways. Thus, there is not *one* idea of the social world linked to a project throughout its lifetime, and the phases themselves conveniently produce and scale the social world and by doing so not only make it politically manageable, but also frame the political debate.

Phase 1 (investigation)

In the first phase, the social world is disciplined and defined by the major organizing device or technology which is brought into action; namely the Social and Environmental Impact Assessments (whether strategic or not (Hansen 2010)). Greenland chose to pursue a Strategic Environmental Assessment (SEA). A SEA constitutes a process to support political decision-making in order to ensure that questions of sustainability are scrutinized and addressed on an informed basis. These impact assessments investigate a project's future consequences. Therefore, impact assessments are interesting because they imagine and evoke future social worlds through the prism of sustainability.

First, the contemporary and future social world is often given its own domain; its own reports, written by researchers who are experts in the dynamics of 'the social world'. Questions of 'the environment' or 'health' are dealt with by other experts in other reports. A second aspect of the disciplining characteristics of assessments (the technologies of sustainability) relates to the demarcation of 'who belongs to the social world?', or to put it in another way: Who are the relevant people and in what way are they made relevant? If we are to respect and integrate local knowledge and local voices, then who are we to contact? In Greenland, the SEA was primarily working with a demarcation of the social world as the 'local', defined by the vicinity to the smelter (see Jacobsen, this volume). The demarcation of the social world as was related to the idea of 'the local' as a specific social category tames the project in both time and space; the social world is localized and timed in particular histories, so to speak. In that way, sustainability is configured in certain temporal and spatial scales. In Greenland, many persons, families, and communities far, far away from the factory itself would be affected hard by the societal dynamics generated by the processes of industrialization, according to a governmental white paper (Naalakkersuisut 2010). They are, however, not in the optics of the assessments' scaling of sustainability, categorized as 'locals', and thus are not integrated in the assessments of 'local' social consequences. Basically, 'the locals' in these project assessments emerge as a product of the defined project; they do not emerge from a complex understanding of several projects – real or anticipated – that cross-cut each other and impact on people's livelihoods. Thus, the individual projects define 'the locals' – not the other way round. The complex navigation and movement of people between several spheres linked to several extractive industries and political agendas are flattened and targeted towards the particular project that is assessed. The social world is tamed and colonized by the project as it takes up a position as the only and most significant Other that people (have to) relate to.

A closer reading of the SEA (SMV 2010) and the written responses by stakeholders, institutions, authorities, and private persons indicates that the social world takes many forms within an overall configuration. For example, it is stated in the SEA for the environment that increased smelter infrastructure can result in increased hunting pressure (SMV 2010: chapter 1:24). Here, the social world emerges as a wave of particular agents (users of resources) flooding the open landscape with potential negative impacts. The social world configured as 'users' in this report has to be controlled and monitored, which makes legislation emerge as the proper tool. In the same report, the social world is evoked as people who have to live with consequences of pollution (waste, noise, and dust, for example). Here, the social world is constructed on the theme of the harmful consequences for consumers entangled in the flows of pollution stemming from the smelter. In some cases, high-level pollution is mentioned (e.g. persistent organic pollutants (POPs) and mercury) but 'victims' are not singled out and scaling is moreover built around the concept of the ecosystem. In the SEA's health section, the social world is produced along the concept of demography (age groups, gender, residence, occupation, consumption patterns, illnesses, etc.) and ideas of status and vulnerability are predominant. Here, concern is aired about the potential consequences for different demographic groups. Furthermore, the health section investigates the consequences of the working environment at the smelter facility, and thus the report makes 'the vulnerable worker' a part of the social world.

In the SEA's section on the regional and economic consequences of the smelter, the social world is scaled in such a way that the nation and its interconnected regions emerge. A point of departure for this section is that '[a] decision on the location of a smelter is in reality a choice of a regional, irreversible and path-dependence creating development strategy regardless of the fact that there is no national development strategy for the Greenlandic regions' (SMV 2010: chapter 5:7, author's translation). This report produces a scaled social world defined on the basis of *regional* economic status and sustainability, which will appear in the wake of the new national flows of resources (financial and human) required by the smelter. In this perspective, the social world is, among other things, the Greenlandic workforce. The report explicitly up-scales the project and the social world onto the national level by asking the question: 'What is best for Greenland?' (SMV 2010: chapter 5:31).

These different kinds of configurations of the social world lay the groundwork for the written responses submitted by the public during a mandatory round of consultations. Most of the responses subscribe to the social world already mapped out and add a few elaborations and details that they consider important to be taken into consideration. However, one of the most interesting responses came from Jørn Hansen (2008) – a citizen of the town Sisimiut – who bypassed the orchestration of the social world in the SEA by formulating the following question:

> It is not evident how an increase of a city with 6–700 families (1500–2000 people) will influence the social environment, well-being, childcare institutions, educational environment, language, and the inter-relations with foreign peoples. Wherever the smelter may be located it must be foreseen that only a few local or Greenlandic employees are available to work at the smelter. Consequently, outside workforce including families can be expected in great numbers. I find an analysis of this fact an important element to be integrated in a SEA and encourage this to be taken into consideration. The many people will become a very important contribution to everyday life – resulting in both positive and negative events, wherever the smelter is located.
>
> (Author's translation)

He points out that the social world can be scaled to everyday life in the town. His concern illustrates that the SEA's multiple configurations of the social world leave out, hide, or silence other takes on the social dynamics and consequences for other understandings of sustainability.

Phase 2 (construction)

The next phase in the project narrative is 'construction'. Of course, Greenland has neither the expertise nor human resources to pursue the construction of such a gigantic project on its own. The construction of hydroelectric dams, harbour, factory, and infrastructure can only be done by inviting in thousands of guest workers and foreign entrepreneurs. For a long time, the Chinese were mentioned as the most probable group to do the work (Sejersen 2015). 'Foreign workers' as a special new category on the labour market soon became a heated topic; not because Greenland had animosity towards these workers, but because Alcoa insisted that the foreign workers were to be seen as belonging to a *new* social category within the Greenlandic social world. The project was, according to Alcoa, to be what they termed 'internationally competitive'; otherwise, it did not make sense financially to build the factory in Greenland. Alcoa was arguing that the economic sustainability of the project was dependent upon the implementation of a new wage structure and, by this suggestion, Alcoa in fact constructed a new social category ('cheap foreign workers'). To legitimize this, Alcoa up-scaled the social world by internationalizing the labour market. In this perspective, national control and the legislation of the social world were seen as hindrances to the global competitiveness of the company and thus a challenge to the project's financial sustainability. Put differently, the aluminium factory should be detached from the locality (Greenland) because it had to be internationally feasible economically. Primarily, the aluminium company wanted to reduce the rights and wages of foreign workers. This met a lot of criticism from all corners of Greenland and fear of social dumping was aired. The dumping of demarcated social categories has a particular history in Greenland, where some Greenlandic workers were given lower salaries than Danes until the 1990s. For many in Greenland, the introduction of equal salaries among different ethnic groups for equal work was a hallmark in the process of decolonialization (Janussen 1995).

However, in 2012, the Inuit-run regional Government of Greenland passed new legislation ('storskalaloven'), which allowed for cheaper labour to be used in the construction of mega-projects. Of interest here, among other things, is that the use of cheaper labour is demarcated to a particular phase of the project: the construction. As soon as 'the construction' is over, this social category does not exist any longer. It is argued that, as soon as the factory starts producing aluminium, a new social reality will emerge and the groundwork for a sustainable society has been established. This brings us to the production phase.

Phase 3 (production)

During the production phase – estimated to be about 60 years – the intention is to have Greenlanders employed at the factory. The expectation is that a newly educated Greenlander, competent in navigating the requirements of an international firm, will emerge on the social scene. This future imaginary infuses promise and hope for a socially sustainable Greenland, where Greenlanders are not to become bystanders to the industrial development in their own country. Thus, seen from the position of the Greenlandic Government, one may say that the primary product of the factory is not aluminium. The true and most interesting product of the factory is to create and maintain this new social world of employees. It is supposed to be a more economically sustainable situation, where an increasing number of Greenlanders are paying taxes and fewer people are reliant upon state support. The new future Greenlander is an industrial worker, and thus a new social category will emerge in this phase and be perceived (and celebrated) to be productive for the continuation of the welfare of the nation. Because this new social category is at the centre of the project, several research projects and public assessments looked into how mobile the population is, and thus if it is possible to establish a new labour market (e.g. HS Analyse 2009; Nordregio 2010). In fact, if Greenland is not able to deliver this new social world of mobile industrial Greenlandic workers, the company will have to employ foreign workers to produce aluminium. In that case, the factory will not produce what the Greenlandic Government wants the most: jobs for Greenlanders.

Phase 4 (closure)

At some point in the future, the factory will close down. Often, in large projects like this one, a closure phase is included and an exit strategy developed. However, this has not been the case here. How are we to understand this in relation to a future post-factory social world? This predicament became apparent when in 2014 a research committee published a report on the feasibilities of large-scale projects in Greenland (The Committee for Greenlandic Mineral Resources to the Benefit of Society 2014). The committee argued that long-term social benefits could be obtained *only* if the politicians applied a long-term sustainability perspective, where coming generations were taken into consideration as well. The committee advised an understanding of the social world on the basis of an up-scaled temporality (a longer time frame). The advice was: do not spend all the money now (even though it is tempting); diversify capacity building and invest in the future; otherwise, mega-projects are like peeing in one's own pants. Again, the promotion and mobilization of a future imaginary of the social world is used to reflect on the organization of the contemporary social world. The committee spelled out that under the production phase the social setup has an expiry date. First, the committee pointed out the dynamics of the social world (they will not stay the same when the smelter is closed); and second, it pointed out the political dimensions of the organization of the social world (it is a political decision how society prepares for a future without a smelter). The committee argued that Greenland should diversify its industrial strategy to work with flexible categories of jobs and to make investments in the future if the large-scale industries were to be economically and socially sustainable for the nation in the long run. By doing so, the committee challenged the authorities' containment and stabilization of the social world on the basis of project phases where the closure phase was not addressed. The committee pursued a spatial and temporal scaling of the social world that exceeded the time–space horizon of the authorities' phases. One could say that the scaling of the imagined future social world by the project-makers and authorities into a better national future was creating the *direction* and *momentum* of the aluminium project, *here and now*. The committee, on the other hand, used the image of the future social world to rethink the *purpose* of the project, *here and now*. In any case, it shows that the evocation of a future social world is a powerful and productive organizing device for contemporary action and imagination related to sustainability.

Conclusion

In mega-projects, as well as in any other project, we see the disciplining of sustainability by acts of scaling and how these scaling activities are used to make up people. Indeed, the scaling of sustainability in time and space is intimately linked to scalings of social worlds. Future imaginaries of the social world are used productively to organize, mobilize, and legitimize, as well as to infuse hope, direction, and promise, but also to hide and silence voices. The scaling of the temporality and spatiality of the social worlds affects the knowledge regime that is constructed, and this not only gives voices legitimacy as well as a time and place, but in fact produces the voices and articulations that can enter the projects. The application of project phases as tools to make sustainability configurations has created categories of people as 'locals', 'citizens', 'stakeholders', 'workers', and 'coming generations', etc. who become the protagonists in different ways by the political identities that are made possible. The idea of *project phases* can be used as an organizing device for the location and framing of the social worlds, and the infusion of productive but disciplined agency into sustainability discussions. By scaling the social world, people and authorities can put forward claims for particular scales of sustainability.

Project plans and project organizing tools are, by their very extrapolatory nature, focusing on the future and shifting ideas of sustainability and its scale. These future imaginaries are not only imperfect, incomplete, and unfinished, but under continuous construction, negotiation, and productive manipulation. In these processes, people-categories are constructed and used politically. However, these imperfect *futures*, which lay the horizon of potentiality and expectation, also affect how we understand, construct, and govern the social world of *today*. For Inuit trying to navigate a postcolonial and neo-liberal world, driven by a desire to live out increasing levels of self-determination and to carve out new meaningful avenues for economic, cultural, and political sustainability, these imaginaries of the future social world are used productively. They scale activities and plans to give present decisions momentum, legitimacy, and direction. The chapter's analysis of the specific mega-project in Greenland shows how scalar narratives of future sustainability provide the framings and problematizations for the construction of discourses of both the contemporary and the future social world, and how the scale-related configurations of sustainability make different political positions possible. Thus, the configurations and use of sustainability as a frame of reference in political debates are highly productive to construct and promote future imaginaries, to infuse moral perspectives, and to demarcate social worlds. The scalability of sustainability offers all stakeholders in any political controversy quite flexible manoeuvring room to put forward mouldable and critical arguments, anticipations, and concerns. Sustainability as a concept opens up and offers a creative social and moral space, which is often hidden and forgotten in the technical and economic discourses that dominate many project processes.

References

AHDR (Arctic Human Development Report) (2004). Akureyri: Stefansson Arctic Institute.

Biersack, A. and Greenberg, J.B. (2006). (eds). *Reimagining Political Ecology*. Durham, NC: Duke University Press.

Committee for Greenlandic Mineral Resources to the Benefit of Society (2014). To the benefit of Greenland. Available from: http://greenlandperspective.ku.dk/this_is_greenland_perspective/background/report-papers/To_the_benefit_of_Greenland.pdf (accessed 30 June 2017).

Danmarks Nationalbank (2017). *Grønland udfordret trods stærkt fiskeri*. Available from: www.nationalbanken.dk/da/publikationer/Documents/2017/08/Analyse_Gr%C3%B8nland%20udfordret%20trods%20st%C3%A6rkt%20fiskeri.pdf (accessed 13 October 2017).

Hacking, I. (1986). Making up people. In: Heller, T.C., Sosna M., and Wellbery, D. (eds), *Reconstructing Individualism: Autonomy, Individuality, and the Self in Western Thought*. Stanford, CA: Stanford University Press, pp. 222–236.

Hansen, A.M. (2010). SEA effectiveness and power in decision-making: a case study of aluminium production in Greenland. Ph.D. dissertation, University of Aalborg.

Hansen, J. (2008). Høringssvar til SMV rapport. *Samtlige skriftlige høringssvar til SMV rapport*, 30 January, pp. 29–30.

HS Analyse (2009). *Kendskab og holdning til aluminiumsprojektet – efteråret 2009*. Nuuk.

Janussen, J. (1995). Fødestedskriteriet – og Hjemmestyrets ansættelsespolitik. *Grønland*, 2, pp. 73–80.

Ling, T. (2000). Contested health futures. In: Brown, N., Rappert, B., and Webster, A. (eds), *Contested Futures*. London: Routledge, pp. 251–271.

Lynge, M. (2011). Kuupik Kleist: Alle skal være med til at løfte opgaverne. *KNR.gl*. Available from: www.knr.gl/da/nyheder/kuupik-kleist-alle-skal-v%C3%A6re-med-til-l%C3%B8fte-opgaverne (accessed 13 July 2011).

Naalakkersuisut (2010). *White Paper on the Aluminium Project Based on Recent Completed Studies, Including the Strategic Environmental Assessment (SEA)*. Nuuk: Greenland Self-Government.

Nordregio (2010). *Mobilitet i Grønland. Sammenfattende analyse*. Stockholm: Nordregio.

Nuttall, M. (1992). *Arctic Homeland*. Toronto: University of Toronto Press.

Økonomisk R. (2010). *Økonomisk Råds Rapport 2010*. Nuuk: Naalakkersuisut.

Ortner, S. (1984). Theory in anthropology since the sixties. *Comparative Studies in Society and History*, 26, pp. 126–166.

Osborne, T. and Rose, N. (1999). Do the social sciences create phenomena? The example of public opinion research. *British Journal of Sociology*, 50 (3), pp. 367–396.

Robinson, T.D. (2015). Reform og den anden I-D-I-O-T. Om konkurrencestatens monopol på tid. *Tidsskriftet Antropologi*, 72, pp. 61–55.

Rowse, T. (2009). The ontological politics of 'closing the gaps'. *Journal of Cultural Economy*, 2 (1–2), pp. 33–49.

Sejersen, F. (2012). Mobility, climate change and social dynamics in the Arctic: the creation of new horizons of expectation and the role of the community. In: Hastrup, K. and Olwig, K. (eds), *Climate Change and Human Mobility: Global Challenges to the Social Sciences*. Cambridge: Cambridge University Press, pp. 225–254.

Sejersen, F. (2015). *Rethinking Greenland and the Arctic in the Era of Climate Change*. London: Routledge.

Selvstyrekommissionen (2003). *Betænkning afgivet af selvstyrekommissionen*. Nuuk: Grønlands Hjemmestyre. Available from: http://naalakkersuisut.gl/da/Naalakkersuisut/Selvstyre/Selvstyrekommissionen (accessed 28 June 2017).

Skatte- og Velfærdskommissionen (2011). *Vores velstand og velfærd – kræver handling nu*. Nuuk: Naalerkkersuisut. Available from: http://naalakkersuisut.gl/~/media/Nanoq/Files/Attached%20Files/Finans/DK/Betaenkning/Skatte%20og%20Velf%C3%A6rdskommissionens%20bet%C3%A6nkning%20Marts%202011.pdf (accessed 30 June 2017).

SMV (2010). *Strategisk Miljøvurdering*. Nuuk: Udarbejdet i forbindelse med aluminiumsprojektet af Grønlands Selvstyres SMV arbejdsgruppe.

Swyngedouw, E. (1997). Neither global nor local: 'glocalization' and the politics of scale. In: Cox, K.R. (ed.), *Spaces of Globalization: Reasserting the Power of the Local*. New York: Guilford Press, pp. 137–166.

Søndergaard, B. (2013). Hammond: Jeg håber på et selvstændigt Grønland. *Kristeligt-Dagblad.dk*. Available from: www.kristeligt-dagblad.dk/danmark/hammond-jeg-h%C3%A5ber-p%C3%A5-et-selvst%C3%A6ndigt-gr%C3%B8nland (accessed 30 June 2017).

Turnbull, D. (2000). *Masons, Tricksters and Cartographers: Comparative Studies in the Sociology of Scientific and Indigenous Knowledge*. Amsterdam: Harwood Academic Marston.

7 Same word, same idea?

Sustainable development talk and the Russian Arctic

Elana Wilson Rowe

Introduction

In the northern coastal city of Arkhangelsk in spring 2017, President Vladimir Putin appeared on the stage with the presidents of Iceland and Finland, with ambassadors, academics, and civil society from all over the world listening raptly from a crowded conference hall. The event was Russia's biannual flagship Arctic conference. The conversation provided an opportunity to state – once again – Russia's main interests in the Arctic as seen from Moscow. Early in the panel, Putin echoed the dual framings of the Russian strategy documents explored in this chapter, characterizing the Arctic as both a 'treasure chest of unique nature' and a territory of 'colossal economic possibility'. Later, in summarizing Russia's Arctic objective, he noted:

> Our goal is to facilitate sustainable development of the Arctic, and this means the establishment of a modern infrastructure, the utilization of resources, the development of commercial bases, the improvement of quality of living of the indigenous peoples of the Arctic, preservation of their unique cultures, traditions … And these tasks must never be considered divorced from questions of preservation of biodiversity and robust Arctic ecosystems … Defence of the polar environment is one of the key priorities of international cooperation in the region, including scientific cooperation.
>
> (Kremlin.ru 2017a)

Minister of Environment Sergey Donskoy's intervention at the conclusion of the Canadian chairmanship of the eight-country Arctic Council in 2015 rang similar notes:

> Russia is open to collaboration and joint implementation of large-scale projects in the Arctic, particularly in the Arctic region of the Russian Federation. This entails not just extraction of natural resources or energy, but also use of the Northern Sea route as the shortest route for transportation of goods between Europe and Asia … Climate change and technological

> breakthroughs make the Arctic, its wealth and resources, accessible for commercial development … this should happen only in accordance with the highest environmental requirements and with due respect to the people living in the region and their traditional ways of life … the cost of failure in the fragile and unique Arctic environment is too high.
>
> (Ministry of Environment of the Russian Federation 2015)

Broadly summarized, one could say that the goal of sustainable development as defined in these statements is nothing less than the colossal task of succeeding in protecting the integrity of a treasure chest while carefully removing its interior riches. These statements also reflect the three main aims of Russia's oft-repeated Arctic policy trilogy: defence and pursuit of national interests, including the economic prospects relating to development of Arctic resources; the welfare of Arctic residents, in particular indigenous peoples; and environmental protection (Laruelle 2014; Sergunin and Konyshev 2015; Wilson Rowe 2009; Wilson Rowe and Blakkisrud 2014).

As explored in the introductory chapter to this volume, sustainability/sustainable development has become an 'obligatory' concept that can encompass many kinds of policies and practices (in other words, when was the last time we heard a political leader or company declare they were going to pursue 'unsustainable development'?). In general, however, the term does usually imply an attempt to consider simultaneously, mitigate, or prioritize the competing concerns of different actors (and the causes for which they speak). Attempting to realize such a balancing act is not a new concern for any Arctic state, including Russia. Russia inherited a set of 'home-grown' science-policy vocabularies and practices relating to environmental risk and a strong focus on protected areas/national parks from the Soviet Union. Likewise, a preoccupation with questions of equality – particularly in response to obvious economic inequalities generated by natural resource extraction projects – is another trademark of the post-Soviet era in local debates (see Kelman *et al.* 2016; Wilson Rowe 2017a). Therefore, while it is an easy assumption to make that 'sustainability talk' functions primarily to appeal to international financial institutions, mirror the Arctic policies of other Arctic states, and/or mitigate the reputational risks of Russian and international extractive companies, these historical factors alone suggest that it is worth taking a look at the rhetorical work the concept does in a Russian policy-making context.

There is a strong and growing scholarship examining how sustainability-related questions are negotiated in practice in various specific locations in the Russian Arctic, particularly as it relates to the extractive industries (oil, gas, and mining). However, there is little research conducted at the national level to examine what kind of high-level political work the concept of sustainability is doing in Arctic policy-making in Moscow. This chapter contributes in this regard by first presenting the long policy lines of Russia's relation to the Arctic and its resources, followed by a look at Soviet-era environmental thinking relevant to understanding contemporary sustainability discourse. The chapter then presents an analysis of Russian policy documents and high-level political statements to explore how sustainability discourse is used in policy discussion of Arctic development questions. The statements of RAIPON, the organization for the indigenous peoples of the Russian North, are also analysed as a point of contrast. This chapter concludes with a discussion of if and how, in national debates, a 'sustainably developed Arctic' differs from simply a 'resource rich' or 'developed' Arctic.

While the chapter does not seek to cover the realization (or not) of sustainability in practice in the vast and differentiated Russian Arctic, it is worth noting that the balance between economic, social, and environmental concerns is negotiated in a variety of ways with different sorts of outcomes at local and regional levels in the Russian Arctic. This may depend on community specifics (capacity, strength of relationship between relevant actors), project specifics (companies involved, type of industry) and the specificities of the physical landscape. Some highly natural-resource dependent areas have found ways of co-existence that are free from ongoing or intense conflict. Other Russian regions, dealing with different companies or a different local starting point at the community level, have seen significantly more conflict over resource-rent distribution and environmental destruction (Wilson and Stammler 2016).

Environmental and developmental legacies in the Russian Arctic

In today's Russian Arctic, there are four main issues that provide the backdrop for Russian policy-making around Arctic development questions and are thus briefly rehearsed here (Wilson Rowe 2017b). First, there is a tension between traditions of and continued need for large state subsidies of Arctic infrastructure and social services and the desire to have the Arctic be primarily a source of profit for the entire country. This is particularly prominent in large infrastructure projects, such as rejuvenation of ports along the Northern Sea Route (NSR) or renewal of the icebreaking fleet (Moe 2014).

A second issue is the locus of decision-making power. Putin's recentralization of power from the regions to the federal level contrasted sharply with the widespread decentralization of the 1990s (Blakkisrud 2004). Despite the difficulties and inefficiencies that a strong centralization in such a large and diverse country may bring about, the leading role of the federal centre in Arctic issues does not seem to be open to question – at least not in Moscow. While regions and cities and communities may not have a large bearing on policy statements from Moscow, they can have a significant influence over large-scale resource development projects initiated by companies in their region, municipality or community, and the region's abilities to capture benefits from resource extraction relates also to capacity to handle questions of equity and social development (Kelman *et al.* 2016). Third, a balancing act – between an 'open' and 'closed' Arctic – also characterizes the region. Specifically, Russia's evolving relationship to its north entails a tension between the securitization of northern space and the nationalization of northern resources working against more international and market-driven orientations (Khrushcheva and Poberezhskaya 2016; Laruelle 2014; Wilson Rowe 2009).

Finally, and of particular relevance to this book's emphasis on problems of sustainability, a fourth core issue for regional policy-making is navigating between commercial ambitions and environmental concerns. The demise of the Soviet Union in 1991 left Russia with serious environmental issues, as the Soviet regime had largely failed to protect the environment from the negative consequences of industrial development (Bruno 2017; Oldfield 2005; Oldfield and Shaw 2016; Ostergren and Jacques 2002; Rowe 2013). Industrial pollution was seen as a problem for the industry to solve by itself, and often ended up quite low on the list of industry priorities, well below the all-important production targets (Rowe 2013:17).

Despite being primarily subservient to industrial concerns, the Soviet regime developed environmental monitoring infrastructure and environmental expertise and practices (Oldfield 2005). This included home-grown environmental ideas, such as Aleksandr Fersman's notion of 'complex utilization', which saw pollution problems as stemming from the suboptimal use of resources and, therefore, a difficulty to be resolved within the production process itself through the application of improved technology (Bruno 2017; Rowe 2013). In the post-Soviet period, environmental issues have largely failed to capture the imagination of the broader Russian public (Crotty and Hall 2012; Henry 2010), although Russia has proceeded to participate, sometimes decisively, in international environmental negotiations (Kotov and Nikitina 1998; Victor *et al.* 1998:24; Wilson Rowe 2013). A highly noticeable outlet for Soviet and immediate post-Soviet environmental interest was, however, a movement that argued for protecting significant tracts of land from industrial development – the zapovednik system. Such a focus on 'pristine' nature was more acceptable to the Soviet leadership, in part because it upheld a division between industrialized areas and wilderness areas (Weiner 1999).

The sum of legacies suggests that there is indeed a number of discursive repertories and debates around ongoing core issues for the region available for Russian policy formulation when it comes to questions of 'sustainable development' in the Arctic. However, as will be explored in greater detail below, the scale within which competing concerns are to be negotiated is a national/regional one where zones of usage and protection can be balanced against each other, rather than seeking to achieve sustainability on a global or broader scale.

Arctic sustainable development in policy discourse

So, when it comes to achieving sustainability on the national scale, what sorts of ideas and arguments are forwarded? In this section, we gain three perspectives on how key actors in Moscow define sustainable development and frame the 'referent object' of sustainability by asking 'sustainability of what?' and 'for whom?'. The three different 'high-level' perspectives from Moscow (governmental strategies,

Kremlin (the political statements of the Russian presidential apparatus) and the Russian Association for Indigenous Peoples of the North (RAIPON)) were selected to explore the possibility of different understandings and purposes of sustainable development discourse. The two official political voices are complementary. The government site contains Arctic sustainability documents of a more formal nature (official papers, strategies), whereas by contrast the Kremlin site contains transcripts and summaries of meetings and, thus, gives us insight into how the concept is put into rhetorical practice in everyday politics.

Government strategies

The development and release of long-term planning strategies for many aspects of sociopolitical life is a prominent feature of Russian politics. Whether these strategies hit the ground in a significant way depends upon the programmes that are subsequently developed, and the extent to which these programmes attract sufficient state, but sometimes private, financing over time. Nonetheless, the strategies can give us a sense of how the region is represented for policy use broadly (and the extent to which there is broad consistency indicating political agreement on these framings). Interestingly, as we shall see, sustainable development is often the overarching outcome for which the strategies are meant to strive, but much more specific policy vocabularies and policy solutions are put to use in describing how one might get to that sustainably developed future.

The *Strategy for Socio-Economic Development of Siberia Towards 2020* (Government of Russia 2010) states the overall aim of ensuring sustainable development of the region (which involves achieving a level of welfare comparable to that in central Russia) and to meet national security/foreign policy objectives. The policy document lists challenges to realizing the policy's goals, include the region's reliance on natural resources (and, by extension, vulnerability to global price changes), capital flight, lack of economic diversity, deteriorating infrastructure (the words 'restore' and 'resurrect' are used frequently), low wages, and 'social depressiveness'. Nonetheless, the policy document attempts to recast some of Siberia's geographical challenges (vastness, remoteness) as positives – forwarding the idea of Siberia as a 'transport bridge' between the countries of Western Europe, North America, and East Asia. Existing and expanded railway connections, a revived Northern Sea Route with updated port facilities, use of the region's rivers, and even ambitious plans to create a rail-link across the Bering Sea and into Alaska are named as regional possibilities to achieve this more geopolitically prominent bridge position.

The *Strategy for Arctic Development* (Government of Russia 2014) can be read as a kind of zone-specific follow-on policy to the *Strategy for Siberia*, although this strategy also covers the European Arctic that was excluded from the Siberian strategy. Regardless, the form and formulations are much like the broader Siberian strategy. The policy opens by rehearsing the standard statement on Russia's overall aims for the Arctic – the realization of national interests and the pursuit of development. This is followed by a quite detailed and sobering recitation of the challenges facing the Arctic region relating to declining population, challenging climatic conditions, distance from major centres of trade, low-quality social services, low capacity in search and rescue, an ageing icebreaker fleet, and many more issues. The list of policy aims is equally broad – covering aspirations for social issues, economic development (tourism and natural resource development in particular), maritime safety/development of the Northern Sea Route, and sovereignty concerns. The protection of indigenous lifeways – as well as a stated aim of ensuring that indigenous youth are educated to be able to take part in 'modern' careers relating to Arctic development – occupies a more prominent position in this document than it does in the Siberia strategy. The 'minimization of environmental damage' is also mentioned briefly, with greater attention devoted to questions of environmental monitoring.

The specific notion of sustainable development pops up a few times. The strategy notes:

> Sustainable socio-economic development of the Arctic zone of the Russian Federation rests on the basis of systematic cooperation between government, commercial and non-commercial organizations and civil society, from utilization of mechanisms of public–private partnership for the realization of key investment projects … the resolution of social problems, as well as the establishment of economic mechanisms to stimulate domestic activity.
>
> (Government of Russia 2014:7)

Here we see that the concept of sustainability was not connected to sustainability standards or visions of sustainable outcomes, but rather to suggest what the actor picture and forms of interaction around such a project would look like. This echoes the framing and definitions forwarded by the small sample of statements on sustainability from RAIPON, as we shall see below. By contrast, the protection of the natural environment of the Arctic and measures for its defence are listed as the fifth specific goal (Government of Russia 2014:21), but divorced from a sustainable development conceptual framework. This includes nature protection and the elimination of the ecological consequences of development activities in conditions of growing economic activity and global climate change.

Interestingly, the part of the document specifying goals for the Sakha Republic (Yakutia) uses the term *sustainable development* in a fashion closer to the context we might expect from sustainability debates elsewhere in the Arctic, in which sustainability less frequently refers to issues of process as noted above, but rather ties into achieving multiple socioeconomic and environment-related goals simultaneously. For example, the authors of the section on the Sakha Republic mention that they would like to use the natural wealth of the Republic to increase the quality of life for their population, while following the conditions of sustainable development of regional ecosystems (Government of Russia 2014:32). The thing to be sustained in sustainability is thus both the population's well-being and the regional environment.

In the *Concept for the Sustainable Development of the Indigenous Peoples of the Russian North, Siberia and the Far East* (Government of Russia 2009), the target of sustainable development is, as the title suggests and as used elsewhere in the *Concept*, the indigenous peoples of the region themselves. The document discusses and celebrates the unique cultures of the indigenous peoples of Russia in the broader context of Russia's historical multiculturalism and also comments on the rights guaranteed them in the Constitution of Russia and in international law. More specifically, the sustainable development of indigenous peoples is defined in the concept as follows:

> the sustainable development of the small-numbered peoples of the North refers to the strengthening of their socioeconomic potential, the preservation of their ancestral areas of land use, traditional ways of life and cultural values on the basis of the full support of government and the mobilization of the internal resources of these peoples in the interests of today's and future generations.
>
> (Government of Russia 2009:2)[1]

Specific tasks identified by the document in achieving this goal include securing indigenous peoples' access to and areas of traditional land use, further developing traditional activities, increasing the quality of life of indigenous peoples to the Russian average levels of well-being, improving health care (and demographics), cooperative development of traditional activities, collectives, and other forms of self-government (self-management) and, finally, preservation of the cultural heritage (Government of Russia 2009:8–13).

Kremlin

On the Kremlin website, there were 17 hits for 'sustainable' combined with 'Arctic' in a ten-year time period (2007–May 2017). Of these, only five included a substantive discussion of Arctic issues. In the other texts, the Arctic was only mentioned briefly in passing and the referent object of the sentence indicated that the term was being used more in connection to the idea of 'can be maintained' than in reference to broader sustainability objectives.

One substantive example from this sample of texts is Putin's statements at the major Arctic conference, which are recounted in the opening of this chapter. The continuation of his remarks turned the topic to climate change in ways that clearly illustrated how the pursuit of sustainable development was scaled (Sustainable at which scale? Balancing factors from how many stakeholders?).

Putin did indeed describe global warming as a 'fact' that is proven and documented. However, he notes that climate change also 'supports our optimism' for the region. Among other things, climate change improves prospects on the Northern Sea Route and creates more conducive 'conditions for utilizing the region for economic goals'. He also raises uncertainty about the causes of climate change, particularly the impact of anthropogenic sources, so the 'main question is simply how to adapt as global climate change may also be cyclical' (Kremlin.ru 2017a). Considering Putin's clearly articulated uncertainty over the role of carbon in global climate change (which ties in clearly with broader Russian policy and societal debates – see Wilson Rowe 2013), the link between choices made at the Russian Arctic regional level about exploiting petroleum resources and the global scale of sustainable development (including the challenge of mitigating climate change) is fundamentally undermined.

Another substantive example from these texts comes from a meeting of the Security Council on Arctic policy questions. Sustainable development was mentioned as well by Putin, who was chairing the session:

> I would like to underline, that our country is interested in sustainable development of the region on the basis of cooperation and absolute respect for international law. In pursuit of this goal, we will seek constant exchange of opinion with our partners, fully follow international regulations relating to ecological security in the region.
>
> (Kremlin.ru 2014)

On a similar note, then Prime Minister Medvedev spoke about the importance of the Arctic for wealth and the importance of 'sustainable development of the region on the basis of cooperation and, without question, respect for international law' as being of primary importance (Kremlin.ru 2012). In these statements, sustainability's referent object may be vague or changeable, but, as in the strategy documents reviewed above, the concept is associated with a particular kind of process for going forward. In this case, the emphasis is on the involvement of international actors and adherence to international practices.

RAIPON

RAIPON's interventions and usage of the term *sustainable development* was widespread in their written documents available online. In fact, in the same ten-year time window that the kremlin.ru site generated only 17 hits covering both the Russian Arctic and sustainable development, RAIPON resulted in 154 hits. Of these, this chapter has analysed the first 20 hits as a sample to better understand how sustainable development is being used by Russia's primary organization for representing Arctic indigenous concerns at the national and international levels. This gives us an important point of contrast to the documents and statements reviewed above, while leaving as questions for further research and a broader study the striking prevalence and likely varied range of usages of the term sustainable development by RAIPON.

Out of the 23 mentions of sustainable development in the sample of RAIPON texts, 11 instances were tied directly to the sustainable development of indigenous peoples (with indigenous peoples being the direct object of sustainable development), which is in keeping with the final government strategy reviewed above. For example, the 'sustainable development of cultural diversity of national traditions of the indigenous small-numbered peoples of the North' is mentioned (RAIPON, 2017a). This is meant to be achieved through good function of ethno-cultural centres. In a broader sense of sustainable development of the peoples themselves, a statement made by Nina Vesailova, a Russian indigenous politician, is illustrative:

> The peoples themselves need to determine their path. No one besides us can know what is important to us. The government can act as our helpers. The government of Russia is working on trying to consolidate the mechanisms of partnership with indigenous peoples. In Russia, the native small numbered peoples of the North are part of the composition of public bodies … but the adopted measures are not sufficient.
>
> (RAIPON, 2017b)

Here again we can identify an echo of what the strategy documents hinted at above. Sustainable development may not be a set of standards against which development can be measured, but indicates a process wherein key voices are heard (or at least, present at the table).

Nine references were to the region more broadly (sustainable development of a geographic space) and three references were using sustainable development in a quite general sense (or, in one instance, a very specific sense with the sustainable development of reindeer herding). In a more general statement about sustainable development of the region itself, specific elements, such as the northern delivery system via the Northern Sea Route, are identified as especially important for creating sustainable development in the region, as it is an element that 'directly affects the interests and fate of everyone living in the region' (RAIPON, 2017c).

Concluding discussion

So, how does a 'sustainably developed' Russian Arctic differ in rhetorical shape from simply a 'resource rich' or 'developed' regional treasure box? In the above, we can see that there are two main 'referent objects' of sustainable development. In other words, the sources analysed here suggest that the answer to the question of 'what' is to be sustained in sustainable development is twofold in Moscow policy rhetoric – the Russian Arctic as a whole for the benefit of Russia more broadly, and indigenous peoples and their traditional ways of life.

First, indigenous peoples themselves are to be developed sustainably and, to a large extent, through process rather than a journey through a predefined/predetermined sustainable development outcome. This comes across very clearly in RAIPON's (the Russian organization representing indigenous interests at the federal level) presentations and discussion of sustainable development. This can be done through specific efforts, such as functioning cultural centres, but primarily through appropriate political processes. As discussed above, a sustainably developed Arctic (in rhetoric, if not practice) requires certain forms of process and the inclusion of relevant stakeholders, including indigenous and local populations, at the table to speak on behalf of their interests and aims. This emphasis on process is important as the criteria for or outcome of sustainable development are rarely envisioned or detailed explicitly. Rather, in both RAIPON's documents and in key Russian strategy documents, sustainable development seems to be an outcome that is to be achieved through the participation of affected parties, which would arguably speak on behalf of competing interests and issues to be balanced in pursuit of sustainable development. The extent to which this attention to the process of reaching sustainable development actually empowers/serves weakly positioned interests, such as those of indigenous peoples' organizations, remains a question to be examined empirically in a systematic way. The general impression from scholarship on indigenous affairs in Russia is that inclusive processes are far from the default mode of operation (see Sidorova *et al.* 2017; Wilson 2017).

Second, the Arctic region itself is also a frequently designated object to be sustainably developed. We can see from the representations of the Arctic articulated by high-level politicians (reflected in Kremlin sources) and by the consensus-based strategy documents that the Arctic is an important zone for economic development and is meant to benefit the entire country. At the same time, the fragility of Arctic ecosystems and the importance of addressing regional social development challenges are also frequently mentioned.

Here, it is important to note that the sustainably developed Russian Arctic is likely not envisioned as a homogeneous space or continuous expanse for policy-makers, but an internally demarcated one. As discussed in the background section on Soviet environmental practices, there is a strong tradition of setting aside parts of nature to remain pristine and untouched by industrial practices. Russia continues this tradition of conservation via zoning of Arctic space today, with the recent establishment of the 'Russia Arctic' park, covering several island chains along the Northern Sea Route as one noticeable example. Russian policy-makers present the new park as a win for biodiversity and an extension of their broader efforts to clean up Soviet-era waste from the Arctic (Kremlin.ru 2017b). In this way, a sustainably developed Arctic is a space of zones as well, which allows extractive projects and conservation concerns to occupy the same policy terrain. For example, in a discussion with Putin, Minister of Natural Resources and Environment in Russia Sergei Donskoi reviewed both improvements to expedite and simplify offshore Arctic licensing procedures and, in the same meeting and in response to a question from Putin about especially fragile natural territories, can extol efforts to protect the Arctic environment through the expansion and strengthening of Arctic nature reserves (Kremlin.ru 2017b).

While international standards and regulations and partners are frequently mentioned in conjunction with sustainable development (one could even suggest that the usage of the term sustainable development is used to mirror the language of such international counterparts), the analysis of costs and benefits entailed in any development project remain within – at most – a systemic form of thinking around a nationally-scaled, yet zoned, Arctic. The firmness of the scaling of sustainability to the Russian Arctic itself comes across clearly in relation to questions of oil and gas development. The relation of oil and gas development extraction to questions of the global carbon balance is not mentioned in any of the documents reviewed. While this is not unlike the petroleum politics of Norway (see Steinberg and Kristoffersen this volume; and Medby, this volume), in which extraction and consumption are seen as two separate policy challenges, the same kind of discursive wall in Russia is shored up by the fact that the top leadership of the country retains a form of climate scepticism. As we saw above, while Putin acknowledges that the climate is warming, he retains uncertainty as to whether this warming is caused by greenhouse gas emissions (or is caused by humans at all). This climate scepticism comes in handy and, in tandem with an anchoring of the scale of Arctic sustainable development at the national level, allows the concept of sustainable development to be applied to nearly all activities in the Russian Arctic, including petroleum development.

Note

1 The direct translations from Russian are given in brackets here, whereas the words that had a more English-language ring are used above: 'ancestral [aboriginal] areas of land use [habitat].'

References

Blakkisrud, H. (2004). What's to be done with the North?. In: Blakkisrud, H. and Hønneland, G. (eds), *Tackling Space: Federal Politics and the Russian North*. Lanham, MD: University Press of America, pp. 25–52.

Bruno, A. (2017). A Eurasian mineralogy: Aleksandr Fersman's conception of the natural world. *ISIS*, 107 (3), pp. 518–539.

Crotty, J. and Hall, S.M. (2012). Environmental awareness and sustainable development in the Russian Federation. *Sustainable Development*, 22 (5), pp. 311–320.

Government of Russia (2009). *Concept for the Sustainable Development of the Indigenous Peoples of the Russian North, Siberia and the Far East*. Moscow: Government of Russia.

Government of Russia (2010). *Strategy for Socio-Economic Development of Siberia Towards 2020*. Moscow: Government of Russia. Available from: www.rg.ru/2010/11/20/sibir-site-dok.html (accessed 13 February 2018).

Government of Russia (2014). *Strategy for the Development of the Arctic Zone of the Russian Federation*. Moscow: Government of Russia. Available from: http://government.ru/news/432 (accessed 13 February 2018).

Henry, L.A. (2010). Between transnationalism and state power: the development of Russia's post-Soviet environmental movement. *Environmental Politics*, 19 (5), pp. 756–781.

Kelman, I., Loe, J., Wilson Rowe, E., Wilson, E., Poussenkova, N., Nikitina E., and Fjærtoft, D. (2016). Local perceptions of corporate social responsibility for Arctic petroleum in the Barents region. *Arctic Review on Law and Politics*, 7 (2), p. 152.

Khrushcheva, O. and Poberezhskaya, M. (2016). The Arctic in the political discourse of Russian leaders: the national pride and economic ambitions. *East European Politics*, 32 (4), pp. 547–566.

Kotov, V. and Nikitina, E. (1998). Implementation and effectiveness of the acid rain regime in Russia. In: Victor, D.G., Raustiala, K., and Skolnikoff, E.B. (eds), *The*

Implementation and Effectiveness of International Environmental Commitments: Theory and Practice. Cambridge, MA: MIT Press.

Kremlin.ru (2012). Uchastnikam I gostyam mezhdunarodnoi vestrechi vysokikh predstavitelei gosudartv – clenov Arkticheskogo soveta. Available from: www.kremlin.ru/events/president/letters/14989 (accessed 11 June 2017).

Kremlin.ru (2014). Zasedanie Soveta Bezopasnosti på voporosu realizatsii gosudarstvennoi politiki vi Arktike. Available from: www.kremlin.ru/events/president/news/20845 (accessed 11 June 2017).

Kremlin.ru (2017a). Mezhdunarodniy forum 'Arktika – territoriya dialoga'. Available from: www.kremlin.ru/events/president/news/54149 (accessed 11 June 2017).

Kremlin.ru (2017b). Soveshanie po voprosu kompleksnogo razvitiya Arktiki. Available from: www.kremlin.ru/events/president/news/54147 (accessed 11 June 2017).

Laruelle, M. (2014). *Russia's Arctic Strategies and the Future of the Far North*. Armonk, NY: M.E. Sharpe.

Ministry of Environment of the Russian Federation (2015). Statement by Minister S.E. Donskoy at the Arctic Council Ministerial Meeting, 24 April. Arctic Council Archive. Available from: http://hdl.handle.net/11374/909 (accessed 29 April 17).

Moe, A. (2014). The Northern Sea Route: smooth sailing ahead?. *Strategic Analysis*, 38 (6), pp. 784–802.

Oldfield, J. (2005). *Russian Nature: Exploring the Environmental Consequences of Societal Change*. Aldershot: Ashgate.

Oldfield, J.D. and Shaw, D.J.B. (2016). *The Development of Russian Environmental Thought: Scientific and Geographical Perspectives on the Natural Environment*. Abingdon: Routledge.

Ostergren, D. and Jacques, P. (2002). A political economy of Russian nature conservation policy: why scientists have taken a back seat. *Global Environmental Politics*, 2 (4), pp. 102–124.

RAIPON (2017a). Krugliy stul: 'Etnokulturniy tsentry korennikh malochsilennikh narodov Severa, Sibiri I Dalnego Vostoka RF: kontsepti sokhraneniya traditzionnoi kul'tury I praktiki realizatsii strategii gosudarsvennoi natzionalnoi I kul'turnoi politiki RF. Available from: www.raipon.info/info/news/2468/?sphrase_id=1272969 (accessed 11 June 2017).

RAIPON (2017b). Nina Visalova: pravitel'stvo Rosii stremitsya k sovershenstvovaniyu mekhanizmov parnerstva s korennimi narodami. Available from: www.raipon.info/info/news/2539/?sphrase_id=1272969 (accessed 11 June 2017).

RAIPON (2017c). Ustoichivoye razvitiye udalennikh severnikh territorii obsudyat na pervoi konferentzii po Severnomy zavozy v Nar'yan-Mare. Available from: www.raipon.info/info/news/2583/?sphrase_id=1272969 (accessed 11 June 2017).

Rowe, L. (2013). Pechenganikel: Soviet industry, Russian pollution, and the outside world. Doctoral dissertation, University of Oslo.

Sergunin, A. and Konyshev, V. (2015). *Russia in the Arctic: Hard or Soft Power?* Stuttgart: IBIDEM.

Sidorova, L., Ferguson, J., and Vallikivi, L. (2017). Signs of non-recognition: colonized linguistic landscapes and indigenous peoples in Chersky, northeastern Siberia. In: Fondahl, G. and Wilson, G.N. (eds), *Northern Sustainabilities: Understanding and Addressing Change in the Circumpolar World*. New York: Springer.

Victor, D., Raustiala, K., and Skolnikoff, E.B. (eds) (1998). *The Implementation and Effectiveness of International Environmental Commitments: Theory and Practice*. Cambridge, MA: MIT Press.

Weiner, D.R. (1999). *A Little Corner of Freedom: Russian Nature Protection from Stalin to Gorbachev*. Berkeley, CA, University of California Press.

Wilson, E. (2017). Rights and responsibilities: sustainability and stakeholder relations in the Russian oil and gas sector. In: Fondahl, G. and Wilson, G.N. (eds) *Northern Sustainabilities: Understanding and Addressing Change in the Circumpolar World*. New York: Springer.

Wilson, E. and Stammler, F. (2016). Beyond extractivism and alternative cosmologies: Arctic communities and extractive industries in uncertain times. *The Extractive Industries and Society*, 3 (1), pp. 1–8.

Wilson Rowe, E. (ed.) (2009). *Russia and the North*. Ottawa: University of Ottawa Press.

Wilson Rowe, E. (2013). *Russian Climate Politics: When Science Meets Policy*. Basingstoke: Palgrave.

Wilson Rowe, E. (2017a). Promises, promises: Murmansk and the unbuilt petroleum environment. *Arctic Review on Law and Politics*, 1.

Wilson Rowe, E. (2017b). The Arctic in Moscow. In: Orttung, R.W. (ed.) *Sustaining Russia's Arctic Cities: Resource Politics, Migration and Climate Change*. New York: Berghahn Books, pp. 25–41.

Wilson Rowe, E. and Blakkisrud, H. (2014). A new kind of Arctic power? Russia's policy discourses and diplomatic practices in the circumpolar north. *Geopolitics*, 19 (1), pp. 66–85.

8 The right to 'sustainable development' and Greenland's lack of a climate policy

Lill Rastad Bjørst

Introduction: thanks on behalf of the citizens of the Maldives

More than 400 participants are attending the Future Greenland 2015 business conference in Nuuk. It is the fourth time that the employers' association Business Greenland is hosting the conference, and again this year, there is a full house in Katuaq (Culture venue in Nuuk). The theme of this year's conference is 'Growth and welfare – scenarios for the development of Greenland'. The debate has been centred on regional development, mining, tourism, and the economic situation of the country and the need for investments. Today is the second day of the conference, and the mayors of the country's four municipalities are attending a panel together with the chairman of Business Greenland, Henrik Leth. It is one of the last sessions, and the auditorium is packed. The floor is open for questions, and from the audience I ask the panel (Figure 8.1):

> What about the global responsibility? I am aware that Greenland is no longer accountable to the Kyoto protocol, but if the industrial project of large-scale mining and tourism [that has been mentioned at the conference] is underway, it will double the CO_2 emissions from Greenland – a region which is warmed faster than any other region on earth according to science.

Leth answers my question decisively:

> about climate change. Yes, it is true that we will double our CO_2 emissions – but not developing Greenland because the rest of the world has been polluting is not an option (laughing and clapping from the audience). Personally, I do not feel guilty for developing new industries. Of course, we will do it as environmentally responsibly as possible … but not starting a mine to save the people living in the Maldives – this is not going to happen.[1]

All four mayors smile consentingly while nodding their heads. The conference chair, Martin Breum, says in an ironic tone, 'Well, then I say thanks on behalf of the citizens of the Maldives'. The 400 participants respond with laughing and clapping.

Figure 8.1 The author queries a panel at the Future Greenland 2015 business conference in Nuuk about climate change.

Source: © Carina Ren.

The 2015 International Panel on Climate Change report states that greenhouse gas emissions are accelerating despite reduction efforts and that emissions grew more rapidly between 2000 and 2010 than in each of the three previous decades. Greenland and the Arctic environment are subject to profound change, and the international research community is speaking about 'a new Arctic reality' (SWIPA 2011). The message from science is that 'we need to move away from business as usual' as substantial emission reductions are needed to avoid dangerous levels of interference with the climate system (IPCC 2015). So why do the politicians and the business communities still think that this does not apply to Greenland?

'Sustainability' and 'sustainable development' in particular were mentioned often at the Future Greenland conference, whereas climate change was absent in most of the conversations. Looking back at the event, Arctic sustainability was primarily understood as a developmental doctrine and less as an environmental doctrine. When considering the relationship between *climate change* and *sustainability*, climate change is often considered a challenge to 'achieving sustainable development', which, in turn, is often described as the end goal. A UNESCO publication from 2009 points out that 'two environmental problems in particular will be crucial constraints to Arctic sustainable development: climate change and loss of biodiversity' (Funston in Nakashima 2009:298). Critical reading of reports and assessments about changes in the Arctic reveals the inconvenient truth about the blurred relationship between *climate change* and *sustainability*. The connection is presented as so obvious that climate change seems to act as a proxy for sustainability in the texts and ends up being absent/present in many discussions related to the future of the Arctic (and the future of Greenland).

In this chapter, I aim at bringing climate change explicitly back into the conversation by confronting both the absence and presence of the concept, asking what should be sustained when discussing climate change and the future development in Greenland. A first task, however, is to find out what Greenland's official climate policy is. As the anecdote from the Future Greenland Conference illustrates, Greenland's approach to climate change has been framed by industrial development plans. These are plans which still need to be implemented since no large-scale extractive projects are up and running as expected only a few years ago (Inatsisartut 2012). To learn more about climate change and sustainability as political concepts, this chapter analyses how, since the turn of the millennia, political actors in Greenland have actively and strategically positioned themselves towards local and global climate regimes. Why is it that Greenland's climate commitments have not really been properly formulated or consolidated? Why has 'The new Arctic reality' presented by science, in which Greenlanders are cast as potential victims of climate change, not been implemented into Greenlandic politics? How and where discussions about climate change and sustainability relate Inuit/Greenlanders to the global ecosystem is not as straightforward as it seems. Something else is at stake. In unpacking this puzzle, this chapter seeks inspiration from STS scholars such as Sheila Jasanoff (2010), who has been investigating the tensions that arise when the imaginary of climate change projected by science come into conflict with subjective, situated, and normative imaginations and agendas of human actors who engage with nature and day-to-day politics.

After briefly introducing the analytical lenses through which the chapter approaches these conflicts, I will begin the analysis by taking the reader back to 2000–2001, when Greenland, without any major discussions or resistance, ratified the Kyoto Protocol with Denmark. Working its way up through the decades, the analytical work is concentrated around a few communicative events related to the slow development of Greenland's climate policy. The data analysed are a combination of governmental documents (both internal and external), newspaper articles (Greenlandic, Danish, and a few from the international press), and not least field observations and interviews conducted in Greenland 2008–2016. Most of the interviews took place in 2008–2011;[2] the global gaze on the Arctic and Greenland as such was centred on climate change (Bjørst 2011), and the Danish chairpersonship of COP15 used it to frame the event and stabilize their position, which I will explain in the analysis. Since then, the global gaze has changed to generally becoming centred on mineral resources, fuelled by very fluid discussions about pathways to sustainable Arctic futures. With the recent focus on the UN's Sustainable Development Goals (United Nations 2015), Arctic sustainability talk in Denmark and Greenland also reflects a very distinct business agenda (see Ministry of Foreign Affairs, Denmark 2017).

Strategic positions, scale, and sustainability

My focus will be on language and its capacity to make and shape politics, particularly by constructing and privileging scales when discussing climate change and sustainability in Greenland. Such analytical work on the becoming of Greenland's climate policy can offer a quick glimpse into the tension around the various potential referent objects of sustainability. This approach will expose some of the close relationships between sustainability talk and climate change talk by revealing how the congruence between the two is part of a political struggle in which certain elements appear to be fixed and important, while other elements appear problematic or absent (such as climate change in most Greenlandic political discussions about the future of Greenland).

To this analytical approach, it is not random who or what is positioned discursively as a winner, loser, witness, or victim in the climate change debate. The discursive practice of being positioned or to position oneself will also have moral implications (Harré and van Langenhove 1999:23), and such positionings are a product of constant negotiation and competition. Žižek (1990) asks for a critical reading of the ideological illusions that are embedded in available subject positions in a debate. When researching the strategic positions of the Greenlandic and Danish Governments, it becomes clear how both are playing games and are trying in various ways to occupy the most privileged position in any given round of negotiations (be it as a climate symbol, a climate-responsible country, a change agent, indigenous peoples, Arctic citizens in a developing country, primary victims or witnesses to climate change, or as citizens in an up-and-coming independent state). The ideologies embedded in the discursive practices and positions are most effective when they achieve a common-sense status (Fairclough 1992:87–88), even if this may lead to social change in the long run. In the following, the focus will be on the strategic positions of the Greenlandic politicians in the climate change debate and to what exact end they are using the concept of sustainability.

In addition to the analysis of strategic positions, an analytical focus on the scale-making processes related to sustainability will be deployed. As emphasized by Bruno Latour, scale is what actors achieve by 'scaling, spacing and contextualizing each other' (Latour 2007, pp. 183–184). In other words, scales are not just there; rather they are made up and as such constitute part of the analytical object which needs to be studied. Sejersen (2014) wants Arctic researchers to be aware of down-scaling and up-scaling mechanisms because these can easily make us blind, also as researchers (Sejersen 2014: 67–68). Thus, according to Sejersen, understandings of climate change and, by extension, sustainability are determined by the scale from where the phenomena are constructed and perceived (Sejersen 2014:74).

What has puzzled me is what kind of climate policy was formulated while the government of Greenland was preparing for what it characterized as 'sustainable mineral resource development' as mentioned in the latest oil and mineral strategy (Government of Greenland 2014:90). This is a development the Government of Greenland has been expecting to witness for some time now.

Analysis: climate change and the global gaze on Greenland

In the summer of 2000, Denmark worked towards the ratification of the Kyoto Protocol and, simultaneously, a framework agreement between Denmark and Greenland was negotiated. In a memo from the Greenlandic Ministry of Environment and Nature to its government minister,[3] the current situation was described in this way: 'Greenland's contribution is relatively limited. This means that a CO2-producing industry can make the reduction go in the wrong direction' (Direktoratet for Miljø og Natur 2000:3). In the same memo, an unofficial request made in the Department of Mineral Resources is mentioned which indicated that developments in the oil industry would cause limited or no emissions. Moreover, the memo mentions a general lack of statistics and plans for adaptation and reduction of CO_2 emissions. Climate policy was not the central matter, and industrial visions and future plans were unclear. It is even predicted that the 8 per cent reduction in CO_2 emissions to which Greenland was committed under the agreement had more or less already been achieved (Direktoratet for Miljø og Natur 2000). In other words, signing on to the agreement along with Denmark would not imply any substantial risk. Hence, after signing the agreement in 2001 (Grønlands Hjemmestyre 2001), Greenland was legally part of a global climate regime. Enquiries into the Danish and Greenlandic media at the time show that the ratification of the Kyoto Protocol with Denmark took place with nearly no media coverage. At the time, this event and agreement were hardly recognized by anyone. Home Rule Premier Hans Enoksen's primary priorities back then were revitalizing settlements and small-scale fisheries, self-determination, and *Greenlandization*. Neither foreign policy nor environmental policies were great priorities for the Government of Greenland.

'The Greenlandic case' and 'the Danish case'

In Denmark, then Prime Minister Anders Fogh Rasmussen (Lib., 2001–2009) did not consider the environment or the climate as important policy areas until 2004. That year saw a notable shift in the official discourse when Connie Hedegaard (Con.) was appointed a Danish minister, first Minster for the Environment and later, in order to concentrate on preparing the COP15 to be hosted in Copenhagen, Minister for Energy and Climate. She was convinced that Greenland was an important place to visit and showed an interest in an Arctic science assessment that was recently presented to policy-makers (ACIA 2004). In 2005, she arranged the first so-called 'Greenland dialogue' meetings in Ilulissat to showcase the ice melting firsthand. From the city centre of Ilulissat, it is approximately 2 km to the mouth of the Ilulissat Icefjord. The Ilulissat Icefjord at the end of the Sermeq Kujalleq glacier is one of the fastest moving and best-studied glaciers in the world (Mikkelsen and Ingerslev 2002). The meetings in Ilulissat were set up to facilitate an informal dialogue about actions to be taken to alleviate climate change impacts. At first, she invited Ministers for the Environment whom she identified as key players in the negotiations under the United Nations Framework Convention on Climate Change (UNFCCC). The dialogue meetings were a success and she recurrently invited more politicians, royalty, journalists, artists, etc. to the Ilulissat Icefjord.

The Ilulissat Icefjord materialized as a very strong symbol of climate change, not least in the Danish branding of climate change as an urgent political topic prior to COP15 (Hedegaard 2008; 2010). As Hedegaard explained in an interview with the Greenlandic press:

> It helps to show the Greenlandic case. Greenland has become a symbol of climate change, and it affects you when you see the terrible consequences of global warming with your own eyes, as it is possible in Ilulissat.
>
> (Holmsgaard 2007:15)

The Greenlandic venue facilitated the contrasts within the discussion at which Hedegaard's dialogue meetings were aimed, forcing the political participants to contemplate the worst-case scenarios. Douglas and Wildavsky (1982) emphasize that such a 'mirror of nature' is very efficient if you want to forbid new things, because it hints at the delicate balance of nature (Douglas and Wildavsky 1982:69). In Hedegaard's view, 'the Greenlandic case' tells a visitor about the consequences or effects of global climate change. In Greenland, Danish politicians were able to up-scale their arguments about the responsibility for the global climate and at the same time down-scale visual proof by localizing climate change in the Ilulissat Icefjord. In effect, this was a scale-making exercise which positioned local people as primarily victims and witnesses to climate change. The efficiency of 'the Greenlandic case' is an example of the structure of discursive authority at the time: Hedegaard and the UNFCCC occupied privileged positions from which they were able to articulate climate change as a threat. In their narrative, what was to be sustained was the global climate (and hopefully the Icefjord), but most importantly, the referent object of climate change was the fate of humankind (Paglia 2017).

There was also a parallel agenda, however, since the Greenlandic case was aligned with what Hedegaard called the 'Danish case'. In March 2009, Danish Prime Minister Anders Fogh Rasmussen was interviewed by the magazine *TIME* and commented on the good example of Denmark: 'We try to make Denmark a showroom. You can reduce energy use and carbon emissions, and achieve economic growth' (Walsh 2009:31).

Stories about Denmark as well as Greenland were published internationally prior to COP15 (Mason 2008). Denmark branded itself as part of the solution; in contrast to this, 'the Greenlandic case' was a place to witness the effects and read the barometers of global climate change. These were two very different but complementary 'showrooms', incidentally confirming a victim position for post-colonial Greenland well-known to the Danish public (Bjørst 2008). A climate-crisis discourse performing a specific combination of down-scaling and up-scaling, and in effect victimization of Arctic citizens, is also found in the discussions about climate adaptation taking place in academia (Sejersen 2015). Neither the academic climate crises discourse nor the political framing of the 'Greenlandic case' left Greenlanders or Greenlandic politicians with much room for manoeuvring. However, what Hedegaard had clearly not expected was the emerging wish from the Government of Greenland (as COP15 moved closer) for local industrial development and their insistence on not becoming party to any binding global climate deals accompanying this.

A new strategic position for Greenland

When I interviewed Kim Kielsen, the then Minister for the Environment and Infrastructure, in June 2008, he claimed that Greenland 'was not a developed country yet' and continued to conclude that 'Now that we [Greenland] have to be self-sustaining, it is hard not to do something which will emit more CO2 – then again, it has to be as little as possible' (Bjørst 2008:28).[4] Kielsen did not recirculate Hedegaard's 'Greenlandic case' as promoted in Ilulissat that very same summer. On the contrary, he was slightly provoked by Hedegaard playing host to foreign visitors to Greenland. At that time, Greenland was negotiating what would become the Act on Greenland Self-Government (2009) with the Danish state. In contrast to Hedegaard's climate-crisis discourse, development and economic 'self-sustainability' in Greenland as an up-and-coming independent state was at the top of the agenda in Greenlandic politics. In effect, Greenland decided not to prioritize its commitment to the Kyoto Protocol, in spite of having signed it in 2001. Only at the 2009 autumn session of the Greenlandic parliament did the Government of Greenland finally present a plan for the reduction of CO_2 emissions to meet the Kyoto Protocol goal for 2012. The delay left less than three years to reduce CO_2 emissions by 8 per cent compared to 1990 level (Grønlands Selvstyre 2009).

The plan was only just formalized as a group of Greenlandic government officials travelled as members of a Danish delegation to a UN climate negotiation meeting in Bonn at the beginning of June 2009. This was to become a brief trip for the Greenlandic participants. The Greenlandic position was pitched before the negotiations began by government official Michael Petersen. He emphasized: 'We are a climate responsible country, but we still rely on being able to develop the extractive industrial sector. And this development demands the right to higher emissions of CO2' (Nyborg 2009). This position did not sit well with the Danish delegation. On one of the first days in Bonn, Steffen Ulrich-Lynge, Permanent Secretary in the Home Rule Department for Infrastructure, received the response below from Thomas Egebo, Permanent Secretary in the Danish Ministry for Climate and Energy. He wrote in a very direct tone:

> it is not easy to come up with a model that meets both with Greenland's desire for significant growth in CO2 emissions and which is also in harmony with the basic premise for the work before Copenhagen (COP15), with the aim of creating a new legally binding and ambitious climate agreement.
>
> (Egebo 2009:1)

Because of this disagreement, a position on Greenland was not presented at the meeting in Bonn (or at COP15, for that matter). In fact, the two parties never really agreed and Greenland is only mentioned in a footnote in the Copenhagen Accord (2009). As Denmark realized that COP15 would not deliver the result they had expected anyway, the negotiation between the two parties became much smoother. Basically, as the host of COP15, Denmark froze their position before COP15 and silenced the Greenlandic positions. During the days of COP15, Premier of Greenland Kuupik Kleist began emphasizing the importance of 'common but differentiated responsibilities' as a new strategy for Greenland. Later at COP16 in Mexico, Kuupik Kleist added a new dimension to his argument, advocating special provisions for developing countries (identifying Greenland as one of these) and indigenous peoples. To the Greenlandic newspaper *A/G* he said that 'we are especially interested in having rights that enable development. You have to clarify which countries can reduce their enormous use of energy and give room to others' (Schultz-Lorentzen 2010:7). By reintroducing the embedded inequalities in the UN system (Greenland not being recognized as an independent state), the setup for CO_2 reduction (differentiated between developed and developing countries), and the importance of indigenous peoples' rights, he hinted at what was to be sustained as the indigenous peoples' right to (sustainable) development. This argument is still in line with the advocacy for a 'common but differentiated responsibility' but is now positioned as an indigenous peoples issue. This position came to be endorsed by Danish politicians. Hence, in Denmark's national presentation at COP16, Lykke Friis (Lib.), the new Minister for Climate and Energy, once again mentioned Greenland as *the* place to localize climate change, posing this request to the audience: 'let's not forget that indigenous peoples are on the frontline of climate change and experience environmental changes first-hand' (Friis 2010). However, in the Danish presentation at COP16, the argument about the right to development was left out, and in its place the logic of the 'Greenlandic case' was reintroduced for the purpose of stabilizing the Danish position, once again with the side-effect of positioning Greenlanders as witnesses and victims to climate-induced changes. Suddenly, this discursive practice was no longer a problem for Greenland since the prospects of being included in new binding climate deals had been eliminated. Advocating for rights derived from being an indigenous people was not a position available for Greenland prior to COP15, as it was not congruent with the current policy in Greenland. However, at COP16 Greenland was free to take up this well-known and somewhat 'privileged' position in the debate as an indigenous people. The position was privileged in the sense that it enabled the Greenlandic politicians to speak with an independent voice in the negotiations, and simultaneously silenced the Danish voice on the topic of Greenland. The slow becoming of a Greenlandic climate policy illustrates Greenland's limited possibility of formulating wishes and priorities and representing themselves via the UN system. This situation repeatedly triggers old disagreements with Denmark.

For most people I interviewed during my fieldwork, climate change was not a looming catastrophe; rather, it was considered part of everyday life and simply something to cope with (Bjørst 2011; 2012). The same calmness can be identified on the Government of Greenland's own website about climate change in Greenland, www.ClimateGreenland.com. Concerning Greenland's stance on the commitment to bringing down emissions of CO_2, the government explains:

> Naalakkersuisut, the Greenlandic government, recognises that the country might increase its emissions in the future as investments in largescale industrial projects and mining take off ... Naalakkersuisut confronts this challenge by minimising its emissions where it is possible. The fact that Greenland annually has invested 1% of GDP in renewable energy throughout the last decade is an example of this.
>
> (Government of Greenland 2014)

The government proceeds to present what they call 'A Greenlandic Reality', which describes Greenland as challenged by: a small population; long distances; weak infrastructure; problems with maintaining a well-connected society; and a cold climate causing 81 per cent of the country to be covered by a shield of ice. Based on this information, the website asks the reader to understand why Greenland 'compared with many other countries, has a relatively high emission per capita'. Greenland's ability to take on future emission reduction commitments should therefore be seen in this light (ClimateGreenland.com 2014). In other words, following the descriptions at ClimateGreenland.com, Greenland's wish for 'increasing political independence' in combination with the special 'Greenlandic reality' determines and scales down its present commitments to global climate regimes. Whereas the global climate regime (presented in documents like the Kyoto Protocol, from which Greenland is territorially exempted) is concerned with sustaining global nature, the promotion of the 'Greenlandic reality' reflects Greenland's concern to sustain a Greenlandic economy in order to prepare for independence. Not even sustaining local nature plays a significant role. This is a 'reality' which again speaks to an understanding of sustainability as 'economic self-sustainability for Greenland' (as formulated in the Coalition Agreement 2014–2018:3) and is not linked to the environment, climate change health issues, or CO_2 reductions. Moreover, it is a logic ('reality') which in effect has caused the last five Governments of Greenland (Liberal as well as left-wing coalitions) to pursue the establishment of new industries in order to support the ambitions of increased political and economic independence from

Denmark. What, according to Greenlandic politicians, should be sustained is Greenland's national development project towards becoming its own state and its right to development. Denmark, in contrast, is/was keen to demonstrate global leadership in the climate negotiations and to sustain Denmark as a 'showroom' for green technology and economic growth.

The analysis of the climate debate in Greenland reveals that talking about 'climate' or 'climate change' is a relatively new concept within Greenlandic politics. For almost two decades, Greenlandic politicians have been trying to maintain a double climate strategy (Bjørst 2008) as a way of dealing with climate commitments: on the one hand, politicians have encouraged a global reduction of CO_2 emissions, while, on the other hand, not letting Greenland be limited by global standards on a local scale (Bjørst 2008). Greenland did not meet its obligations under the Kyoto Protocol to reduce emissions by 8 per cent. Rather, in agreement with the Danish Ministry for Climate, Energy and Building, Greenland's fulfilment of the first commitment period was secured through the purchase of CO_2 credits in 2012. For the second commitment period (2013–2020), Greenland was exempted from international reduction commitments through territorial exclusion (Government of Greenland 2014). Events leading up to COP15 and the following year's commitments confirm the pattern: the Greenlandic climate policy can be best described as a policy that never really existed. Greenland's recent decline to join the COP21 agreement (The Paris Agreement 2015) testifies to the durability of the double strategy (underwriting the lack).

Summing up, the analysis of the slow becoming of Greenland's climate policy so far may be described as follows: since 2001, Greenland has been following a double climate strategy that argues for growth and industrial development in Greenland, while advocating for stopping global warming and supporting CO_2 reduction on a global scale. The double climate strategy should be understood as part of an ongoing nation-building process with a growing self-image of Greenland as being on the path to full sovereignty and independence (Gad 2014:99). Particularly the strategic positioning performed by Greenlandic politicians allows me to argue that the lack of a formulated climate policy in Greenland is an effect of a nation-building process that has been underway for decades, involving a growing ambition to be economically independent and *sustainable*. Professor in Anthropology Mark Nuttall, who has researched how the Arctic is approached as a resource frontier, noted how 'Far from being mere victims of the impacts of industrial development, indigenous peoples are participants in, and increasingly beneficiaries of, the development of the Arctic resource frontier' (Nuttall 2010:23). Since 2001, Greenlandic politicians have actively been assuming a variety of subject positions in the local/global climate change debate as victims, as witnesses, as indigenous peoples, and as change agents arguing for their right to development. In contrast, Danish climate diplomacy has been trying to limit Greenlandic positioning to 'victims' and 'witnesses' to global climate change, most markedly during Connie Hedegaard's staging of the Greenlandic case. In the Danish climate discourse and in global arenas, these positions have achieved common-sense status, assisted by representations of Arctic peoples and communities dominant in academic climate change assessments (Martello 2004). Striving for positions that would give access to an independent voice and agency, Greenlandic politicians have probed how a variety of narratives worked in contrast and concert with Danish and global discourses. As a surprise to some, Greenlandic climate change also includes a political question, which this analysis confirms.

Scale-making and sustainability dreams

This study shows that the concept of sustainability in a Greenlandic context privileges the national scale and focuses on the discussions around issues that matter to the coalition government of the moment. The Greenlandic people consider the national scale (the national budget of an envisioned sovereign state) to trump any global responsibilities. Despite being positioned in an iconic role as global climate victims (by Hedegaard and others), when focus is on the right to 'sustainable development', Arctic citizens, in this case the Greenlanders, assume value and become central players in future-making (Sejersen 2015). Climate change discourses, in contrast, position Arctic citizens as vulnerable and challenged and, most of all, potentially restricted in their development (e.g. by CO_2 quotas). As illustrated in the analysis of the climate negotiations leading up to COP15 in 2009, Greenland cannot give a proper 'state response' to its stand on climate change as they are limited in the UN system by not (yet) being a state and therefore must be represented via the Danish agenda and ask for an exception. In effect, Greenland often ends up presenting itself as more indigenous (and less national) as a result of the power structure of the UN climate negotiations (Dahl 2013).

Despite my ambition to bring climate change and climate policies back into the conversation, this study indicates that as climate agendas disappear from Greenlandic politics, the concepts of sustainability and sustainable development seem to evolve (see Jacobsen, this volume; and Sejersen, this volume) to an extent that it is possible to create a scale in which for example the development of extractive industries is the only road to sustainable development *in* and *for* Greenland (Bjørst 2016). This means that today actors can even invent their own scale of sustainability (depending on the referent object) (Bjørst 2016). As the referent object for Greenland is often its national budget, the scale-making of sustainability is limited to creating local jobs, ensuring capacity building and a better living standard.[5] These observations indicate that sustainability talk (and related scale-making) works very well with Greenland's development plans as long as what is to be sustained is the national project, while climate change or even climate talk is absent in most of those future planes.

This scale-making elegantly combines three different sustainability dreams for Greenland: (1) a sustainable economy (a national budget independent from Danish subsidies, based on private investments and the use of own resources); (2) a sustainable community (jobs, growth, regional development, and a new infrastructure); (3) a sustainable polity (a referendum leading to a sovereign Arctic nation state, able to represent itself in the UN and the Arctic Council). The political establishment in Greenland is aware that Arctic statehood comes with responsibilities, and that ensuring sustainability is one such responsibility. Today, Greenland is still in the waiting room to become its own state, and the position as an indigenous people is often a way of gaining a voice and creating room for manoeuvring in international climate politics. With his discursive practice of combining the indigenous people's position with the wording of the Kyoto Protocol about the 'common but differentiated responsibilities', Kleist elegantly reintroduced Greenlandic development dreams into the negotiations. Ever since, this practice has inspired Greenland's manoeuvring in international climate politics.

Conclusion: a policy and a commitment that never really were

I think it is important to return to my question posed to the panel at the Future Greenland 2015 conference. Climate change was not a matter of concern or uncertainty at the conference. I invited the panel and the audience to be part of a problem that no one would or could see. The politicians did not even bother to try to practice their skills in climate talk, but rather legitimized their list of priorities to the audience by placing 'Growth' at the top and 'Climate' at the bottom of their list (if this was even deemed worthy of entry to the list).

Hence, this chapter has presented an analysis of a climate policy and a commitment that never really were. The result of this absence is that the Greenlandic politicians can legally allow as much pollution as they like, and can even brand such a development as sustainable (for Greenland). This means that the Greenlandic politicians and business community can ensure a good 'investment climate' and argue in favour of a sustainable development without committing to climate targets or CO_2 reductions. Greenlandic politicians do not seem to consider sustainable development a showstopper if the words *climate* and *environment* are deleted. As mentioned, they are even talking about the prospects of 'sustainable mining' (Bjørst 2016).

Both at the Future Greenland Conference and at the international climate events, the imaginary of climate change projected by science came into conflict with day-to-day politics and the national project. Thus, the Greenlandic politicians found a way to play games, and kept changing their positions in the ongoing climate dialogue. This chapter has shown that it is certainly not random who and what is positioned discursively as a winner, loser, witness, or victim in the climate change debate. This is a product of constant negotiation, in which elements of competition can appear, and positions are far from stable. Climate is a multiple object, but it does not necessarily connect to everyday life; not even for the people who are said to be at risk (Bjørst 2012). Whereas the local was constructed inside the global and inside the minds of Connie Hedegaard and others, it did not relate Inuit/Greenlanders to the global ecosystem in local politics. In the down-scaling and up-scaling mechanisms I encountered at the Future Greenland conference and during my fieldwork, it was difficult to determine whether Greenland and sustainability were going from big to small, or vice versa. Rather than one global perspective, there rather appeared to be a series of situated globalities (see Law and Urry 2004). Among these, the 'economic self-sustainability for Greenland' offered one very local and situated perspective with Greenland at the centre of the world and with Greenlanders positioned not as victims or witnesses, but as potentially marginalized citizens of the world fighting for their right to sustainable development.

Notes

1 Quotes have been translated by the author from the Danish.
2 As part of my PhD project (Bjørst 2011).
3 Alfred Jacobsen (IA).
4 Today, Kielsen is the Premier of Greenland.
5 From studies of mining discussions in Greenland (Bjørst 2016) I have learned that being sustainable (in the mining sector) is primarily a question of creating local jobs, ensuring capacity building and a better living standard. In the same vein, tourism development is staged as the more sustainable alternative to mining. This means that tourism is often (in discourse) turned into a strategic tool to achieve a more sustainable future for Greenland with permanent local jobs and development (Bjørst and Ren 2015:97).

References

ACIA (2004). *Arctic Climate Impact Assessment*. Cambridge: Cambridge University Press.

Bjørst, L.R. (2008). Grønland og den dobbelte klimastrategi. *Økonomi og Politik* 4, pp. 26–37.

Bjørst, L.R. (2011). Arktisk Diskurser og klimaforandringer i Grønland. Fire (post) humanistike klimastudier. Unpublished doctoral dissertation, Det Humanistiske Fakultet, Københavns Universitet.

Bjørst, L.R. (2012). Climate testimonies and climate crises narratives: Inuit delegated to talk on behalf of the climate. *Acta Borelia*, 29 (1), pp. 142–157.

Bjørst, L.R. (2016). Saving or destroying the local community? Conflicting spatial storylines in the Greenlandic debate on uranium. *The Extractive Industries and Society*, 3 (1), pp. 34–40.

Bjørst, L.R., and Ren, C. (2015). Steaming up or staying cool? Tourism development and Greenlandic futures in the light of climate change. *Arctic Anthropology*, 52 (1), pp. 91–101.

ClimateGreenland.com (2014). 'A Greenlandic reality'. Available from: www.silappissusaa.gl/om-climategreenland.aspx (accessed 4 April 2014).

Coalition Agreement (2014). Siumut, Demokraatit and Atassut. Coalition Agreement, Election Term 2014–2018 'Fellowship – Security – Development'. Available from: http://naalakkersuisut.gl/~/media/Nanoq/Files/Attached%20Files/Naalakkersuisut/DK/Koalitionsaftaler/Koalitionsaftale%202014-2018%20engelsk.pdf (accessed 20 January 2014).

Copenhagen Accord (2009). UNFCCC. Copenhagen Accord 2009. Decision -/CP.15 (advance unedited version). Available from: http://unfccc.int/files/meetings/cop_15/application/pdf/cop15_cph_auv.pdf (accessed 22 February 2010).

Dahl, J. (2013). The United Nations and the indigenous space. In: Dahl, I.J., and Fihl, E. (ed.), *A Comparative Ethnography of Alternative Spaces*. Basingstoke: Palgrave Macmillan, pp. 19–40.

Direktorat for Miljø og Natur (2000). 'Notat til Alfred Jakobsen vedr. Kyoto-Protokollen.' Grønlands Hjemmestyre, p. 3.

Douglas, M. and Wildavsky, A. (1982). *Risk and Culture: An Essay on the Selection of Technological and Environmental Dangers*. Berkeley, CA: University of California Press.

Egebo, T. (2009). 'Brev til departementschef Steffen Ulrich-Lynge.' Klima- og Energiministeriet. Copenhagen.

Fairclough, N. (1992). *Discourse and Social Change*. Cambridge: Polity Press.

Friis, L. (2010). 'National statement COP 16'. In: FN's Klimakonference 2010 or COP16 (Conference of Parties 16). Available from: www.kemin.dk/national_statement_cop_16_copenhagen_accord_energy_let.htm (accessed 8 March 2018).

Gad, U.P. (2014). Greenland: a post-Danish sovereign nation state in the making. *Cooperation and Conflict*, 49 (1), pp. 98–118.

Government of Greenland (2014). Vores råstoffer skal skabe velstand [Our raw materials have to create prosperity]. Government of Greenland, Departementet for Erhverv, Arbejdsmarked og Handel. Available from: http://naalakkersuisut.gl/~/media/Nanoq/Files/Publications/Raastof/DK/Olie%20og%20Mineralstrategi%20DA.pdf (accessed 6 June 2014).

Grønlands Hjemmestyre (2001). Rammeaftale mellem landsstyremedlem Alfred Jacobsen og Miljø- og energiminister Svend Auken om ratifikation af Kyoto-protokollen. Miljø- og Energiministeriet og Grønlands Hjemmestyre.

Grønlands Selvstyre (2009). *Redegørelse for virkemidler til reduktion af udledningen af drivhusgasser*. Grønlands Selvstyre, EM 2009/127.

Harré, R. and van Langenhove, L. (1999). Refleksive Positioning: Autobiografi. In: Harré, R. and van Langenhove, L. (eds), *Positioning Theory: Moral Contexts of Intentional Action*. Oxford: Blackwell Publishers, pp. 60–73.

Hedegaard, C. (2008). *Da klimaet blev hot*. Copenhagen: Gyldendal.

Hedegaard, C. (2010). *Kampen om klimaet*. Art People.

Holmsgaard, E. (2007). Miljøminister: Det hjælper at vise den grønlandske case. *Sermitsiaq*, 30 July, p. 15.

Inatsisartut (2012). Inatsisartutlov nr. 25 af 18. December 2012 om bygge- og Anlægsarbejder ved storskalaprojekter. Available from: http://lovgivning.gl/lov?rid={6D7F52B4-6893-4BDC-A943-601817D309A0} (accessed 11 December 2017).

IPCC (2015). *Climate Change 2014: Mitigation of Climate Change* (Vol. 3). Cambridge: Cambridge University Press.

Jasanoff, S. (2010.) A new climate for society. *Theory, Culture & Society*, 27, pp. 233–253.

Latour, B., (2007). *Reassembling the Social: An Introduction to Actor–Network-Theory*. Oxford: Oxford University Press.

Law, J. and Urry, J. (2004). Enacting the social. *Economy and Society*, 33, pp. 390–410.

Martello, M.L. (2004). Global change science and the Arctic citizen. *Science and Public Policy*, 31, pp. 107–115.

Mason, G. (2008). Call me an optimist, but if Denmark can go green, Canada can too. *The Globe and Mail*, 8 December, A6.

Mikkelsen, N. and Ingerslev, T. (2002). *Nomination of the Ilulissat Icefjord for Inclusion in the World Heritage List*. Copenhagen: GEUS.

Ministry of Foreign Affairs, Denmark (2017). SDGs in the Arctic: local and global perspectives – key insights. Available from: http://um.dk/en/foreign-policy/the-arctic/the-sdgs-in-the-arctic/key-insights (accessed 29 December 2017).

Nakashima, D. (2009). Climate change and Arctic sustainable development: scientific, social, cultural and educational challenges. Available from: http://unesdoc.unesco.org/images/0018/001863/186364e.pdf (accessed 8 March 2018).

Nuttall, M. (2010). *Pipeline Dreams: People, Environment, and the Arctic Energy Frontier*. Copenhagen: IWGIA.

Nyborg, K. (2009). Avataq savner politisk uafhængig information om industrielle projekter. *KNR*, 18 May.

Paglia, E. (2017). The socio-scientific construction of global climate crisis. *Geopolitics*, 23, pp. 1–28.

Paris Agreement, The. (2015). Available from: https://unfccc.int/resource/docs/2015/cop21/eng/l09r01.pdf (accessed 10 May 2016).

Schultz-Lorentzen, C. (2010). 'Grønland til klimatopmøde,' AG, 1 November, p. 7.

Sejersen, F. (2014). Klimatilpasning og skaleringspraksisser. In: Sørensen, M. and Eskjær, M.F. (eds). *Klima og mennesker: Humanistiske perspektiver på klimaforandringer*. København: Museum Tusculanums Forlag, pp. 59–79.

Sejersen, F. (2015). *Rethinking Greenland and the Arctic in the Era of Climate Change: New Northern Horizons*. London: Routledge.

SWIPA (2011). *SWIPA: Snow, Water, Ice and Permafrost in the Arctic (Executive Summary)*. Arctic Monitoring and Assessment Programme (AMAP). Oslo: Arctic Council.

United Nations (2015). *2030 Agenda for Sustainable Development and its 17 Sustainable Development Goals*. Available from: www.un.org/sustainabledevelopment/development-agenda (accessed 26 December 2017).

Walsh, B. (2009). Denmark's wind of change: How Denmark's green energy initiatives power its economy. *Time*. Available at: http://content.time.com/time/magazine/article/0,9171,1883373,00.html.

Žižek, S. (1990). Beyond discourse analysis. In: Laclau, E. (ed.), *New Reflections on the Revolution of Our Time*. London: Verso.

9 Building a Blue Economy in the Arctic Ocean

Sustaining the sea or sustaining the state?

Philip Steinberg and Berit Kristoffersen

Introduction

As sustainability discourses extend to the Arctic, planners and politicians are faced with the fact that the Arctic is fundamentally a maritime region. The oceans cover the majority of the planet's surface, and the Arctic is no exception. Although precise definitions of the Arctic vary (Arctic Human Development Report 2004:18), by any definition the Arctic remains unified by a central ocean that takes up the majority of its expanse (Steinberg 2016). As such, any strategy for developing the Arctic's economies or securing its environmental future must be, necessarily, a maritime strategy. And thus it should come as no surprise that calls for sustainable development in the Arctic are linked with the global turn to a Blue Economy.

In this chapter, we explore this turn through an investigation of Norwegian ocean policy. Norwegian officials from both the right and the left have systematically argued for 'sustainable growth' in the ocean as essential for the country's future. In a 2015 lecture celebrating the tenth anniversary of the Norwegian High North Strategy, Jonas Gahr Støre, leader of the Labour Party, told the audience at the University of Tromsø ('The Arctic University of Norway'):

> If I had the chance to define the High North Strategy for the years ahead, then the connecting link for the next chapter in the story about Norway, all of Norway, which would sum up climate change, nature, environment, knowledge, people, challenges, expectations, [and] geopolitics, is the emphasis on the sea.
>
> (10 November 2015)

Two years later, Per Sandberg, Norwegian Minister of Fisheries for the right-wing Progress Party, explained to the European Commission for Fisheries that 'The Norwegian government's new ocean strategy is all about increasing opportunities for sustainable growth, creating more jobs and exploiting the potential of our oceans' (24 April 2017).

By extending this Blue Economy discourse from its origins in fisheries to the breadth of ocean uses, Norwegian leaders from across the ideological spectrum are opening up political space to redefine the country's Arctic policy. At the same time, we argue, Norway is engineering a new future for its all-important offshore oil and gas industry by placing it within an emergent 'blue' ideal for integrated ocean management. In effect, then, by locating the Norwegian state's primary mission – sustaining itself – within a more conventional referent object of the sustainability discourse – the ocean environment – the Norwegian state is proposing a future where its governmental authority is both enabled and exemplified by its rational management of ocean resources.

To develop this argument, the remainder of this chapter proceeds in three parts. First we consider the literature on Blue Economy discourse and development programmes, drawing in particular on the work of Young Rae Choi (2017) who, in her study of China's ocean development strategy, emphasizes that the Blue Economy is neither simply a well-intentioned programme for sustainable development nor a cynical attempt at 'green-grabbing'. Rather, Choi argues, promotion of a Blue Economy should be seen as part of the effort to apply state governmentality in an emergent space of sovereign interest. We then apply this analysis to the Arctic through a study of Norway's Blue Economy initiative that is shaping state policy with reference to its northern resource frontiers. We conclude by suggesting that in its effort to group together numerous competing ocean uses within a single, sustainably managed, ocean industry, the Norwegian state constructs the ocean as a space that is beyond politics and therefore appropriate for state intervention. This finding, in turn, suggests that when the sustainability discourse is integrated with a post-political managerial agenda, the ultimate referent object of sustainability may be the manager itself: in this case, the state.

The Blue Economy as a sustainability strategy

In their comprehensive review of calls for a Blue Economy at the 2012 United Nations Conference on Sustainable Development (the 'Rio + 20' conference), Silver *et al.* (2015:137) locate the term within the broader promotion of a Green Economy: a 'managerial ontology of natural capital' in which a state's natural resources become valued as capital stock, with its value typically incorporating its ecosystem services as well as its extractive resource potential. In contrast with 'brown' development models in which resources are costed without regard to their externalities, in the Green Economy (and, by extension, its maritime, Blue, version), the rational management of these resources (including through internalization of externalities) is to be taken up by partnerships of states, global finance bodies, and environmental NGOs, with measured liquidation of these resources to be undertaken by private capital (see also, Winder and Le Heron 2017).

Green (and Blue) Economy programmes locate themselves within the broader sustainability discourse, but then take that discourse in a specific direction that favours the economization of nature's capital (Corson and MacDonald 2012; Onesti 2012). Silver *et al.* (2015) are generally critical of this economization programme. It effectively reduces the environment to use value while justifying the combination of public management with private accumulation that is the hallmark of neoliberalism. In the process, the concern for equity that frequently features within the rhetoric, if not the actual practice, of sustainable development initiatives is subsumed within an overarching agenda for growth and accumulation.

Nonetheless, Silver *et al.* caution that when Green Economy discourse is applied to the ocean – through references to the Blue Economy – there is more at hand than the 'green-grabbing' characterized by the Green Economy's critics. In their analysis of references to the Blue Economy at Rio + 20, Silver *et al.* identify four distinct elements: a 'natural capital' discourse that values the ocean for its long-term ecosystem services and that typically calls for the enrolment of large financial institutions (e.g. the World Bank) and global conservation groups (e.g. The Nature Conservancy) to facilitate leveraging of this value; a 'good business' discourse that promotes and commends private enterprise that operates in a relatively environmentally friendly manner; a 'SIDS' (small island developing states) discourse that highlights the ocean's economic and environmental significance in order to facilitate the economic development of these states; and a 'small-scale fishers' discourse wherein appeals to the Blue Economy are used to promote the survival of small-scale fishers against the threats posed by overfishing by large-scale, heavily capitalized fishing fleets. While aspects of each of these elements can support each other, the four can also work at cross-purposes, and thus Silver *et al.* suggest that the Blue Economy discourse provides grounds for negotiation rather than simply being a tool of 'green-grabbing' hegemony.

While Silver *et al.* shed light on the way in which the Blue Economy concept is applied in the framework of global governance initiatives and how it is being utilized by SIDS, their analysis appears less well suited for interpreting the ways in which the term is being applied in the Arctic. Notably absent from Silver *et al.*'s article is any mention of the ocean resource most prevalent in discussions of the maritime Arctic's resource potential: offshore oil and gas. To an extent, this omission is understandable as oil and gas are rarely considered 'green' (or, in the ocean, 'blue') industries, although, as is discussed below, in Norway in particular the state often takes credit for 'greening' oil and gas extraction and thereby integrating it into its Blue Economy strategy (see also Kristoffersen 2015). In addition, the gap may be the result of Silver *et al.*'s focus on international conferences. Oil and gas extraction is presently restricted to areas within a single state's sovereign jurisdiction or, in some instances, in areas where two states have bilaterally established a joint development zone. Therefore, global regulation of the industry (including its 'greening') might best be achieved through corporate take-up of best practices rather than through spatial management tools. Nonetheless, even if hydrocarbon extraction is rejected as an industry that can never be 'greened', its persistence in the ocean must be integrated into any understanding of the Blue Economy agenda. As of 2015, nearly 30 per cent of total global oil output came from offshore production. Maintaining a focus on oil and gas is particularly important when one turns to

Norway. Norwegian fields accounted for 7 per cent of global offshore production, and Norway's Statoil is engaged in production around the world (US Department of Energy 2016).

The inapplicability of Silver *et al.*'s framework to the Arctic goes beyond its omission of oil and gas, however. Silver *et al.* focus on parts of the world where management by international institutions (UN system organizations, bilateral aid bodies, environmental NGOs, etc.) is typically seen as beneficial for advancing one or more of the four goals that the authors identify (achieving full valuation of the ocean's resources, promoting good business practices, enhancing the development of small island states, or promoting small-scale fisheries). By contrast, as was articulated in the Ilulissat Declaration (2008), the five Arctic Ocean littoral states prefer a unilateral approach where each state has uncompromised sovereignty over the development of 'its' ocean. While a limited role may be given to cooperatively formed regional organizations (e.g. the Arctic Council, the Barents Euro-Arctic Council), the Ilulissat Declaration makes it clear that there is no role for higher global governance bodies, since none is sanctioned by the United Nations Convention on the Law of the Sea, which reigns supreme as the spatial management framework for the Arctic Ocean. Reflecting on Arctic Blue Economy initiatives, then, it becomes apparent that the problem with Silver *et al.*'s framework is not just that it largely focuses on the Global South while the Arctic is in the Global North (although parts of the region would generally merit classification as 'Global South' when using standard development indicators). Rather, the ill fit stems from the fact that in the Arctic Blue Economy strategies are more often associated with national and regional 'sustainable development' initiatives than with attempts to marshal global conservation capital, secure global development funds, or support small-scale fisheries.

Given the differences between applications of Blue Economy discourse in the Global South and Global North, one might assume that a better entry point for understanding Arctic Blue Economy initiatives would be Winder and Le Heron's (2017) study of Blue Economy references in European Union workshops and policy documents. Winder and Le Heron understand ocean-space as an assemblage of multiple practices and elements that intersect, inform each other, and congeal into institutions in varying and unstable ways. Winder and Le Heron suggest that the Blue Economy assembles the ocean in one specific way (particularly with reference to its biological and economic properties) while forestalling other possible assemblages. Significantly, though, they also note that once one understands the ocean (and its rhetorical-institutional frameworks) as assemblages, one frees oneself to consider alternate rationalities for ocean management, and they develop this final point through a consideration of ocean management initiatives in New Zealand.

Winder and Le Heron's assemblage approach has been thoroughly considered and critiqued from numerous angles,[1] and we will not rehearse all of these critiques here. However two bear further elaboration. One concerns the relatively weak version of assemblage theory employed by Winder and Le Heron and the suggestion, articulated by Bear (2017), that to truly understand the ocean as an assemblage one must also accommodate the mosaic of temporalities and spatialities (rhythms, repetitions, stabilizations, and destabilizations) that constitute the ocean's materiality (see also, Steinberg and Peters 2015). Although we do not belabour this point in the analysis of Norway's ocean development strategy that follows, elsewhere we have demonstrated how Norway's policies for oil and gas extraction in its icy waters seek to stabilize and rationalize the intersection of climatological, biological, and oceanographic forcings that vary over time and space and that, in their very complexity, defy stabilization into fixed, essential categories (Steinberg *et al.* 2018).

Of greater relevance to our concern with the role of sustainability narratives within Blue Economy programmes, however, is the critique by Choi (2017), who stresses that an explicit consideration of governmentality is largely missing from Winder and Le Heron's analysis. In her analysis of China's Blue Economy policy, Choi writes,

> It is on the one hand an expansion of capitalist space driven by the state to the oceans, which are perceived as underdeveloped frontier spaces through which infinite possibilities of 'better' uses are imagined, institutionalized, and invested. On the other hand, it is intrinsically a spatial intervention that rearranges people and resources so as to avoid waste and to achieve their economic use. In other words, *infinite possibilities collide with finite space.*
> (Choi 2017:39, emphasis added)

The need to foster 'infinite possibilities' in 'finite space', in turn, requires far-seeing management of spaces and the practices that occur within them. Proactive management is required to prevent environmental decline (e.g. from overfishing, unregulated pollutant outflow, etc.), to stave off conflict (e.g. between seascape-dependent coastal tourism industries and unsightly offshore wind farms), and to foster emergent industries that will require state-sponsored investment before becoming profitable (e.g. deep-sea bioprospecting). Thus, in China, the Blue Economy agenda has been used to justify 'a comprehensive sea governance system through which the country's sovereign sea space is imagined and managed in its entirety' (Choi 2017:38).

Significantly, although this 'embedding [of the] marine economy in national development planning … has made sovereign sea space increasingly visible and legible and effectively conceptualized it primarily as economic space' (Choi 2017:39), it has not been accompanied by the reduction of the ocean to a denatured spatial abstraction. Rather, the Blue Economy is constructed as particularly sensitive to nature's variations and interdependencies. Substantial resources are devoted to expanding knowledge of the sea, not just to increase extractive potential but also to inform policy that can sustain its numerous environments. The ocean's future – China's future – is defined as one that necessarily must be sustainable. This is understood as requiring comprehensive management, as well as cutting-edge science to support that management.

However, this knowledge is always carried with an eye towards defining and separating the ocean's numerous processes, physical states, functions, and uses. The sustainability of the sea and the sustainability of the state are thus seamlessly interwoven: a vibrant sea economy is required to sustain the state and a strong, managerial state is required to sustain the sea as an environment of resources and riches.

The pathway towards the goal of sustainability is thus characterized by the institutionalization of spatial rationalities that, in the interest of regulation, simplify the very processes that they attempt to sustain. To quote Choi again:

> The natural attributes have not changed; it is the new relations between the natural world and the economy, assembled in the soil of the desire for economic growth and technological optimism that justify particular uses in particular geographic places.
>
> (Choi 2017:39)

In our previous work on Norway's efforts to manage oil and gas activities in the Southeast Barents Sea, we similarly illustrated how ocean management, although informed by an attentiveness to the complex biological processes of ecoclines, resulted in a reduction of dynamic surfaces to static binary categories that neither adequately reflected temporal-spatial variability nor accounted for the complex processes whose vulnerability initiated the planning process in the first place (Steinberg and Kristoffersen 2017; Steinberg *et al.* 2019). In this chapter, as we move our focus to Norway's broader effort to construct a Blue Economy, we find that the effort to achieve sustainable development in Norway's maritime Arctic is accompanied by a mixture of optimism and concern similar to that which characterizes the Chinese effort: Once the ocean is established as a space of 'infinite possibilities in finite spaces', one requires comprehensive management to maximize potential, minimize harm, and ensure a sustainable future.

Norway: the Arctic Blue Economy as maritime manifest destiny

As leaders of a self-declared ambitious maritime nation, Norway's political elite has embraced the Blue Economy as the cornerstone of the country's future. Illustrating Choi's assertion that the Blue Economy agenda institutionalizes governmentality, Norway's leaders go beyond simply appealing to the ideal of sustainability. Rather, Norwegian officials assert that in the hostile, complex, and interconnected environment of the ocean frontier, the intersection of what Choi calls 'infinite possibilities in finite spaces' requires proactive management. Thus, for sustainability to be achieved in the ocean, environmental sensitivity and a thirst for growth must be informed by the application of knowledge. Norway claims a long history of understanding, extracting resources from, and sustainably managing, its northern waters; indeed the encounter with the Arctic as a fundamentally maritime region plays a central role in Norwegian state identity (Medby 2015; see also Leira *et al.* 2007).

Thus, Norway's institutional *knowledge* and *history* of the Arctic marine environment (which might be joined together via the concept of *expertise*) are heralded by Norwegian officials as justifications for Norway's leadership. This fusion of history with knowledge is buttressed in turn by appeals to geographic destiny. Labour Party leader Jonas Gahr Støre has stressed that Norway will 'conquer ocean space' by becoming a 'global knowledge hub' akin to Silicon Valley (Støre 2015), while also noting that 'for Norway [being] North in the world, North – towards the Arctic, nobody is better positioned than us'. In a similar vein, the Norwegian Government's 2017 Ocean Strategy, *New Growth, Proud History*, boasts of Norway's leading presence across a range of ocean sectors:

> Every day hundreds of thousands of Norwegians go to work in the ocean industries, which together represent about 70 per cent of our export income. Norway is one of the world's largest producers of oil and gas. We are one of the world's largest and most advanced seafaring nations. We are the world's second largest exporter of fish and seafood. In addition, we have a world class service and supply industry. Norway is also at the forefront of marine research and responsible management of marine resources.
>
> (Ministry of Trade, Industry and Fisheries/Ministry of Petroleum and Energy 2017:4)

The strategy, which builds upon a previous strategy, *Maritime Possibilities: Blue Growth for a Green Future* (Ministry of Trade, Industry and Fisheries 2015), emphasizes the embeddedness of the marine economy across sectors and stresses how the marine sector is at the forefront for developing economic knowledge and practice. The Ministry of Foreign Affairs has taken up the programme as well. Stressing the Blue Economy strategy's alignment with UN Sustainable Development Goal 14 (which calls for conservation and sustainable use of the oceans), the ministry publicizes Norway's commitment to ocean development to evidence the country's environmentalist credentials (Aas 2017; Ministry of Foreign Affairs 2017).

In short, Norway is positioning itself as the Arctic Ocean's rightful and natural steward, a country with the capacity, presence, and vision, as well as the appropriate geographic location and cultural-economic history, to manage its northern waters. Norway thus claims to be poised to maximize production and conservation across a range of sectors, while minimizing conflict. Norway advocates its maritime manifest destiny[2] through the management of a single, integrated plan for extracting value from the ocean that will facilitate 'blue growth through green restructuring' (Ministry of Trade, Industry and Fisheries/Ministry of Petroleum and Energy 2017:8). Norwegian officials have often described this as a single, multi-sector ocean industry, where technology and knowledge are smooth and mobile, floating across and among the various sectors that extract value from the sea. Thus the Conservative Party's press release announcing the strategy declared that the '[n]ew ocean strategy will ensure more jobs in the ocean industry' (Høyre 2017). The same year, the Labour Party's Jonas Gahr Støre launched his campaign for prime minister by noting that 'the first thing I will do as Prime Minister is to establish a value creation program for the ocean industries' (Norwegian Broadcasting Agency 2017). This vision of a single, multi-sectoral ocean industry (or, in Støre's terms, a linked family of industries) is illustrated in a poster produced for a 2016 workshop co-organized by the Ministry of Trade, Industry and Fisheries and the Ministry of Petroleum and Energy (Figure 9.1). Wind power, oil drilling, fish farming, and trawling live in peaceful co-existence, aided by the gaze of a satellite and the order of a geodetic map. The material properties of the ocean – the presence of sea ice, the topography of the seabed, the power of currents, the variability of depth – not to mention the

Figure 9.1 Poster from a conference on the Norwegian Blue Ocean policy (Bergen, 30 May 2016), sponsored by Ministry of Trade, Industry and Fisheries, and the Ministry of Petroleum and Energy.

Source: © Ministry of Trade, Industry and Fisheries.

conflicts that persist among multiple users, are elided amid the ideal of peaceful co-existence under the watchful eye of the state.

Blue Economy, brown oil

Not evident from this poster, however, is that this managerial approach towards an ocean of multiple, compatible uses united by the pursuit of a Blue Economy has its origins in the remaking of the north in the wake of an energy crisis. As Norway's oil production peaked in 2001, the oil industry and its state allies increasingly turned towards northern waters as the country's next hydrocarbon frontier. However, the extension of hydrocarbon exploration to the Norwegian Arctic opened a new space of political conflict, as the potential for oil and gas drilling sat uneasily amid efforts to sustain local and national fishing economies as well as protect rich ecosystems in fragile Arctic environments. To meet its various goals, the Norwegian state proposed an integrated approach based on the zoning of ocean space. The *Lofoten-Barents Sea Management Plan* sought to 'facilitate the coexistence of different industries, particularly the fisheries industry, maritime transport and petroleum industry' (Ministry of Environment 2006:7–8). The plan established guidelines for hydrocarbon extraction in the country's northernmost waters while calling for the protection of a number of environmentally sensitive regions, including the fish-rich Lofoten islands in the southern part of the Norwegian Arctic and the areas around the marginal ice zone (the area with seasonal ice cover) in the north.

While the strategy of constructing order through allocating specific activities to individual regions is a common technique of the managerial state, including in ocean regions (Steinberg 2011), it has the effect of depoliticizing what are, in fact, inherently conflictual situations. Choi notes that the Blue Economy agenda often celebrates the co-existence of various groups 'that may be incompatible in other venues' (Choi 2017:40), and this would certainly seem to be the case with Norway's advocacy of a single 'ocean industry' that unites numerous means of extracting value from the sea. In fact, a close examination of Figure 9.1 reveals that its vision is dependent on a number of cartopolitical oversimplifications. Although spatial differentiation can be employed to prevent the co-location of incompatible economic activities, many ocean activities have impacts that affect other, adjacent areas. Thus, for instance, although the image shows the oil rig and the school of herring co-existing by occupying different areas of ocean-space, in actuality the fish could swim under the rig or, conversely, pollution from the well could contaminate adjacent fish habitats. Additionally, all spatial management strategies have outer bounds, both for defining the limits of the space being managed and for defining the extent of managed activities. In the case of the Norwegian 'ocean industry' graphic, the focus is exclusively on the deep sea, which means that both coastal regions and coastal activities are, quite literally, off the map. In actuality, some of the most heated contests in Norwegian ocean management in the past decades have involved coastal residents who have contested further expansion of the 'resource frontier', whether through nearshore oil and gas drilling, mining (dumping waste), or fish farming. The image thus smooths over potential dispossessions, or 'new extractivism' in the 'name of sustainable ocean management' (Winder and Le Heron 2017:20).

The Norwegian oil industry has a long history of co-opting opposition by proclaiming its environmentalist credentials (Kristoffersen 2014), and the repositioning of that industry as the cornerstone of an emergent Blue Economy would seem to be a continuation of this trend. Indeed, while the commitment to develop new marine technologies may be seen, in one sense, as part of a programme for advancing Norway's ocean economy beyond oil and gas, these technological developments can also permit oil and gas extraction to reach into new frontiers. The technological fusion thus goes in both directions. As the Minister of Fisheries told the European Commission, 'Testing new technologies are also central points to our ocean strategy. Norway has a long-standing tradition of transferring knowledge between industries. And we would like to see more of this' (Sandberg 2017). The implication here is that out of this technological fusion Norway will become the leader in a new multi-sector ocean industry that can set a high-technology, low-environmental-impact model for countries around the world as they seek to engage the Blue Economy in pursuit of sustainable development.

Conclusion: contesting the Blue Economy

As the quotations presented at the beginning of this chapter suggest, Norway's turn to a Blue Economy strategy crosses, and claims to transcend, political divides. It thus echoes the concept of the post-political, in which contesting practices and priorities are subsumed within questions of 'management'. A number of scholars, most notably Eric Swyngedouw (2007), have associated this celebration of the post-political with the turn to sustainable development discourse. Within the sustainable development discourse, the omniscient state asserts its authority in the name of efficiency and productivity, securing a future for itself and its citizens and thereby elevating itself seemingly above politics.

As our analysis of Norway's Blue Economy strategy suggests, Norwegian ocean management policy is directed less towards the sustainable management of the marine environment than towards the expansion of oil and gas production in a manner that accommodates other ocean uses. The state thus emerges as both the initiator and beneficiary of these management activities. Indeed, its institutions, as well as its authority, are deemed necessary for keeping politics *out of* the sea. And yet despite the best efforts of politicians from both the left and the right, ocean policy remains a central focus of political conflict. For instance, in the national election in September 2017, the future of the offshore oil and gas industry and its central role in the Norwegian economy were key points of debate (Holter 2017; Lahn 2017). The ensuing discussion over the petroleum economy spilled over to related questions in marine management, including the government's ambition to quintuple production from aquaculture as well as concerns regarding the rights of coastal fleets to fishery resources. In short, the integrated, spatially differentiated, multi-sector 'ocean industry' depicted in Figure 9.1 emerged less as a model of post-political consensus than as a comprehensive list of areas for public contestation.

In conclusion, the Blue Economy agenda, as a state programme that seeks to blend the sustainable development discourse with a turn to post-political management, is itself the site of political conflict. As such, it should be understood less as a means for rationalizing the longevity of ocean resources than as a means for legitimizing and maintaining state power, as well as extending that power to emergent resource frontiers. As the world's oceans present new opportunities for the realization of state power, the case of Norway's Blue Economy agenda reminds us that efforts to sustain the seas may be rooted in efforts to sustain the state.

Notes

1 The issue of *Dialogues in Human Geography* in which Winder and Le Heron's (2017) article appeared (vol. 7, no. 1) also included five critical responses.

2 Manifest destiny is a term that originated in the United States in the 1840s, when expansionist journalists and their allies in office asserted that, due to the confluence of historic, geographic, and cultural factors, the United States was naturally destined to expand across the extent of the North American continent.

References

Aas, K. (2017). The oceans: a challenged resource. *Huffington Post.* Available from: www.huffingtonpost.com/entry/the-oceans-a-challenged-resource_us_59d7a2cbe4b0cf2548b33674 (accessed 17 October 2017).

Arctic Human Development Report (2004). Akureyri, Iceland: Stefansson Arctic Institute.

Bear, C. (2017). Assembling ocean-life: more-than-human entanglements in the Blue Economy. *Dialogues in Human Geography*, 7 (1), pp. 27–31.

Choi, Y.R. (2017). The Blue Economy as governmentality and the making of new spatial rationalities. *Dialogues in Human Geography*, 7 (1), pp. 37–41.

Corson, C., and MacDonald, K.I. (2012). Enclosing the global commons: the convention on biological diversity and green grabbing. *Journal of Peasant Studies*, 39 (2), pp. 263–283.

Holter, M. (2017). Forget peak demand, Norway's oil chief wants more crude to pump. Available from: www.worldoil.com/news/2017/10/19/forget-peak-demand-norways-oil-chief-wants-more-crude-to-pump (accessed 22 October 2017).

Høyre [the Conservatives] (2017). Skal skape flere blå jobber [Will create more blue jobs]. Available from: https://hoyre.no/aktuelt/nyheter/2017/jobb-arbeid-hav-naering (accessed 31 July 2017).

Ilulissat Declaration (2008). Arctic Ocean Conference, Iluilissat, Greenland, 27–29 May. Available from: www.oceanlaw.org/downloads/arctic/Ilulissat_Declaration.pdf.

Kristoffersen, B. (2014). 'Securing geography': framings, logics and strategies in the Norwegian High North. In: Powell, R. and Dodds, K. (eds), *Polar Geopolitics? Knowledges, Resources and Legal Regimes*. Aldershot: Edward Elgar, pp. 131–148.

Kristoffersen, B. (2015). Opportunistic adaptation: new discourses on oil, equity and environmental security. In: O'Brien, K. and Selboe, E. (eds), *The Adaptive Challenge of Climate Change*. Cambridge: Cambridge University Press, pp. 140–159.

Lahn, B. (2017). Oljedebatten er kommet for å bli [The oil debate is here to stay]. Commentary in *Dagbladet*, 24 September. Available from: www.dagbladet.no/kultur/oljedebatten-er-kommet-for-a-bli/68718558 (accessed 28 September 2017).

Leira, H., Borchgrevink, A., Græger, N., Melchior, A., Stamnes, E., and Øverland, I. (2007). *Norske selvbilder og norsk utenrikspolitikk* [Norwegian self-images and Norwegian foreign affairs]. Oslo: Norwegian Institute of International affairs (NUPI). Available from: www.files.ethz.ch/isn/46181/200704_Norske_selvbilder.pdf (accessed 15 June 2017).

Medby, I.A. (2015). 'Big fish in a small (Arctic) pond: regime adherence as status and Arctic identity in Norway'. In: *Arctic Yearbook 2015*. Akureyri, Iceland: Northern Research Forum, pp. 313–326.

Ministry of Environment (Norway) (2006). Meld. st. 8 (2005–2006): Helhetlig forvaltning av det marine miljø i Barentshavet og havområdene utenfor Lofoten (forvaltningsplan). Published in English as: Report to Parliament (2005–2006) 8: Integrated management of the marine environment of the Barents Sea and the sea areas off the Lofoten Islands. Available from: www.regjeringen.no/globalassets/upload/md/vedlegg/stm200520060008en_pdf.pdf (accessed 26 July 2017).

Ministry of Foreign Affairs (Norway) (2017). The place of the oceans in Norway's foreign and development policy. White Paper 22 (2016–2017) to the Storting. Available from: www.regjeringen.no/contentassets/1b21c0734b5042e489c24234e9927b73/en-gb/pdfs/stm201620170022000engpdfs.pdf (accessed 26 August 2017).

Ministry of Trade, Industry and Fisheries (Norway) (2015). *Maritime muligheter – blå vekst for grønn framtid* [Maritime Possibilities – blue growth for a green future]. Available from: www.regjeringen.no/contentassets/05c0e04689cf4fc895398bf8814ab04c/maritim-strategi_web290515.pdf (accessed 7 March 2018).

Ministry of Trade, Industry and Fisheries/Ministry of Petroleum and Energy (Norway) (2017). *New Growth, Proud History: The Norwegian Government's Ocean Strategy*. Available from www.regjeringen.no/en/dokumenter/the-norwegian-governments-ocean-strategy/id2552610/ (accessed 28 September 2017).

Norwegian Broadcasting Agency (2017). Her er Jonas bredside mot Høyre og FRP [Here is Jonas' broadside against the Conservatives and the Progress Party]. Available from: https://nrk.no/norge/her-er-jonas-bredside-mot-hoyre-og-frp-1.13481187 (accessed 31 July 2017).

Onesti, M. (2012). Latin America and the winding road to Rio+20: from sustainable development to green economy discourse. *Journal of Environment & Development*, 21 (1), pp. 32–35.

Sandberg, P. [Norwegian Minister of Fisheries] (2017). Talk for the European Parliament's Fisheries Committee, 24 April. Available from: http://web.ep.streamovations.be/index.php/event/stream/170424-1500-committee-pech (accessed 15 June 2017).

Silver, J.J., Gray, N.J., Campbell, L.M., Fairbanks, L.W., and Gruby, R.L. (2015). Blue economy and competing discourses in international oceans governance. *The Journal of Environment & Development*, 24 (2), pp. 135–160.

Steinberg, P.E. (2011). The Deepwater Horizon, the *Mavi Marmara*, and the dynamic zonation of ocean space. *Geographical Journal*, 177 (1), pp. 12–16.

Steinberg, P.E. (2016). Europe's 'Others' in the Polar Mediterranean. *Tijdschrift voor economische en sociale geografie*, 107 (2), pp. 177–188.

Steinberg, P.E. and Kristoffersen, B. (2017). 'The ice edge is lost … nature moved it': mapping ice as state practice in the Canadian and Norwegian North. *Transactions of the Institute of British Geographers*.

Steinberg, P.E. and Peters, K. (2015). Wet ontologies, fluid spaces: giving depth to volume through oceanic thinking. *Environment and Planning D: Society & Space*, 33 (2), pp. 247–264.

Steinberg, P.E., Kristoffersen, B., and Shake, K. (2019). Edges and flows: exploring legal materialities and biophysical politics at the sea ice edge. Under review for publication in *Ocean Legalities* (Irus Braverman and Elizabeth Johnson, eds.). Durham, NC: Duke University Press.

Støre, J.G. [Labour Party leader] (2015). *Nordområdene – 10 år etter – ved starten av et nytt kapittel i fortellingen om Norge* [The High North – 10 years later. At the beginning of a new chapter in the story about Norway], UiT-breakfast, UiT – the Arctic University of Norway, 10 November, transcribed by the authors. Available from: https://mediasite.uit.no/Mediasite/Play/d0d4758a9c7140d8b9fccf4f0d7d42be1d (accessed 15 June 2017).

Swyngedouw, E. (2007). Impossible 'sustainability' and the postpolitical condition. In: Krueger, R. and Gibbs, D. (eds), *The Sustainable Development Paradox: Urban Political Economy in the United States and Europe*. New York: Guilford Press, pp. 13–40.

US Department of Energy (2016). Offshore production nearly 30% of global crude oil output in 2015. Available from: www.eia.gov/todayinenergy/detail.php?id=28492 (accessed 25 July 2017).

Winder, G.M. and Le Heron, R. (2017). Assembling a Blue Economy moment? Geographic engagement with globalizing biological-economic relations in multi-use marine environments. *Dialogues in Human Geography*, 7 (1), pp. 3–26.

10 Saving the Arctic

Green peace or oil riot?

Hannes Gerhardt, Berit Kristoffersen, and Kirsti Stuvøy

In July 2017 Greenpeace's *Arctic Sunrise* ship again sailed to the Barents Sea to protest oil drilling. As before, there were activists from all over the world voicing their concern about state and corporate efforts to extract profit from undersea fossil fuels. Featuring themselves in front of Statoil's *Songa Enabler* oil rig in social media, they continued their efforts to delegitimize Arctic oil and gas activities in their campaigns to 'Go Beyond Oil' and 'Save the Arctic' (Greenpeace 2017d). This was a continuation of the more aggressive efforts to board oil rigs in the territorial waters of Greenland, Russia, and Norway between 2011 and 2014. These previous actions pushed the limits of legality, but the campaigns went viral on social media and garnered strong support. Clearly, Greenpeace was seeking an audience that reached far beyond the Arctic, with the aim of bringing to light the environmental risks that link the Arctic to the rest of the world. Yet when we consider the reactions by the various states targeted, we find an inherent clash with this largely global focus, as the states jostled to reaffirm their own national priorities in the region.

The confrontation between Greenpeace and the three states in these interventions reveal an underlying struggle over the broad question of what should be sustained in the Arctic, or perhaps more accurately, what has priority with regard to 'sustainability'. In the case of Greenpeace, we find a regional focus on the natural systems endangered by oil rig operations as well as a global focus on climate. This priority, in turn, is linked to the relationship of this sustainability to the inherent value of these natural systems and, more generally, the well-being of humanity. In this sense the Arctic is seen as the heritage and responsibility of all, for both its natural value and its role in the evolving climate threat. It is this view of sustainability in the Arctic context that also informs Greenpeace's approach to a solution, namely a global movement to establish a moratorium on Arctic oil/drilling and the establishment of an internationally recognized marine sanctuary for non-coastal parts of the Arctic Ocean. This focus on international collaboration to achieve environmental governance is here seen as an attempted embodiment of the organization's namesake, achieving a green peace.

When turning to the states targeted by Greenpeace's interventions, however, we find that they did not directly respond to the sustainability demands being made. Instead, Russia, Norway, and Greenland focused on turning Greenpeace's global perspective back to the national, territorial scale; in this case it is first and foremost the dominance of the nation state that needs to be sustained. This underlying political 'sustainability' is further associated with other state priorities in the Arctic. In the case of Greenland, sustaining the nation state is linked to self-determination and future sovereignty; in Russia to domestic and geopolitical authority; and in Norway to a desire to maintain its technological ('environmental') supremacy in the realm of Arctic oil exploration. In all three cases, however, the Greenpeace actions are decried as the work of a fringe group of irresponsible eco-rebels engaging in little more than oil riots.

In this chapter we explore the details of this struggle over sustainability imperatives. The analysis draws on interviews with Greenpeace activists in each of the different settings of Greenland, Russia, and Norway,[1] as well as reactions by each of the states involved as expressed in publicly available texts – e.g. press releases, reports, and government statements. Our analysis proceeds in the following steps. First, we consider how Greenpeace's Arctic campaign is built around scalar disruptions and moral geographies that seek to prioritize an environmental sustainability in the Arctic that would fundamentally depart from the current business-as-usual approach. We then spell out how the underlying geographic imaginaries of Greenpeace connect localities around the world to this specific region, thereby challenging the taken-for-granted dominance of the national scale. Next we explain the responses to Greenpeace's actions by Greenland, Russia, and Norway by considering the contexts within which they operate. More specifically, we demonstrate how the activist interventions ('riots') trigger separate sustainability prioritizations by the targeted states that are ultimately based on maintaining a particular conception of the nation state into the future. We conclude that any demands for environmental sustainability that challenges the scalar assumptions inherent in these state visions is quickly sidelined.

Sustainability in the Arctic: business as usual

The concept of environmental sustainability in the Arctic is not new, but it is only in the last few years that it has become normalized. Today it is practically impossible to find a government or corporate report on resource extraction in the Arctic that does not make some mention of the need to ensure sustainable practices. The fact that the term sustainability is so readily used by the actors engaged in Arctic oil and gas drilling is itself revealing, as it uncovers an important and even unassailable assumption that this activity is no way incompatible with sustainability. This is an assumption that the three state entities in this case study all have in common.

But what does sustainability in this context actually mean? In broad terms the use of the term sustainability in the corporate and state discourses on natural resource extraction in the Arctic emphasize environmental safety as a precondition for the larger aim of economic development. This safety, in turn, is assumed to be achievable given adequate research and study. Hence, there is continuous reference made to Environmental Impact Assessments, studies, and consultancy. Take, for instance, *Greenland's Oil and Mineral Strategy 2014–2018*, which states: 'The current development of the … resources sector must be based on sustainability. It is important that the development takes place in an environmentally safe manner' (Government of Greenland 2014:66). Sustainable development here means economic growth that is done with the requisite safety protocols being followed.

Yet for environmental groups such as Greenpeace this sustainability discourse has become predictable, generic, and increasingly vacuous. More specifically, the case is made that 'sustainable development' has lost sight of exactly what in the environment is to be sustained. As one leader from Greenpeace stated:

> If you look at the Arctic Council, every time they make a statement they put the word 'sustainable' in there, even though it would take me (only) a minute to explain why the rest of the statement is not sustainable.
>
> (Interview with Greenpeace)

And later: 'I've heard oil companies say in one sentence they need to be sustainable, prevent climate change, and drill in the Arctic.' From this perspective, it is clear that the term 'sustainability' has lost its referent and rather has become a public relations nicety that means nothing more than business as usual.

Related to this sceptical view on the ground, Erik Swyngedouw has made the case that the sustainability discourse has become an integral part of the post-political condition. Swyngedouw (2009; 2010) thus identifies 'a particular regime of environmental governance that revolves around consensus, agreement, participatory negotiation of different interests and technocratic expert management in the context of a non-disputed management of market-based socio-economic organization' (2010:227). Within this regime it is possible to have disagreements, but only with regard to 'the choice of technologies, the mix of organizational fixes, (and) the detail of the managerial adjustments' (Swyngedouw 2009:611). A more fundamental questioning of 'the system' is not tolerated. For Greenpeace, then, a major effort in its oil rig occupations in the Arctic was fundamentally to challenge this vacuous notion of sustainability. Yet, as we shall see, it was the way they went about this task that drew most of the states' attention.

Not just ecocentric: Greenpeace's Arctic sustainability focus

Greenpeace commenced its oil rig occupations in the Arctic under the campaign 'Go Beyond Oil', which was broadly focused on the perceived need to move away from dependency on fossil fuels. The oil rig occupation off the Greenlandic coast was performed with banners advertising this campaign in 2010. Since 2013, however, Greenpeace has switched its focus to a more regionally focused campaign, 'Save the Arctic', which bases its vision in part on the accomplished Antarctic-Environmental Protocol. This treaty, which entered into force in 1998, essentially makes Antarctica and its associated ecosystems a protected nature reserve off-limits from any commercial exploitation. The 'Save the Arctic' campaign goal is to ban all oil rig activities in the Arctic and to set up a marine sanctuary, free of industrial fishing, in the parts of the Arctic currently not under state control.

For Greenpeace, a major effort then is to again assert a clear referent to the idea of sustainability, which in this case is both the fragile ecosystem of the Arctic and a 'natural' climate for the globe. Underlying this sustainability emphasis, however, lies a relation to what Greenpeace identifies as its primary goal: 'to ensure the ability of the earth to nurture life in all its diversity' (Greenpeace 2017a). It is this goal that has prompted the characterization of Greenpeace as pursuing an ecocentric environmental ethics, in which nature is viewed to have value beyond its economic utility (Milton 2013).

There are certainly ecocentric strains in Greenpeace's discourse and history. For instance, Greenpeace's reconceptualization of the whale – transformed from being conceived as a resource to be harvested into an inherently valuable species to be saved – played off a juxtaposition of people and nature that has become a trademark of the organization (Kramvig *et al.* 2016). In this way the fundamental relation to the environment in terms of sustainability is not first and foremost development, but rather one of preserving ecosystems and climates for the sake of the intrinsic value of the biodiversity that they enable.

Yet it would be misinformed to assess Greenpeace's approach as purely ecocentric. In the last several years Greenpeace has made significant efforts to dampen the man vs nature dichotomy in its Arctic discourse as it tries to incorporate and reach out to the peoples of the Arctic. Hence, the prospect of an oil rig blowout, for instance, is increasingly also framed as impacting local communities and livelihoods. The focus on sustaining a natural climate is another good example where linkages exist to both the ecocentric roots of Greenpeace and the increasing attempt to incorporate the human dimension that makes the Arctic quintessentially different than the Antarctic. In their own words, 'Climate change threatens to unravel Earth's complex web of life, and puts livelihoods, coastal cities and food production at risk. It's the gravest threat to life we know' (Greenpeace 2017b).

In the Arctic, climate change becomes a particular focus in Greenpeace's campaign due to the region's unique position within this unfolding event. The Arctic is here presented in a threefold way: first, as a ground-zero for climate change effects; second, as an integral part of the current natural climate system (i.e. the world's air conditioner); and last, as a flashpoint for increased exploration and extraction of carbon-based fuels.

Greenpeace's Arctic campaign is thus based on a clear sense of the acute need for intervention, which came in the form of the oil rig occupations. These occupations, in turn, succeeded in bringing global attention to the dangers posed to the Arctic's natural systems and the well-being of the local peoples that depend on them. As one Greenpeace activist reported after the occupation of Cairn Energy's *Leiv Eiriksson* rig off the coast of Greenland in 2011:

> We stopped this rig from drilling for four days, which was four days in which a Deepwater Horizon-style blow-out couldn't happen in this beautiful and fragile environment. Our climbers are in jail now, but this won't stop us opposing the madness of drilling for oil that we can't afford to burn and in a region where a spill would be almost impossible to clean up. This isn't over. We must keep on pushing till the oil companies get out of the Arctic.
>
> (Greenpeace 2017c)

This type of reporting, which communicated a David vs Goliath dynamic between Greenpeace and the fossil fuel industry, ultimately resulted in worldwide sympathy for the Greenpeace cause with nearly 8.5 million people signing their support for the 'Save the Arctic' campaign.

Saving the Arctic: scale and geographical imaginaries

From a scalar perspective, Greenpeace's sustainability focus can be said, on a surface level, to be regional and global. The emphasis on the uniquely Arctic ecosystem and the demand for a marine preserve stretching across a large swath of the Arctic Ocean reveals this regional focus. Significant aspects of Greenpeace's Arctic discourse on climate change are also related to this regional lens, as the point is repeatedly made that the Arctic region is particularly vulnerable to rising temperatures. Yet there are also global scalar dimensions involved, particularly with regard to the oil/gas exploration and exploitation that is ongoing in the region.

The demand for a moratorium on offshore oil rig activity on the part of Greenpeace is not limited to the Arctic, but is a global demand linked to address the potentially negative effects that global warming has everywhere. Consequently, the direct actions taken to temporarily halt the oil rig operations in Arctic waters were not undertaken and publicized only for Arctic residents. Rather, they were conducted on behalf of the global citizenry, who, on the one hand, are conceived as having a right to and a responsibility for the heritage of the Arctic's intrinsic natural value, while, on the other hand, having a direct stake in the potentially devastating environmental outcomes that may arise from Arctic development.

Yet when digging into this moral geography a little more, it becomes necessary to add some nuance to the claim that Greenpeace's understanding of the Arctic and it's call to action neatly fits into a regional or even global scalar compartmentalization. Greenpeace's vision is essentially a deconstruction of the national scale by positing an inherently networked moral geography that manages to connect localities around the world (global) to this specific region. Thus, for Greenpeace, a key strategy is to show how we are all connected to the oil rigs through their continued supply of fossil fuels. The oil rigs are entwined in our everyday worlds through our dependency on oil, and their impacts on us in our specific locales, potentially far away and across many borders.

This view reminds of Doreen Massey's (2004) call to see place as a node of interconnections, a power geometry that we must take personal responsibility for. In this sense, the oil rigs, for instance, are not necessarily seen as isolated dots on a territorialized map; rather, they are seen more as emergent phenomenon resulting from a plethora of forces ranging from the pressures and ambitions felt in far-off board rooms and carbon-based desires of people across the globe. For Massey, our place identities need to recognize these networks that we are entwined in, which link the places we know and identify with, usually at a local and national scale, to distant nodes across the globe.

Such a sentiment is well captured in a Save the Arctic campaign photo (Figure 10.1), which shows a woman from behind, dressed in modern city attire, standing in what appears to be the Arctic Ocean, pointing to the oil rigs in the distance. The image declares 'Be Part of the Generation That Ends Oil', which also seems to be saying that we are part of a generation that currently sustains oil. Yet, to move us in a more responsible direction, the advertisement seems to say, requires that we internalize our personal connection to these far-off sites: only in recognizing our power geometry, the spaces of interconnection beyond territorialized scale, will we be able to save the Arctic, and indeed our planet.

The perception of the Arctic as the heritage and responsibility of all humanity and the call for establishing a globally shared marine sanctuary in the non-coastal Arctic Ocean in many ways creates the Arctic as a non-territorialized space that flies in the face of the various seabed claims currently being submitted via the UN Convention on the Law of the Sea. As we will see in the remainder of this chapter, the regional and global focus inherent in Greenpeace's environmental sustainability aims, especially when conceived in their networked form, leads to a stiff reaction from the targeted states whose own

Figure 10.1 Poster from Greenpeace campaign 'Save the Arctic'.

Source: © Greenpeace.

political sustainability imperatives are perceived to be challenged by this geographical imaginary. We now turn to a comparison of the state responses to Greenpeace's campaigns in Greenland, Russia, and Norway.

Greenland: sustaining the dream of postcolonial sovereignty

In the case of Greenland, the occupation of oil rigs in Baffin Bay in 2010 and 2011 was met by a heated response by the Greenlandic leadership, with the protesters seen as infringing on Greenland's hard-won economic rights over their subsoil. Then Prime Minister Kuupik Kleist presented an impassioned statement in response to the second oil rig occupation:

> This constitutes an obvious illegal act that disregards the democratic rules … Greenpeace has once again succeeded in impeding Greenland's opportunities to secure the economic foundation for its people's condition of life … The Greenland Government regards the Greenpeace action as being a very grave and illegal attack on Greenland's constitutional rights.
>
> (Government of Greenland 2010)

Greenland's focus on its sovereign right to extract its own resources indicates both a geo-economic and geopolitical lens through which the protests are viewed. On the one hand, there is the clear demand for the right to access and profit from the potential oil reserves that lie off of their coast. On the other hand, and this is somewhat hidden in the statement above, is the extrapolated view that blocking oil exploration is a direct obstruction of Greenland's aspiration to gain full independence, as the hoped-for oil profits are seen as a key path to an aspired full independence from Denmark, an aim held by both major political parties in Greenland.

These concerns on the part of Greenland, furthermore, must be understood within the specific context that Greenland is operating in, particularly it's colonial relationship with Denmark. This relationship has gone through various iterations, but the general trend has been one of Greenland demanding and receiving ever greater forms of self-determination. The most recent breakthrough came with the passed referendum on Self-Government in 2008, which essentially placed all domestic governance decisions in the hands of the Greenlandic Government, including the rights to the subsoil.

The referendum passed with 75 per cent approval and was heralded and celebrated with great fanfare. Then Prime Minister Kuupik Kleist summed up the mood: 'This morning we awoke with new hope in our heart … From today we are starting a new era in the history of our country, a new era full of hope and possibilities' (George 2009).

It is apparent that there is a sense of being weighed down by the colonial rule of Denmark, the perception of a skewed relationship in which Greenland is subservient. In a recent study of Greenland, Adam Grydehøj (2016) argues precisely this point, indicating that there is little economic reason to pursue independence; the desire is instead rooted in the pained perception of an unequal relationship.

It is in this broader context that we must place the defensiveness with which Greenland treats its right to explore and exploit its subsoil. Not only is this right still new, and the result of a long history of activism, but the very fruits of this right, the possibility of a large oil find, would lay the foundation for bringing this activism to its culmination, i.e. full independence. Interfering with this right, as Greenpeace has done, is thus seen as a postcolonial and patronizing attempt to once again take the right to self-determination away from the Greenlandic people. Exacerbating this sentiment is the particular history that Greenpeace has had in Greenland with regard to the seal hunt. Greenpeace's imposition of western norms and values on an indigenous culture is still blamed for crashing the sealing market and severely damaging the livelihood of Inuit across the Arctic (see Graugaard, this volume).

Thus, in the case of Greenland, the government chose to villainize Greenpeace and the imperialism that it viewed as being inherent in the organization's global focus. Attention was instead redirected to the imperative of sustaining Greenland's self-determination and it's dream for full independence. This was a response that was widely lauded by Greenlanders. Ironically, concerns over the environmental dangers of oil rig drilling pushed by Greenpeace was possibly even less on the agenda after this confrontation as Greenlanders rallied around postcolonial sentiments of non-interference. Environmental sustainability was, after this, largely subsumed by the more pressing demand to back off. As Aleqa Hammond (2014) stated in a speech while serving as Prime Minister of Greenland, 'We do not need to be reminded by others of the preciousness of nature's wealth, because it feeds us, clothes us and sustains us every day.' In this way, environmental sustainability is revealed to be an issue that Greenland refuses to have defined from the outside.

What the governmental response by Greenland discloses in this case is a strong commitment to the preservation of the national scalar lens in approaching Arctic governance. The issue of environmental sustainability is sidelined and replaced by declarations of illegal interventions and constitutional rights. The quasi-state of Greenland is thus essentially seeking to sustain its own vision of what it is to become, a fully independent state, yet in order to do so the geographical ordering of the Arctic needs to be maintained as well. In concrete terms this means avoiding interference from non-national entities, and preserving the all-important national right to the subsoil.

Russia: forceful posturing for home and abroad

In 2013, Greenpeace activists, who later became known as 'the Arctic 30', illegally boarded the *Prirazlomnaya* platform in the Pechora Sea in Russia's exclusive economic zone (EEZ). The activists had climbed the platform and released a banner saying 'Save the Arctic', which from the Russian perspective justified the intervention of enforcement troops who arrested the activists and seized the

Greenpeace vessel (MOFRF 2015).[2] The activists were originally charged with various offences, including piracy and terrorism, but in the end the state chose 'hooliganism' ('rioters') as the key offence to prosecute. The activists, however, were eventually given an amnesty in conjunction with the Sochi Winter Olympics, and were able to leave Russia after more than three months of detention.

Later, the Permanent Court of Arbitration formed under the UN Convention on the Law of the Sea (UNCLOS) ruled that the charges against the activists had no basis and ordered Greenpeace's ship to be released with a compensation payment of 5.4 million euro to the Netherlands, under whose flag the ship was sailing. Russia, although party to UNCLOS, abstained from participating in this arbitration and then refused to recognize its verdict, insisting instead on their sovereign right to safely explore and develop their resources.

To better understand how Greenpeace's efforts were quickly overshadowed by questions of sovereign rights and responsibilities it is essential to understand the overarching Russian political context. On the one hand, Russia's overall outlook must be seen as shaped by a general sense of competitiveness with the West, culminating in a deep malaise about NATO's expansion 'right to our doorstep'. On the other hand, there is the related sense of an increasing challenge to the legitimacy of the political regime. There have been serious anti-Putin protests in the winter of 2010/2011, anti-corruption demonstrations since 2015, and a host of smaller demonstrations around the country, including ones focused on environmental issues such as waste, forests, and the breach of environmental legislation (Gabowitsch 2017; Henry 2010; Newell and Henry 2016).

Given these conditions, there is an underlying sense in the Russian state of being beleaguered on both domestic and international fronts. Thus, when confronted by Greenpeace activists in its EEZ, the Russian Government quickly and defensively focused on the sovereign rights and jurisdiction of the state and what it believed to be its self-evident responsibility to preserve security and public order. The rational and justified strong hand of the state is here pitted against the Greenpeace activists' 'irrational' sabotage of Russian interests in the Arctic, making them, in the logic of this discourse, little better than terrorists. Greenpeace activists were thus 'othered' as rogue and illegitimate outsiders. The subsequent unfavourable ruling by the Permanent Court of Arbitration was then dismissed as Western bias, with Russia upping the ante by emphasizing their perceived victimhood by blaming the Netherlands for not taking responsibility for the Greenpeace vessel's actions. Apart from the brief mention of safely managing their resources in the report cited above, the question of environmental sustainability was quickly forgotten.

Again, we find the challenge to the national scalar imaginary that is insinuated, if not integral, to the Greenpeace interventions as triggering a response that is very much focused on preserving the nation state. It is worth pointing out that the Arctic region has deep ties to Russian identity and is in many ways inextricably bound to the constructions of the Russian nation (Hønneland 2016). While this imaginary may be playing on the minds of Russian officials, it is also exploited by them with regard to the broader public. One Greenpeace activist we interviewed commented on Russia's rhetoric in the Arctic as follows:

> it is totally nationalistic ... The authorities try to use the Arctic offshore race to unite (the nation), to find (an) enemy outside of these countries, to capitalize their political power ... Because in Russia this story is sold as the Russian political frontier, where NATO is coming, want to take all our Arctic Seas, that's why we have to spend more money here, unite around the Kremlin. It's pure, hundred percent propaganda.
>
> (Interview, Greenpeace)

As this quote clearly shows, the Arctic is here very much maintained as a space that is defined by its national jurisdictions. Indeed it is the forceful defence of the Russian sector in this Arctic division that is used as a way to sustain the Russian nation state in the face of domestic strife and global geopolitical competitions.

Norway: there's nothing to see here

In May 2014 Greenpeace activists were able to board and postpone the drilling of the Transocean *Spitsbergen* platform for almost two days. At 74 degrees north, they protested the northernmost exploration ever conducted, 175 km south of the bird reservoir at Bear Island. In contrast to Russia, however, the rig was both in motion *and* outside of the exclusive economic zone, and with Greenpeace sailing under a Dutch flag the Norwegian authorities were forced to wait in order to create a security zone that would allow them to intervene. Further complicating the situation was the fact that Russian authorities protested any Norwegian drilling activities in this area as it was considered to be within the 'Svalbard box'; Russia disputes Norway's claim that the Svalbard shelf is subject to Norwegian sovereignty (Pedersen 2006:345). After five days of protest the Norwegian Coast Guard was given the green light to end what had culminated in the occupation of the platform. The activists were peacefully removed, repatriated with their ship, towed away, and then let go without charge.

According to interviews, while the Greenpeace activists had anticipated benefiting from Norway's compromised legal position in this situation, they were nonetheless somewhat surprised by the gentle response (no arrests) by the Norwegian authorities. There were clearly legal and geopolitical conditions that impacted this response, yet beyond these, it is important also to consider the underlying domestic debate that surrounds the oil question in Norway more generally. While Norway has long been a country that identifies with its history of resource extraction (fish, whales, timber), the continuation of that trajectory to include oil has increasingly been fraught with tension. For its part, the government has generally been a proponent of oil exploration and extraction; their continuing vision, which goes hand in hand with that of the two-thirds government-controlled oil company Statoil, has been one of making Norway the leader in global offshore technology (see also Steinberg and Kristoffersen, this volume). It is a vision of industrial prowess, innovation, technological competitiveness, and wealth (including an oil fund that holds one trillion dollars), which explains their eagerness to push into new and unchartered drilling frontiers.

Yet, there are other Norwegian dimensions at play in this debate as well. There is the more traditional self-image, rooted in the Norwegian past, where nature and the landscape are seen to play an important part of the collective Norwegian psyche (Jensen 2016). There is also the outer presentation of Norway to the world as the good global citizen, handing out the Nobel Peace Prize and serving as a model of global citizenship. Yet, these parts of the Norwegian story sometimes clash with the government vision of being an advanced petro state (Kristoffersen 2015). How do you square oil rigs with the value of pristine nature and unsullied landscapes? Or justify the risky pushing of environmental (and diplomatic) limits while posing as an exemplar of global stewardship in ocean governance?

With these contradictions in mind, Greenpeace's direct action in the Norwegian Barents Sea was an attempt to rattle one of the most ardent champions of the business-as-usual discourse on sustainability. Greenpeace's interventions aimed to hit Norway where it hurts by not only treating them as an equal 'oil villain' to Russia (Jensen 2016), but by chastising their 'opportunistic' and 'aggressive' attempt to exploit global warming by drilling so close to the marginal ice zone (Interview, Greenpeace). More broadly, Greenpeace sought to challenge the publicly popularized celebration of Norwegian history as the 'oil fairytale', where oil resources are perceived as the underlying foundation for Norway's wealth and generous welfare state.

In this light, the primary aim of the Norwegian Government, beyond avoiding the sticky legal and geopolitical pitfalls of the situation, was to deflect from any underlying conflicts and contradictions within the national story. Unlike in the Russian case, where actions taken were partly a show for national and international consumption, the Norwegian Government wanted the public and the international community as far as possible from any media coverage of Greenpeace's actions. In attempting to get back on-message, the Norwegian discursive response focused heavily on safety. Furthermore, the Norwegian Government, which was undoubtedly aware of the many potential international snafus that could arise, largely stayed out of the fray, leaving the case to be made by Statoil. Commenting on the unwinding of the Greenpeace action, the oil company explained:

> For Statoil the safety of the people and the environment is the first priority, and we do not want activity that can increase the risk level. Greenpeace has been explained the risk associated with actions against a rig in open waters.
>
> (Statoil 2014)

In other words, now that Greenpeace's reckless stunt was over, a return to the 'safe' status quo – and the dominant sustainability imperatives that it enables – was possible.

In being confronted with Greenpeace's moral geography, which ties the citizenry of the entire globe to the responsibility of the Arctic, the Norwegian state tried to downplay the incident by deferring to Statoil to debunk the moral claims of Greenpeace. The underlying message in this response is that the best governance arrangement for the region is the one that is already in place. Allowing the nation states in the region to decide over their territorial domain of the Arctic is, in this view, the best way to ensure the most rational, efficient, and safe way to govern what Statoil calls the 'workable Arctic'. For the Norwegian state, then, maintaining business as usual, understood as clear territories of state regulated oil/gas markets, is the best way for them to ensure the sustainability of the nation state they envision, a hegemonic petro powerhouse in the Arctic.

Conclusion: can there be a green peace?

Greenpeace has engaged in a number of oil rig occupations and protests as part of their broader 'Save the Arctic' and 'Go Beyond Oil' campaigns. These interventions, or 'riots', were able to bring global attention to Greenpeace's critique of the business-as-usual approach to the environmental sustainability discourse and practice where the referent of this sustainability is seen to be increasingly vacuous. For Greenpeace, however, the aim is to engage the global citizenry in taking responsibility for the Arctic ecosystem and the natural climate. This mobilization is linked to the value of preserving the diversity of life and the well-being of humanity in the hope of achieving a 'green peace'.

This focus, however, must also be understood to have its own particular geographical imaginary. Greenpeace's vision for the Arctic, after all, is the establishment of an international ocean preserve and a ban on all oil and gas extraction. The focus on the Arctic as a global heritage that is interconnected to the rest of the world in its development and impacts works to undermine the assumed division of the Arctic into clearly bounded national territories, many of which Arctic states are seeking to expand into the area that Greenpeace would like to declare as a marine sanctuary. By drawing attention to what Massey would call power geometries in which any place is seen as networked and morally entwined with myriad other places, Greenpeace hopes to create awareness and support from around the world to envision and create a different Arctic.

It is perhaps this vision of a different Arctic that most offends the various states being targeted by Greenpeace. While Greenpeace was able to stage awareness-raising interventions, their actions also triggered reactions by the targeted states, quickly revealing that the question of environmental sustainability is intricately bound up with other state imperatives, and hence other types of sustainability to prioritize. In the case of Greenland, there was a switch in its sustainability focus to self-determination and ultimately the dream of full sovereignty; for Russia, emphasis fell on preserving geopolitical respect and domestic security; and in Norway the desire appeared to be mostly centred on maintaining the geo-economic conditions in place that enable the market edge that Norway (Statoil) maintains in Arctic oil/gas technology. Yet in all three cases these varying imperatives can be linked back to a fundamental commitment to the political sustainability of the image of the nation state that was being embraced by each of these governments.

Recognizing how the issue of environmental sustainability brought to the fore by Greenpeace was quickly trumped by other state interests is revealing when considering the lens with which the state views the Arctic. Greenpeace stated its intention to bring the plight of the Arctic as a whole to the attention of the world. While they were quite successful in this aim, they also succeeded in revealing an underlying agreement between the targeted states, not only that environmental sustainability and Arctic drilling are compatible, but that any starting point for discussion on Arctic governance must start with the assumption that the region is first and foremost organized on the national scale, a territorialized state system within which political sustainabilities take priority. Thus, while Greenpeace's vision of a world beyond oil is controversial, it appears that their geographical imaginary of such a world is even more so.

Notes

1 We have anonymized the interviewees and will therefore not give date or country when materials from these conversations are presented.

2 This was in stark contrast to the Coast Guard's reticent/discreet and defensive response to the captain's plea for action in a similar Greenpeace action in 2012, to which they responded 'no orders to intervene' (Livinov 2012).

References

George, J. (2009). Feature: Greenland takes another step to independence. *Nunatsiaq News*, 24 June. Availbale from: http://nunatsiaq.com/stories/article/features_greenland_takes_another_step_to_independence/ (accessed September 2018).

Government of Greenland (2010). The Greenland Government condemns the action taken by Greenpeace against Greenland. News from the Government. Available from: http://en.mipi.nanoq.gl/sitecore/content/Websites/uk,-d-,nanoq/Emner/News/News_from_Government/2010/08/greenpeace_action.aspx (accessed June 2017).

Government of Greenland (2014). Greenland's oil and mineral strategy 2014–2018. Available from: www.govmin.gl/media/com_acymailing/upload/greenland_oil_and_mineral_strategy_2014-2018_eng.pdf (accessed September 2018).

Greenpeace (2017a). Our core values. Available from: www.greenpeace.org/international/en/about/our-core-values (accessed September 2017).

Greenpeace (2017b). Stop climate change. Available from: www.greenpeace.org/international/en/campaigns/climate-change (accessed September 2017).

Greenpeace (2017c). Navy climbers end our Arctic oil rig occupation. Available from: www.greenpeace.org/africa/en/News/news/Navy-Climbers-End-Our-Arctic-Oil-Rig-Occupation (accessed September 2017).

Greenpeace (2017d). Save the Arctic. Campaign homepage. Available from: www.savethearctic.org (accessed September 2017).

Grydehøj, A. (2016). Navigating the binaries of island independence and dependence in Greenland: decolonisation, political culture, and strategic services. *Political Geography*, 55, pp. 102–112.

Hammond, A. (2014). New Years reception speech by Premier Aleqa Hammond, Brussels, 15 January. Available from: http://naalakkersuisut.gl/~/media/Nanoq/Files/Attached%20Files/Naalakkersuisut/DK/Taler/2014_BXL_New%20Years%20Reception.pdf (accessed June 2017).

Henry, L.A. (2010). Between transnationalism and state power: the development of Russia's post-Soviet environmental movement. *Environmental Politics*, 19 (5), pp. 756–781.

Hønneland, G. (2016). *Russia and the Arctic: Environment, Identity, and Foreign Policy*. London: I.B. Tauris.

Jensen, L.C. (2016). *International Relations in the Arctic: Norway and the Struggle for Power in the New North*. London: I.B. Tauris.

Kramvig, K., Kristoffersen, B., and Førde, A. (2016). Responsible cohabitation in Arctic waters: the promise of a spectacle touristic whale. In: Abram, S. and Lund, K.A. (eds), *Green Ice*. Basingstoke: Palgrave, pp. 25–48.

Kristoffersen, B. (2015). Opportunistic adaptation: new discourses on oil, equity and environmental security. In: O'Brien, K. and Selboe, E. (eds), *The Adaptive Challenge of Climate Change*. Cambridge: Cambridge University Press, pp. 140–159.

Livinov, D. (2012). Taking action from Arctic Sunrise. Blogpost from Arctic Sunrise, 27 August. Available from: www.greenpeace.org/international/en/news/Blogs/making-waves/taking-action-from-the-arctic-sunrise/blog/41903 (accessed June 2017).

Massey, D. (2004). Geographies of responsibility. *Geografiska Annaler B*, 86 (1), pp. 5–18.

Milton, K. (2013). *Environmentalism and Cultural Theory: Exploring the Role of Anthropology in Environmental Discourse*. London: Routledge.

Ministry of Foreign Affairs of the Russian Federation (MOFRF) (2015). On certain legal issues highlighted by the action of the Arctic Sunrise against Prirazlomnaya platform. 5 August. Available from: www.mid.ru/en/web/guest/foreign_policy/news/-/asset_publisher/cKNonkJE02Bw/content/id/1639745 (accessed July 2017).

Newell, J.P. and Henry, L.A. (2016). The state of environmental protection in the Russian Federation: a review of the post-Soviet era. *Eurasian Geography and Economics*, 57 (6), pp. 779–801.

Pedersen, T.R. (2006). The Svalbard continental shelf controversy: legal disputes and political rivalries. *Ocean Development & International Law*, 37 (3–4), pp. 339–358.

Statoil (2014). Greenpeace board Transocean Spitsbergen. Statement by Statoil. Available from: www.statoil.com/en/news/archive/2014/05/27/Greenpeace.html (accessed July 2017).

Swyngedouw, E. (2009). The antinomies of the postpolitical city: in search of a democratic politics of environmental production. *International Journal of Urban and Regional Research*, 33(3), pp. 601–620.

Swyngedouw, E. (2010). Post-political populism and the spectre of climate change. *Theory Culture & Society*, 27(2–3), pp. 213–232.

Tamkin, E. (2017). Russian environmental activists on engaging people under Putin. *Foreign Affairs* 3(17).

11 Sustaining the Arctic nation state

The case of Norway, Iceland, and Canada

Ingrid A. Medby

It was a blustery October's day in Reykjavik – 'typically Icelandic', I was told. I was there to interview a member of the Icelandic Parliament, the Alþingi, about her sense of an Icelandic 'Arctic identity'. We met at a hotel that promised 'fresh ocean air all around', ostensibly striving 'to stay in touch with nature'. The irony of the hotel's location, practically right next to Reykjavik Airport, was left unmentioned.

The parliamentarian enthusiastically answered my questions about her 'Arctic identity'. She explained the Icelanders' pride in being 'green' and clean, powering the nation by sustainable, geothermal energy. We ordered coffee, and soon she was explaining how they were now – equally proudly – considering oil exploration. Once more, the irony was left unmentioned. Such exploration would be conducted in cooperation with the 'experienced' Norwegians (see Steinberg and Kristoffersen, this volume), making sure that – in her words – Iceland is 'part of the making of the future of the Arctic'. While the question of oil remains controversial in Iceland, both sides of the debate are arguing along lines of 'who we are'; political practice may be disputed, but the actor, the 'nation state', never is.

Continuing, the parliamentarian referred to a shared circumpolar identity: different landscapes, but the same atmosphere and mentality. As an Icelander, she felt strongly that she had an Arctic identity, connected to their closeness to nature – 'we are always talking about the weather!' – to natural forces, to the ocean, to glaciers. In her words, 'ice, the glaciers, you know, it's also a part of us. Arctic, glaciers – it's somehow the same. You connect with it somehow.' In the north, we have a 'connection'.

According to her, it is due to this connectedness to nature and what she referred to as a 'special energy' only shared by northerners that Icelanders care so much about the environment. She explained that this connects them to their history all the way back to the settlement of the volcanic island in the Atlantic; they are *part* of nature themselves. As such, sustainability politics is indeed about nature, but a nature that is also seemingly intrinsic to the Icelandic people and identity. Sustaining the Arctic is also about sustaining themselves.

Introduction

As the above anecdote illustrates, politics of sustainability co-constitute ideas of Arctic statehood and identity, past, present, and future. However, while the aim of Arctic sustainability seems to be undisputed and undisputable, it often remains undefined in terms of sustainability of what, where, how, and by whom. For the Arctic states, their privileged position as such spells not only particular rights, but also particular responsibilities – among these, the promotion of this Arctic sustainability. With ever more actors and stakeholders making their presence felt, the performance of these responsibilities becomes demonstrative of authority, credibility, and sovereignty: an active performance of what it means to *be* an Arctic state.

This chapter explores how politics of sustainability become tied to those of identity among policy-makers in three Arctic states; and how enacting sustainability thereby comes to reproduce and reify the idea of the Arctic 'nation state' itself. That is, as state representatives articulate 'sustainable' practices as demonstrative of a seemingly inherent characteristic of the national community, they also indirectly 'sustain' the very ideal from which the practice purportedly flows, namely the 'Arctic nation state's' identity.

The specific focus here is three of the eight Arctic states: Norway, Iceland, and Canada. These are all states whose political leadership share explicit concern with Arctic identity and sustainability, yet whose relationships to the region differ in practice. The chapter draws on interviews with state personnel about their sense of Arctic identities and relations more generally; hence, 'sustainability' entered conversations as reflective of their conceptualizations of what it means to be, perform, and represent an Arctic state. While their articulations of sustainability demonstrate the concept's elasticity, potentially seeking to sustain objects ranging from environments to economies, the subject, i.e. the Arctic nation state, always remains constant. In this manner, Arctic politics is discursively reified as the domain and responsibility of states – that which 'always' has been and 'always' will be.

In what follows, the chapter proceeds in three steps. First, brief conceptualizations of sustainability and statehood are offered, placing the chapter within the wider context of the book. This is followed by, second, a brief introduction of the three Arctic states in question – Norway, Iceland, and Canada – and their respective relationships to the region. Third, articulations of Arctic identity and sustainability by the study's interviewees are presented for each of the three states. Notably, the point here is not that state actors are deliberately attempting to 'sustain' (or even construct) an identity for political gains but rather that this is an effect of their statements, rhetoric, and practices. Here, the state-level deployment of the concept of 'sustainability' effectively reproduces the very idea of 'the national' itself, which is fundamental to both political support and professional purpose. Thus, in the process of performing politics of sustainability, the Arctic state is in effect invested with authority – the bearer of Arctic rights and Arctic responsibilities.

Sustainability and statehood

As demonstrated throughout the book, the concept of sustainability seems only to have increased in currency, use, and meaning in recent years. Particularly since the UN's *Our Common Future* (1987), or so-called 'Brundtland Report',[1] popularized 'sustainable development', the concept has become widely employed as a political virtue worth global effort (see Benson *et al.* 2016). While numerous definitions exist, the etymology of sustainability traces back to the *ability* to *sustain* something, applied to (ecological) systems' indefinite regenerative ability. In the Arctic context too, sustainability has become a near-ubiquitous objective – at least rhetorically and symbolically (Kristoffersen and Langhelle 2017; see also Swyngedouw 2007). However, as soon becomes clear, what this much-lauded aim means in practice is far from straightforward. What is to be made 'sustainable', by whom, where, when, and *how* are all questions open to debate. Highlighted by the aforementioned Icelandic parliamentarian, sustainability may simultaneously be about protecting nature and about protecting an economy through petroleum extraction. Instead of seeing these as irreconcilable, both instances point to the maintenance of a present 'good' for the future: a way of life in and of the Arctic. And indeed, it is the acknowledgement of the concept's political purchase coupled with its inherent ambiguity that guides and unites the wide-ranging enquiries throughout this book.

Here, the interest lies in the conceptual premise of sustainability, or the 'act of sustaining something', and what political role it may play in the Arctic. In other words, the focus is less on what sustainability is said to mean in practical terms and more on what it means as a political *concept* to begin with. Evoking the concept of sustainability is an act in and of itself; state actors speaking of Arctic 'sustainability' is a narrative positioning of the Arctic 'nation state' and the environment in a relation of interdependence, which is (to be) maintained over time (see Gad *et al.*, this volume). On the surface, claims of sustainable action/politics imply that the Arctic environment is that which ostensibly needs to be safeguarded through policy (although, as seen above, these may include practices and nouns seemingly far from, even in opposition to, environmentalism per se). However, more fundamentally, political statements and practices of Arctic sustainability have the unintended consequence of 'sustaining', or reifying, its 'sustainer' – in this case, the nation state and its purported identity. And, conversely, the 'sustained', i.e. the Arctic, is rendered passive and static, objectified, in need of safeguarding (see Swyngedouw 2007). As such, what is of particular interest is how this normative concept percolates through political discourses more broadly, and how it comes to take on added meaning in the question of 'who we are' – indeed, co-constituting Arctic identities.

Of course, state actors are far from the only ones engaged in Arctic politics or 'sustainability' (however defined); the region's governance arrangements having been described as more akin to a 'mosaic' (Young 2005) or 'bazaar' (Depledge and Dodds 2017). However, they do hold a certain status and standing as the voting members of the Arctic Council. For the eight states with territory north of the Arctic Circle, their formal geographically defined title also encompasses notions of homelands and communities, histories and futures. Paying attention to scale, it is in this context striking how the 'national' (or 'state'/'federal') seems only to be strengthened in the face of increasing globalization and multilateral engagements (see, e.g. Dittmer *et al.* 2011; Gerhardt *et al.* 2010). In what is frequently framed as the ever more global, 'new' Arctic, the eight states' particular, privileged position is reiterated and reified (see, e.g. Graczyk *et al.* 2017; Wilson 2016).

Even so, though, defining who or where or what the Arctic state really *is* in practice is not straightforward. Rather than seeing the state as an 'actor', it is here understood as an abstract ideal that 'materializes' as an effect of practice (see Abrams 1988; Mitchell 2006; Painter 2006); the 'Arctic state' comes into being through the numerous practices and performances enacted under its banner (see Butler 2011). Hence, it is enacted into being by a range of *people* (Jones 2007; Kuus 2008), including those interviewed below. Statements of/about sustainability are among the enunciations through which the state emerges; and moreover, they are reflective of how these practitioners understand their own roles as representatives of Arctic states. These statements are simultaneously products and *productive* of political discourse (Foucault 1972), which means that, for example, the above-cited Icelandic parliamentarian's articulations of what it means to 'be' an Arctic state also *make* it such (Butler 1999; see also Butler 2011). Conversely, articulating sustainability otherwise holds the potential to disrupt and challenge much more than specific policies – namely deeper-set convictions, questioning not only 'how' we should act sustainably, but also 'who' should do so and 'why'. That is not to say that state representatives are deliberately seeking to sustain a notion of (Arctic) nation statehood; rather, the concept of 'sustainability' itself is premised on an idea of a future that leads on from a certain interpretation of the present, a continuation and maintenance of a discursive status quo and specific worldview.

Also worth noting, 'nation state' represents an ideological, discursive coupling of a people/society (the nation) with a political-territorial organization (the sovereign state), assuming their congruence (Gellner 1983; Sparke 2005; but see Medby 2014). As will become clear below, Arctic sustainability practices are in these states frequently framed as demonstrative of such an identity – a seemingly homogeneous national community, sharing an identity across time and space (Anderson 1983; Guibernau 2013). Not only does this render it difficult to challenge politics, but also potentially homogenizes how and who may indeed legitimately enunciate and enact them. Thus, in short, claiming that the production of Arctic policy is 'based' on a national identity becomes a practice through which its practitioners indirectly also reproduce and 'sustain', the ideal of a nation state and its identity; one which they themselves represent as well as rely on for their professional purpose. *How* these idea(l)s are discursively wedded – i.e. the convergence of identity, nation statehood, and sustainability – in Arctic states is that which will be further explored below.

A case of three Arctic states

In order to start approaching how the above unfolds in practice, articulations of 'Arctic identity' by Arctic state personnel here offer insights on the state and identity's relation to 'sustainability' (see Medby 2017). Three of the eight Arctic states – Norway, Iceland, and Canada – provide the focus due to their shared state-level rhetoric of Arctic identities; yet, their identities and relationships with the Arctic more generally are of contrasting character (see, e.g., Arnold 2010; Baldwin *et al.* 2012; Dodds and Ingimundarson 2012; Grant 2001; Jensen 2013; Johnstone 2016). On the one hand, these states are all relatively small, historically 'peripheral' states on the international scene; on the other, they have vastly different geographies, histories, and peoples – not least in the Arctic. Nevertheless, in their particular emphases on sustainability as *state* practice, they may all be contrasted with, for example, current US rhetoric, seemingly having jettisoned both Arctic and sustainability (or environment – see Herrmann, this volume) from the immediate policy agenda, or indeed the aspiring-to-be-state Greenland, denouncing responsibility for other states' unsustainable practices (see Bjørst, this volume).

It is worth noting that the semi-structured interviews, most of which were conducted in 2014, centred on Arctic identity and statehood,[2] asking state personnel of their sense of such and potential implications for political practice. Thus, when 'sustainability' entered conversations, this tended to be through respondents' own associations. While not all used the term 'sustainability' per se, they *all* spoke about their sense of a particular duty to protect and steward the region and its environments. The respondents were 49 state personnel in a variety of positions across the state/federal level: some elected politicians, others appointed bureaucrats, ranging from ministers to advisers, Indigenous and not. Importantly, many of these did not work on specifically Arctic-labelled issues; rather, what was of interest was how the Arctic region is identified with (or not) by decision-makers potentially far removed from it, but nevertheless representing a state formally titled as Arctic. Hence, instead of focusing on one specific sustainability strategy, what is here at stake is how such strategies enter political discourses, worldviews, and 'common-sense' practices; and how this in turn may reconfigure relations in and with the region.

As it is the conceptual premise of 'sustainability' that is of main concern, the analysis goes beyond the onomasiology of 'sustainability' per se, to consider also related words and expressions (see Gad *et al.*, this volume). In their deployment, terms such as stewardship, environmental protection, safeguarding for future generations, etc. perform similar (often mutually reinforcing) functions in political parlance. Hence, for the purpose of this chapter they are included, not as syntactically different but as semantically related – forming part of a discourse of political consequence (see Foucault 1972).

In what follows, each of the three states are presented separately, highlighting both their similarities and differences. As soon becomes clear, 'sustainability' remains vaguely defined across them, and among individual personnel, but it is precisely this ambiguity – and yet unquestioned normativity – that leads to the concept's ubiquity. When sustainability becomes a *value*, the question is not only 'What is to be sustainable?', but indeed 'Who defines its currency; and what does it *do*?'.

Articulations of sustainability and identity

Before turning to different manifestations of identity discourses in each state, it is worth noting their shared discursive frame: state personnel in all three described their 'Arctic identity' as simultaneously embedded within a pre-existing national identity – the stable core of 'who we are' – and a new phenomenon, the result of change, that allowed a re-assessment of 'who we *will* be'. In this manner, the geographically based title of Arctic statehood – and with it, sustainability politics – becomes part of the national narrative: a present that is a consequence of the past as well as a promise for the future. However, the ways in which this takes place, i.e. the ways in which politics of Arctic sustainability become embedded in national identity, are reflective of the perceptions each has of their own nation state: enshrined in different imaginations of different national pasts and different national futures – though all, importantly, national.

Norway

In interviews with Norwegian representatives, the topic of sustainability was often brought up as both an example of and justification for not only Norwegian political involvement, but *leadership* in the Arctic region. Environmental protection here ties into a wider narrative of 'good' nationhood, even moral exceptionalism, which feeds into a quest for international status and influence that arguably runs through Norwegian foreign policy more generally (de Carvalho and Neumann 2015; Lahn and Wilson Rowe 2015). One Norwegian politician explained this, as well as the undeniable paradox posed by it:

> Not that I am personally in favour of it, but if there is *one* nation on this Earth that could, perhaps, limit oil extraction based on a climate-reasoning, so to speak, and have the economy to miss that income etc., then that is Norway, right. And also, to show the importance of taking care of the climate and environment – among others, because of the Arctic.

As his quote suggests, there is a tension between Norway's emphasis on environmental protection and their active extraction and export of fossil fuels. However, even here the 'responsible' image is maintained through what Jensen (2012) has termed 'discourse co-optation': claiming that Norway *should* extract hydrocarbons due to their 'cleaner', more 'sustainable' extractive industries in order to avoid the purported inevitable alternative, namely that Russia would do so instead, but in a 'dirtier' way. In short, Norway is 'drilling for the environment' (Jensen 2006; see also Kristoffersen 2015). In this manner, the image of

Norway as a sustainable, responsible, moral actor is sustained in the Arctic, and political practices are not only justified but seemingly necessitated in the process. And indeed, many interviewees linked this leadership to a proud history of Polar exploration (see e.g. Drivenes and Jølle 2004; Wærp 2010): a pioneering spirit that today 'transfers' to environmental protection. Thus, Norwegian politics of sustainability in the Arctic not only sustain a particular self-perception, but also link to patriotism and pride, embedded in 'the' national narrative.

Several interviewees described Norwegian environmental care as not just historically rooted but as environmentally determined: describing how shared experiences of cold and harsh coastal conditions have somehow *shaped* them as 'closer to nature', similar to the opening's Icelandic parliamentarian. Reflecting on national identity's influence on Arctic political practice, one interviewee explained:

> When you meet others, from other countries, who do not have that, sort of, hard nature, right; then you notice that it affects our character, yes, our identity, how we look at things. And I think this closeness to nature means that we have more *respect* for nature.

Here, nature is ostensibly what 'makes' Norwegians, 'makes' political behaviour, and 'makes' it a leader of sustainability, as above. A self-perception of being a small, struggling nation state of fishermen and farmers is re-narrated in the 'new' north (see Stuhl 2013); in the words of one politician: 'Today, the Norwegian identity [is tied to] us being the nation that exploits natural resources but in a *sustainable* way.' Once more, notions of sustainability – here, sustainable development in the arguably unsustainable extraction of fossil fuels – become subsumed in a deeper sense of Norwegian self. Regardless of whether nature has influenced the Norwegian 'character', as long as state personnel *believe* it to be true, it undoubtedly influences *their* behaviour, *their* representation, and enactment of the Arctic 'nation state'.

Iceland

Arctic nature, environments, and their protection were also frequent themes in the Icelandic interviews. However, instead of resulting in leadership, the Icelandic environment was more often described as that which sets the country *apart*, that which defines them as stubbornly independent. In a similarly environmentally determined manner as in Norway, climatic conditions were articulated as providing the Icelanders with a unique character and spirit, and therefore their claim to political independence – to nationhood and sovereignty. In the words of an Icelandic politician, closeness to nature – Icelandic identity – clearly influences political behaviour: 'we are a people who need a lot of freedom. I think that comes from being brought up in an environment where you are totally free … And that also makes a country that is brought up like that.' In other words, the Icelandic nature is what supposedly makes Icelanders, and thus, sustaining nature becomes a question of sustaining much more – the very 'essence' of the nation.

For Icelandic state representatives, Arctic statehood is a title that is explicitly future-oriented – arguably *more* so here than in the other seven states, despite most Arctic engagement being marked by anticipation (Depledge 2016; Dodds 2013; Steinberg *et al.* 2014). Due to Iceland's geographical position mainly south of the Arctic Circle – and until recently, political efforts reaching mainly south, east, and west – the adoption of an Arctic 'brand' is still fresh in Icelandic memory. Of course, Icelanders also look to the past, connecting their Arctic statehood to histories of e.g. Vikings and fishermen (at least rhetorically), but the interviewees were all clear that for Iceland, the Arctic primarily represents economic opportunities: 'Attention to the Arctic and the economic interests are not very close to us now, but of course *we* [political leaders] have to look to the farther future.'

In the process of recovering from a severe economic crisis, also the current Icelandic economy is highly dependent on the successful promotion of their 'pristine' nature. In the words of another parliamentarian: 'It is, of course, very important for Iceland to be a clean country – without any smoke, without any nuclear waste, without any other kind of waste. Clean. That is really a huge issue for Iceland.' Nature is not just something that gives Icelanders a sense of a stable national self, but also attracts interest from external 'others': tourists and investors. And so, the sustainable co-existence in and with nature is something to *display*; for example, presenting geothermal energy use as an item on itineraries of neatly packaged trips and convenient flight layovers. And moreover, sustainable practices are not just laudable achievements, but necessary for sustaining the Icelandic national economy.

While tourism is a relatively recent addition of scale, fishery is deeply embedded in the national, historical narrative as a 'foundation' for an Icelandic identity; and today, autonomy over fishing zones and quotas has come to symbolize the aforementioned struggle for freedom and independence of the Icelandic nation (Bergmann 2014; Robert 2014). As one parliamentarian noted in relation to fishery and control over own (sustainable) quotas: '[I]t's very much at the core of Icelandic foreign policy, identity. And more so than just the economic activity; it's of course an extremely important economic activity, but it's a little bit *more* than that.'

When Iceland was excluded from Arctic Ocean negotiations (Dodds and Ingimundarson 2012), it therefore soon became a question of Icelandic national pride. The demonstration of Icelandic sustainability – from the ways in which they extract energy and transport tourists to how they regulate their fishery – thus also soon comes to be seen as a demonstration of the inherently Icelandic. As such, aiming to sustain the environment matters for much more than 'just' the environment; it matters also to the sense of national self, and indeed, sustains the belief in such.

Canada

In the Canadian case, the concept of sustainability tended to come up under a different banner: that of 'stewardship'. While historically referring to colonized peoples and lands, stewardship is today transferred to the safeguarding of the environment and nature. Indeed, on what it means for Canada to be an Arctic state, one official simply summarized: 'stewardship and sovereignty'. Or, in the words of former Prime Minister Stephen Harper: 'This magnificent and unspoiled region is one for which we will demonstrate stewardship on behalf of our country, and indeed, all of humanity' (2008, cited in C.FATDC 2010:16).

Harper's views on environmental protection or (the lack of) climate change would seem to contradict calls for sustainability (in the environmental sense); and yet, even under his government, it was a topic, however differently framed, differently performed, differently identified with. Perhaps connected to the historical (colonial) connotations of 'stewarding', also 'sustainability' frequently took on a more human/social (even economic) meaning in the Canadians' articulations. Reflecting on Canadian citizens' views of the North, one official explained that, in his view, these were romanticized and:

> mostly environmentally focused, mostly hyper-sensitive of the peoples of the North and their challenges ... What's not part of that conversation, and what's not part of people's thinking, perceptions when they think of the North, is business. You know, how – who is keeping the towns running? Where is the economy coming from? What *sustains* it?

More so than in the other states, the adjective 'sustainable' was applied to communities, economies, and the financial viability of industries. Hence, it is a question of 'sustaining' the nation – indeed, paternally protecting the national community (Dodds 2012) – where the Arctic environment provides the backdrop for the postcolonial state's formation of self.

The environmental determinism seen in the other two states is of a different character in Canada. On the topic of a Canadian Arctic identity, personnel tended to speak of Arctic environments and nature as still foreign, distant, exotic, but something with which Canadians could *connect* through a nature-based identity – that of the south, yet cold, 'winter-y', and *relatively* northern: 'I think it might blur; like, "Arctic" might blur into Canadian vastness and wilderness. A lot of Canadians [are] very proud of the wilderness and the uniquely Canadian – the vastness as well.' While the romanticized Arctic does indeed link to national identity here too, it does so in a far more *symbolic* manner than in Norway and Iceland (where, arguably, the opposite is taking place as national Arctic identities are becoming normalized and domesticated). While most Canadians, including most state personnel, have never physically experienced the Arctic, it is nevertheless a space that *matters* symbolically and politically, not least in separating themselves from, first, Great Britain, and second, the US.

As one federal official explained, 'the changing Arctic' may not affect everyday life in the south, but it does affect, even *define*, how Canadians see themselves:

> as a sort of outdoors people, who are close to nature as well. And ... you know, who have inherited this part of the planet and also want to make sure that it's carefully protected and that we are, basically, stewards of that region.

Although Canadian politics of sustainability are articulated differently than in Norway and Iceland, they nevertheless perform a similar function. What may at first glance seem a unidirectional relation between the Canadian North and south – the latter 'sustaining' the former – upon closer observance proves to be one mutual dependency, where the former sustains the southern 'stewards'' sense of self. Ultimately, identity is always relational and performative – through the practice of 'sustaining' the Arctic, the very notion of the Canadian nation state is too.

Conclusion

Considering state personnel's articulations of Arctic identity and its ties to sustainability, there are many similarities across the three states in question: for all three, sustainability is unquestionably positively laden and constitutes a political aim to strive for in the Arctic and elsewhere. Speaking of Arctic sustainability is uncontroversial (some might even say near-mandatory), and, linking it to identity, a laudable value. Further, in all three it is about sustaining a present object, the Arctic, for the future (or indeed, aspects of it, such as Arctic environments, cultures, economies, etc.); hence, it is a concept of temporal consequence, of governing the future in the present. As the Icelandic parliamentarian at the beginning of this chapter phrased it, political engagement in the region is about '*making*' the Arctic future; importantly, a future based on one interpretation of the present, thereby singularizing the plurality of potential futures. Tied to identity, it is therefore not *only* about who 'we', the so-called national community, supposedly *are*, but also about who we *want* to be – past, present, and future generations in and of the Arctic – seen from the level of the state.

While Arctic sustainability intersects with narratives of national identity in all three states, it nevertheless does so in different ways across the different contexts. Thus, what we may also observe above is the elasticity of the concept; and moreover, how it comes to take on added meaning when coupled with fundamental myths of 'who we are' – justifying certain political practices, and indeed, 'sustaining' the very idea of the cohesive nation state and its identity. For Norwegian state personnel, their politics of sustainability may be seen as demonstrative of their leadership in the Arctic; for Icelandic state personnel, it connects to their sense of unique independence and ties to nature, a making of the Arctic future; and, for Canadian state personnel, it is about stewardship and responsibilities that are inextricably bound up with colonial histories. In short, politics of sustainability come to co-constitute an 'Arctic identity' for these state personnel, albeit weighing past, present, and future differently. In this manner, it is the very ambiguity of the concept that gives it its political power, a universal 'good' under the banner of the Arctic nation state (see Guibernau 2013), investing diverse practices and practitioners with authority.

Returning to the book's guiding questions (as introduced by Gad *et al.*), this chapter has sought to highlight how the introduction of sustainability as a political value supposedly derived from an identity discursively reifies political practices and actors. Through notions of morality and the collective national 'self', state action in the Arctic becomes state action as Arctic – and thereby also becomes naturalized and normalized, uncontested and incontestable. Hence, what is clear from the above is that wedding politics of Arctic sustainability to identity has implications extending beyond the region – to fundamental political beliefs and worldviews. In short, the political implications of the above may constitute a discursive narrowing of what it means to 'be' Arctic and how to be so, and a privileging of the state as the key 'steward' of the region. Applied ever more widely, as indeed the other chapters also demonstrate, sustainability becomes less of a concrete practice and more of a discursive legitimation of authority and rights, in this case those of the state. That is not to say that the term is completely 'emptied' of meaning, that it does not also come with a range of important policy initiatives; Arctic environments and natures, lives and cultures, are indeed articulated as something to be maintained through time. However, importantly, 'to be maintained' implies an Arctic that is passive, static, necessitating the active intervention of a 'stewarding' state. In other words, well-meaning as they may be, these articulations cite a certain worldview and discourse, a certain ordering of relations, through which the idea of the state and state power are reified. In the end, state practitioners' articulations of Arctic sustainability and identity do more than 'sustain' the Arctic per se: they sustain the very idea upon which statements are made, discourses of Arctic 'nation statehood'.

Notes

1 Worth noting here, Gro Harlem Brundtland was also Prime Minister of Norway in 1981, 1986–1989, and 1990–1996.

2 More about the study, its methodology, and results can be found in Medby (2017).

References

Abrams, P. (1988). Notes on the Difficulty of Studying the State [1977]. *Journal of Historical Sociology*, 1 (1), pp. 58–89.

Anderson, B. (1983). *Imagined Communities: Reflections on the Origin and Spread of Nationalism*. London: Verso.

Arnold, S. (2010). 'The Men of the North' Redux: Nanook and Canadian national unity. *American Review of Canadian Studies*, 40 (4), pp. 452–463.

Baldwin, A., Cameron, L., and Kobayashi, A. (eds) (2012). *Rethinking the Great White North: Race, Nature, and the Historical Geographies of Whiteness in Canada*. Vancouver: UBC Press.

Benson, D., Bulkeley, H., Demeritt, D., Jordan, A., Murphy, J., and Selin, H. (2016). Environment and sustainable development scholarship: a celebration. *Environment and Planning A*, 48 (9), pp. 1679–1680.
Bergmann, E. (2014). Iceland: a postimperial sovereignty project. *Cooperation and Conflict*, 49 (1), pp. 33–54.
Butler, J. (1999). *Gender Trouble: Feminism and the Subversion of Identity*. London: Routledge.
Butler, J. (2011). *Bodies that Matter*, 2nd edn. London: Routledge.
C.FATDC (2010). *Statement on Canada's Arctic Foreign Policy*. Ottawa: Foreign Affairs, Trade and Development Canada, Government of Canada. Available from: www.international.gc.ca/arctic-arctique/assets/pdfs/canada_arctic_foreign_policy-eng.pdf (accessed 23 November 2017).
de Carvalho, B. and Neumann, I.B. (eds) (2015). *Small State Status Seeking: Norway's Quest for International Standing*. Abingdon: Routledge.
Depledge, D. (2016). Climate change, geopolitics, and arctic futures. In: Sosa-Nunez, G. and Atkins, E. (eds), *Environment, Climate Change and International Relations*. Bristol: E-International Relations Publishing, pp. 162–174.
Depledge, D. and Dodds, K. (2017). Bazaar governance: situating the Arctic Circle. In: Keil, K. and Knecht, S. (eds), *Governing Arctic Change: Global Perspectives*. London: Palgrave Macmillan, pp. 141–160.
Dittmer, J., Moisio, S., Ingram, A., and Dodds, K. (2011). Have you heard the one about the disappearing ice? Recasting Arctic geopolitics. *Political Geography*, 30 (4), pp. 202–214.
Dodds, K. (2012). Graduated and paternal sovereignty: Stephen Harper, Operation Nanook 10, and the Canadian Arctic. *Environment and Planning D: Society and Space*, 30 (6), pp. 989–1010.
Dodds, K. (2013). Anticipating the Arctic and the Arctic Council: pre-emption, precaution and preparedness. *Polar Record*, 49 (2), pp. 193–203.
Dodds, K. and Ingimundarson, V. (2012). Territorial nationalism and Arctic geopolitics: Iceland as an Arctic coastal state. *The Polar Journal*, 2 (1), pp. 21–37.
Drivenes, E.-A. and Jølle, H.D. (eds) (2004). *Norsk polarhistorie: Ekspedisjonene, Norsk Polarhistorie*. Oslo: Gyldendal.
Foucault, M. (1972 [1969]). *Archaeology of Knowledge*. London: Tavistock.
Gellner, E. (1983). *Nations and Nationalism*. Oxford: Blackwell.
Gerhardt, H., Steinberg, P.E., Tasch, J., Fabiano, S.J., and Shields, R. (2010). Contested sovereignty in a changing Arctic. *Annals of the Association of American Geographers*, 100 (4), pp. 992–1002.
Graczyk, P., Smieszek, M., Koivurova, T., and Stepien, A. (2017). Preparing for the global rush: the Arctic Council, institutional norms, and the socialisation of observer behaviour. In: Keil, K. and Knecht, S. (eds), *Governing Arctic Change: Global Perspectives*. London: Palgrave Macmillan, pp. 121–140.
Grant, S.D. (2001). *Canada and the Idea of North*. Montréal: McGill-Queen's University Press.
Guibernau, M. (2013). *Belonging: Solidarity and Division in Modern Societies*. Cambridge: Polity.
Jensen, L.C. (2006). Boring som miljøargument? Norske petroleumsdiskurser i nordområdene [Drilling for the environment? Norwegian petroleum discourses in the High North]. *Internasjonal Politikk*, 64 (3), pp. 295–309.
Jensen, L.C. (2012). Norwegian petroleum extraction in Arctic waters to save the environment: introducing 'discourse co-optation' as a new analytical term. *Critical Discourse Studies*, 9 (1), pp. 29–38.

Jensen, L.C. (2013). *Norway on a High in the North: A Discourse Analysis of Policy Framing*. Tromsø: University of Tromsø.

Johnstone, R.L. (2016). Little fish, big pond: Icelandic interests and influence in Arctic governance. Nordicum-Mediterraneum – Conference Proceedings from 'No One is an Island: Iceland and the International Community', 11 (2).

Jones, R. (2007). *People/States/Territories: The Political Geographies of British State Transformation*. Oxford: Blackwell.

Kristoffersen, B. (2015). Opportunistic adaptation: new discourses on oil, equity and environmental security. In: O'Brien, K. and Selboe, E. (eds), *The Adaptive Challenge of Climate Change*. Cambridge: Cambridge University Press, pp. 140–159.

Kristoffersen, B. and Langhelle, O. (2017). Sustainable development as a global-Arctic matter: imaginaries and controversies. In: Keil, K. and Knecht, S. (eds), *Governing Arctic Change: Global Perspectives*. London: Palgrave Macmillan, pp. 21–42.

Kuus, M. (2008). Professionals of geopolitics: agency in international politics. *Geography Compass*, 2, pp. 2062–2079.

Lahn, B. and Wilson Rowe, E. (2015). How to be a 'front-runner': Norway and international climate politics. In: de Carvalho, B. and Neumann, I.B. (eds), *Small State Status Seeking: Norway's Quest for International Standing*. Abingdon: Routledge, pp. 126–145.

Medby, I.A. (2014). Arctic state, Arctic nation? Arctic national identity among the post-Cold War generation in Norway. *Polar Geography*, 37 (3), pp. 252–269.

Medby, I.A. (2017). Peopling the state: Arctic state identity in Norway, Iceland, and Canada. PhD thesis, Durham University. Available from: http://etheses.dur.ac.uk/12009/ (accessed 11 September 2018).

Mitchell, T. (2006 [1999]). Society, economy, and the state effect. In: Sharma, A. and Gupta, A. (eds), *The Anthropology of the State: A Reader*. Oxford: Wiley-Blackwell.

Painter, J. (2006). Prosaic geographies of stateness. *Political Geography*, 25 (7), pp. 752–774.

Robert, Z. (2014). Iceland left out of EU mackerel deal. *Iceland Review*. Available from: http://icelandreview.com/news/2014/03/13/iceland-left-out-eu-mackerel-deal (accessed 17 April 2014).

Sparke, M. (2005). *In the Space of Theory: Postfoundational Geographies of the Nation-State*. Minneapolis, MN: University of Minnesota Press.

Steinberg, P.E., Bruun, J.M., and Medby, I.A. (2014). Covering Kiruna: a natural experiment in Arctic awareness. *Polar Geography*, 37 (4), pp. 273–297.

Stuhl, A. (2013). The politics of the 'New North': putting history and geography at stake in Arctic futures. *The Polar Journal*, 3 (1), pp. 94–119.

Swyngedouw, E. (2007). Impossible 'sustainability' and the postpolitical condition. In: Krueger, R. and Gibbs, D. (eds), *The Sustainable Development Paradox: Urban Political Economy in the United States and Europe*. London: Guilford Press, pp. 13–40.

UN (1987). *Our Common Future: United Nations World Commission on Environment and Development*. Oxford: Oxford University Press.

Wærp, H.H. (2010). Fridtjof Nansen, 'First Crossing of Greenland (1890)': bestseller and scientific report. In: Ryall, A., Schimanski, J., and Wærp, H.H. (eds), *Arctic Discourses*. Newcastle-upon-Tyne: Cambridge Scholars Publishing, pp. 43–58.

Wilson, P. (2016). Society, steward or security actor? Three visions of the Arctic Council. *Cooperation and Conflict*, 51 (1), pp. 55–74.

Young, O.R. (2005). Governing the Arctic: from Cold War theater to mosaic of cooperation. *Global Governance: A Review of Multilateralism and International Organizations*, 11 (1), pp. 9–15.

12 'How we use our nature'

Sustainability and indigeneity in Greenlandic discourse

Kirsten Thisted

In the network of signs and symbols that add meaning to the concept of sustainability, indigeneity entails a significant denominator. The two concepts were developed during the same period of time, the 1970s, in adjacent – sometimes interconnected – political environments, which were trying to formulate a counter discourse to the hegemonic discourse of progress (Dahl 2012; Pisani 2006). The oppression of indigenous peoples became a symbol of modernity's oppression of the cohesion between man and nature. From a status as the lowest step on the ladder of human evolution, indigenous peoples were now seen as a lifeline to pre-modern existence and experience, believed to be contained in what was named 'traditional knowledge'. This was endorsed in the United Nations' report *Our Common Future* (the so-called *Brundtland Report*) from 1987:

> These communities [indigenous peoples] are the repositories of vast accumulations of traditional knowledge and experience that links humanity with its ancient origins. Their disappearance is a loss for the larger society, which could learn a great deal from their traditional skills in sustainably managing very complex ecological systems.
>
> (United Nations 1987)

Therefore, when the report states that indigenous peoples should be given 'a decisive voice in formulating policies about resource development in their areas' (United Nations 1987), the point is not only that they have a *right* to decide what happens to the land where they live, it is assumed that by assigning this right to indigenous peoples, a voice is automatically given to nature. The concept 'traditional knowledge' has continuously been developed and theorized under terms such as 'traditional ecological knowledge (TEK)' (Berkes *et al.* 2000), 'indigenous environmental knowledge (IEK)' (Burkett 2013), or 'indigenous science' (Snively and Corsiglia 1998). It forms a theoretical basis for new subjects such as ethnobiology and environmental anthropology (Sillitoe 2007), and it is nowadays enrolled as a parameter of best practice in projects, reports, etc. at all levels. 'Indigenous knowledge' is often used interchangeably with 'local knowledge', but the two concepts do not carry the same meaning in all contexts, since the latter is not necessarily ascribed the same mythical links to 'ancient origins' as the former.

The added value as a guarantee for good governance in environmental and climate issues makes Professor in Philosophy and Community Sustainability Kyle Whyte draw attention to the governance-value of indigenous knowledge also when it comes to *indigenous resurgence and nation-building* (Whyte 2017:5). In Whyte's argument, decolonization and the resurgence of indigenous peoples' former 'place-based and embodied existence' go hand in hand (Whyte 2017:9). Consciousness about the value of indigenous knowledge serves as a unifying and nation-building factor for the indigenous peoples themselves. Indigenous peoples are thus expected to build their community on values derived from 'completely different cultural-linguistic contexts' than Western standards (Whyte 2017:9), even though it is recognized that the two ontologies learn from each other and cooperate. This kind of thinking is common in the literature on indigenous peoples and has also characterized discourses in Greenland, particularly in the years surrounding the implementation of Home Rule in 1979.

However, the debate about nature preservation has a long history in Greenland, which has made the connection between indigeneity and sustainability a lot more complex and contested. This chapter looks into this history in order to better understand the complicated situation of today, where non-Western ontology plays a minor role in Greenlandic environmental politics. First, a section juxtaposes the difficulties of translating the concept of sustainability into the Greenlandic language with the efforts made to claim it as Greenlandic, and a section revisits the literature on the relation between man and nature in precolonial and colonial discourses. Then a section analyses the way knowledge and authority were negotiated in debates on progress and conservation as they unfolded in the Greenlandic press in the early twentieth century, a period crucially reconfiguring life and identities. Finally, a section explicates how local – rather than indigenous – knowledge is presented as that which makes colonialism unsustainable, and how local knowledge is later 'nationalized'. In sum, a brief conclusion finds that the cultural heritage from the Inuit past is central to many Greenlanders' self-identification, as well as to Greenland's brand as a nation. The Greenlanders see themselves as pioneers in the struggle of indigenous peoples for equality and the right to gain independence. However, the approach guiding how the Self-Government works with the sustainability concept affirms modernity and nation, rather than tradition and indigeneity.

Sustainability in Greenlandic

As early as 1992, Greenland expressed its intention that the Home Rule Government would work towards sustainable development. This happened in connection with the signing of the Rio Declaration and Agenda 21.[1] In 1995, *Pinngortitaleriffik*, the Greenland Institute of Natural Resources, was established by a law passed by the Home Rule Government (Law no. 6, 8 June 1994). The institution carries out research and monitoring of animal populations, vegetation, etc. in an international environment of scientists and related institutions. In 1998 Greenland was a signatory on the Nordic prime minister's declaration *Ett hållbart Norden* (A Sustainable Nordic Region), and has been a signatory on Nordic strategies for sustainable development ever since.

'Globalization and sustainable development' was the topic of the Greenlandic *Foreign Policy Statement* in 2001. This examines, inter alia, Greenland's regional and international commitments. In 2003, the Greenlandic Parliament passed the Act on Greenland Nature Conservation (Landsting Act No. 29 of 18 December 2003), and in 2008, Deputy Head of Department Kaj Kleist took the initiative to draw up an overview of Greenland's commitments in the field of sustainability, with particular reference to identifying the shortcomings that Greenland might have in individual policy areas. This resulted in the report *Arbejdet med bæredygtighed og globalisering i Grønlands Hjemmestyre* (The Work on Sustainability and Globalization in Greenland's Home Rule Government; Nielsen 2008). Since then, a number of reports and guidelines have been issued, all under strict consideration of enrolling Greenland in international law and best practice in the field. It is primarily in this sense that sustainability has had a nation-building effect in Greenland. By adopting the concept, Greenland brings itself on an equal footing with other nation states and is able to commit itself at an international level. Thus, the term sustainability has come to Greenland from the outside and has been translated into a Greenlandic context. This is also true in terms of linguistic translation.

The most commonly used term for sustainability in Greenlandic is *piujuartitsineq* or *piujuaannartitsineq*, which is also the term sanctioned by Oqaasileriffik, the Language Secretariat of Greenland. The term has been in use for a while, as for instance in the Coalition Agreement between Siumut and Inuit Ataqatigiit (IA) signed 8 December 2002, and in the above-mentioned Act on Greenland Nature Conservation.[2] Sustainability is only one of many foreign terms that are entering the Greenlandic language at an ever-faster pace. Such terms circulate in English and Danish, while translators are trying to invent Greenlandic terms that people will understand, even if they do not have access to the discourse in Danish and English. This can be a challenging process, and different translators often try out different formulations which then circulate concurrently, until one term is lexicalized and is included in the official vocabulary.[3]

Sustainable is not an easy word to translate. Also in an English context, it is rather ambiguous. According to the dictionary, it means that something is 'able to continue over a period of time', or 'causing little or no damage to the environment and therefore be able to continue for a long time' (*Cambridge Advanced Learner's Dictionary*). The second meaning is probably derived from the Brundtland Report, where sustainable development is defined as 'development that meets the needs of the present without compromising the ability of future generations to meet their own needs' (ch. 2, 1). The term *piujuartitsineq* – or *piujuaannartitsineq* – is a direct translation of this English description. *Piujuar-* means lasting. The affix -inner- is actually not needed, but it adds a durational aspect: lasting for a long time. The affix *-tsipaa* tells that an action has to be taken: *causing* it to last for a long time. *-neq* turns it into a noun or concept. Back-translated into English, the term would mean something like 'sustainabilization'.

Like 'sustainable', *piujuartitsineq* is an abstract term and therefore suited to become a floating signifier: being able to carry a web of different meanings.[4] However, the Greenlandic online dictionary defines *piujuartitsineq* by the much more familiar – and much more concrete – concept *nungusaataanngitsumik atuineq*: using a resource without exhausting it. Unfortunately, this explanation makes translations of English and Nordic texts promoting 'sustainable mining', i.e. of *non*-renewable resources, almost meaningless in Greenlandic. During the 1990s, *nungusaataanngitsumik atuineq* was the most common translation of the sustainability concept. It comes from the classic nature preservation paradigm, where it is estimated at which point a certain species will no longer be able to reproduce itself and survive in the wild. However, when biologists nowadays use the word sustainability in connection with giving advice on the sustainable exploitation of living resources, they work with another rationale, concerning how much it is possible to harvest without jeopardizing the potential future yield. This meaning is not equally well covered by *nungusaataangitsumik atuineq*. Another term, *oqimaaqatigiissaarineq*, to keep the balance, has sometimes been used, but it seems that also in this context *piujuartitsineq* may be a better choice. What is happening right now is that this term is used more often and in different contexts, and thereby a framework of connotations and meanings is being built, like the framework for 'sustainability' in English, and 'bæredygtighed' in Danish. By being introduced to new terms like *piujuartitsineq*, the Greenlandic population is introduced to the international discourses and is simultaneously educated in the ontology expressed in these discourses.

Like elsewhere in and beyond the Arctic, sustainability is thus a keyword in Greenlandic politics and administration, and it seems important for all parties to give the concept a rootedness in Greenlanders' own history. However, not all political parties associate it with indigeneity. Indigeneity is not a basic concept in Greenlandic politics, at least not in all contexts and for all parties. The term is not mentioned in the Act on Greenland Self-Government, and it has been discussed whether Greenlanders can still be defined as an indigenous people after the implementation of this act, since they are now in power in their own territory. The UN Declaration on the Rights of Indigenous Peoples (2007) aims at protecting ethnic minorities from the state. Many Greenlanders hold the opinion that the Greenlanders should now simply call themselves 'a people' (Thisted 2013). However, every time the discussion is opened, it is quickly closed again because the category of indigenous peoples conveys certain special rights, not least with regard to whaling. Indigeneity has even been used as a trump card in the game about CO_2 emissions and the right to industrial development (Bjørst 2011; see also her chapter in this volume).

Especially for the left-wing party IA, indigeneity has, since the party's establishment in the late 1970s, played an important role as a unifying factor that would differentiate the Greenlanders from the Danes and inspire decolonization.

IA politicians have played a leading role in indigenous organizations such as the Inuit Circumpolar Council, and when the IA party came to power in the first election after the implementation of Self-Government in 2009, a small section on 'life view' was inserted in the opening of the Coalition Agreement:

> The Greenlandic life view is based on the consciousness of the 'cohesion of all things' – the unbreakable physical and spiritual connection of human beings with each other, with nature, nature's resources, the universe – a context that extends far beyond the individual.
>
> (Naalakkersuisut 2009)[5]

The Siumut Party, which was in power during the 30 years of Home Rule (1979–2009), has been less oriented towards ontology, and the passage disappeared when Siumut came back into power. However, the following passage has been inserted in the two latest Coalition Agreements, in a paragraph on traditional hunting: 'The principle of sustainability is a Greenlandic invention, where the first provisions concerning sustainable use of animals was formulated in the Thule Acts, and these principles must be maintained and honored' (Naalakkersuisut 2014; 2016).

The Thule Acts were formulated on the initiative of the famous Arctic explorer and author Knud Rasmussen (1879–1933), with the help of the Danish Supreme Court attorney Rudolf Sand, and ratified by the hunters' assembly on 7 June 1929. In 1910, Rasmussen had established a trading station in far northwest Greenland as a private investment, since the Danish state was not willing to extend its sovereignty any further north. The Acts apply to a wide range of areas in society: §§28 and 29 concern nature protection (translated or explained with the words *ima piniagagssat piniartailineqásaput*, 'This is how the game should be hunted'). With the acts, Rasmussen wanted to avoid the sort of problems that he had seen in Danish West Greenland, and would again later on his Fifth Thule Expedition (1921–1924) in Canada and Alaska. With the introduction of firearms, hunting pressure went up and certain species were threatened with extinction. A short text explains the reason and need for animal protection:

> Any free hunter must provide himself and his family food and skins by hunting. But the game is not present in unlimited quantity any more. All over the Earth free peoples have therefore agreed that the animals must be protected at those times of the year where they breed, because the game is less and less in numbers every year. Here with us, it is especially important to protect eider ducks, foxes and walruses from extinction, and any free hunter should gladly agree to such a conservation, since these animals will otherwise have become extinct when the people who now are children become adults … The man who will not understand this and who wishes to assert that such rules should not apply to him, commits a bad act to his fellow human beings and should therefore be punished.[6]

As the very last area to become Christianized and 'civilized', Thule has an aura as the most 'authentic' (pre-colonial Inuit) part of Greenland. Here, some of the old hunting techniques and traditions have been preserved to this day. However, with the reference to the Thule Acts, the concept of sustainability is semantically moved from being associated with the mythical, an essence of 'Greenlandishness', to the contact zone: the period in history when Greenland was in transition from the pre-colonial and pre-modern to the colonial and modern society. Thus, it is the willingness to integrate new ideas and translate tradition into new forms that is set as a benchmark for today's society with the reference to the Thule Acts. As is evident from the extract, the Thule Acts aimed at introducing a whole new way of thinking about nature in the hunting society.

Pre- and colonial discourses about nature

In his book *Grønlands Naturforvaltning – ressourcer og fangstrettigheder* (Greenland Nature Management – Resources and Rights to Hunting), the anthropologist Frank Sejersen distinguishes between two different conceptions of nature that he believes have existed side by side, and sometimes blending, in Greenland during the twentieth century.

The pre-colonial Inuit perceived the animals as fundamentally human. Animals were considered rational beings who made plans for the future and decided which people they wanted to seek or avoid. People and animals were in a close social relationship with each other, and it was the task of human beings to make sure they were in good standing with the animals. In order to achieve this goal, they had to comply with a large number of rules and taboos. If the hunter did not offend, but pleased an animal, it would return to him and let itself be caught again and again, since it was believed that the souls of living beings would reincarnate after death. As Sejersen points out, the Inuit may therefore be said to have been 'sustainable', in the sense that they aligned themselves on the premise that their relationship with the animals should last over time, including future generations (Sejersen 2003:158). This, however, does not mean that they were sustainable in the sense this concept is generally understood today. The anthropologist Ann Fienup-Riordan has convincingly refuted the idea of the Inuit as a kind of 'original ecologists' or 'natural conservationists' (Fienup-Riordan 1990). Fienup-Riordan argues that the way the Inuit perceive animals as an infinitely renewable resource is incommensurable with the Western view of fish and wildlife as finite and manageable 'stocks'. According to the Inuit worldview, it was the *availability* of animals that was within the range of human influence, not their *existence* (Fienup-Riordan 1990:xix, 173). If a hunter did not see any animals, it was not because the animals were not there. It was because they would not come to the hunter and let themselves be seen. Accordingly, if an animal showed itself, it was because it wanted to make or renew a bond between the hunter and itself, and therefore it was an offence to the soul of the animal *not* to kill it.

However, as Fienup-Riordan points out, matters became complicated, since the Inuit were rising to European expectations and started to depict themselves according to Euro-American discourses. Fienup-Riordan's observations pertained to the Yupik Inuit in Alaska in the 1970s and 1980s (1990:xix), but both the description of the pre-colonial worldview and the analysis of how the Inuit took over the contemporary dominant discourses fit very well with Greenland. Also in Greenland, this was a time where natural romanticism and the idea of indigenous people as a sort of 'noble savage' had a renaissance. And, just as in Alaska, it was not a pre-colonial Inuit cosmology that here came to the fore, but a cosmology built upon Western ideals. The Greenlanders were highly familiar with – and co-producers of – the discourse on indigenous sustainability, at least those Greenlanders who engaged in identity politics. This new discourse represented a total break with the previous discourse on nature conservation in Greenland.

Nature conservation and the management of living resources was introduced in Greenland in the early twentieth century, when such arrangements were also made in Denmark, and Danish officials brought the ideas to Greenland (Sejersen 2003:153). Up through the 1900s, more and more restrictions on catch and collection were introduced by the local authorities, who were first and foremost Danes, but who also included elected Greenlandic members. The Danish officials saw the decline in animals and birds as the combined result of the improved hunting techniques (guns and rifles) and the Greenlanders' nature as a primitive people. The discourse about the Greenlanders as impulsive children driven by their instincts was commonplace around Danish administrators, scientists, and politicians at this time. In 1906, Rasmus Müller, a colonial manager – and himself an enthusiastic hunter – wrote the following:

> To try to introduce a hunting act or restriction of the right to hunting here would be as impossible as to thaw the sea dry; The Greenlanders feel free like the bird in the air, they only follow their own will, and they would not really realize what benefit such limitation should be, and therefore, of course, not comply with it.[7]

Müller's tone is a mix of paternalism and fascination. The idea of the free Eskimo hunter was held as an ideal among Danish officials – also because it benefited trade (Thomsen 1998). Likewise, when in 1921 the botanist Morten Peder Porsild, founder and manager of the Danish Arctic Station at Qeqertarsuaq (Godhavn) in Disko Bay, wrote that 'the Eskimo lives like a predator',[8] this was not considered as offensive and racist in a contemporary perspective as it may sound today. Administrators, traders, and scientists all agreed that the Greenlanders' mindset and worldview was alien to European nature conservation. As described by Fienup-Riordan, they may not have been quite wrong in this assertion. But, as the following section will demonstrate, they were wrong in the assertion that the Greenlanders were unable to acquire new perspectives and go against their 'nature', and therefore needed the Danes to protect them from themselves in perpetuity. The conflict was far from a conflict between Danes on the one hand and Greenlanders on the other. The debate quickly became an internal debate between those Greenlanders who believed in the idea of nature preservation, and those Greenlanders who held on to the old cosmology – or had other reasons to reject the idea of restrictions.

In the early 1900s, some Greenlandic hunters opposed the recently adopted restrictions on hunting, on the grounds that man thus made himself master of the animals and took it upon himself to control what was not for human beings to control (Petersen 1965:121–122). These hunters seem to have oriented themselves from a pre-colonial worldview, according to Ann Fienup-Riordan's definitions. It was a basic principle of the pre-colonial Greenlandic Inuit that one was allowed to take as much as one needed, but forbidden to take more than one needed or to let anything go to waste (Petersen 1965:117). However, there seems to have been some disagreement about exactly how 'waste' should be defined. Knud Rasmussen quotes the successful hunter Abia's objections, as the local council convicted him for having shot a large number of reindeer, but bringing only sinew and the best pieces of skin home with him:

> I can not recognize the decision made by the Board of Trustees (Forstanderskabet), which I assume to be formulated by men who have no idea of reindeer hunting ... If I were to adhere to the regulation of the Board, I would be unable to support myself and my family at certain times of the year when there is neither fox trapping nor fishing ... Every year I receive so many orders on skin and sinew thread that I would not be able to overcome everything ... Should I now follow the provisions of the Board, then I should make a long and difficult journey, just in order to catch one or two reindeer, more meat I can not carry on my kayak. And then I should not only suffer from lack of skins and sinew during the winter, but I would not be able to help many as I do now. So, can anyone really come and forbid me, yes punish me for that I leave the meat behind, which does not play any major role for me at this season? Certainly, I feed the foxes with this meat, and it is by them I make my living during the winter.[9]

Before anything else, this hunter is a very pragmatic man. His statement gives us insight into the hunters' difficult situation at a time when the social praxis that used to set the framework for hunting and sharing had been dissolved, and the hunters' challenge was to figure out how to navigate in these brand new waters. Before the arrival and intensification of the trade, reindeer hunting had been carried out collectively and the meat shared so that all got a part. Thus, there were many to transport the killed and butchered animals. Abia lived at a time where work had been divided into various functions. Many people were employed by the colony and therefore could not participate in the hunts. Other people who could not hunt, and who did not have a successful hunter in the close family, now had to buy the things that earlier would have been distributed as gifts. This is why Abia can 'help' so many people – that is: the people who are able to order and buy the products. Likewise, the only reason that Abia could make a living from foxes during the winter is that he sold them to the shop and bought groceries in return. Foxes were a very important export product from Greenland due to the demand in the European market. Therefore, not only hunting techniques, but the whole social practice that the hunt was embedded in was changed at this point in time (for a parallel discussion, see Graugaard in this volume). The new rules and regulations served as a new kind of framework that better matched the new conditions after the gradual dissolution of the old social order.

Knowledge and authority in early twentieth-century debates on conservation and progress

All of this was vigorously debated in the Greenlandic public sphere, gradually coalescing by way of the periodicals *Atuagagdliutit* and *Avangnâmioq* in the early twentieth century. At that time, the so-called 'Danish West Greenland' was divided into North and South Greenland as two units that each corresponded to Copenhagen, not to one another. The two papers were published each in their regional centre, Nuuk (South) and Qeqertarsuaq (North). Issues were irregularly distributed up and down the coast; old volumes were saved and re-read, so many years-old debates were often resumed. Here, the collapse of the old sharing rules and the transition to market mechanisms was a favourite topic. The writers regret how the lack of skins cause prices to rise, so that skins become completely out of reach for people who do not go hunting themselves (see, for instance, Kaspersen 1922). The warmest advocates of preservation regulations appear – not surprisingly – to be the educated Greenlanders. Members of the local councils write articles in which they try to explain the background for the decisions taken by the councils. The point in these articles is that 'the road to the Greenlanders' progress goes through laws and regulations' (Lynge 1923:46).[10] However, many hunters have a diverging idea of progress: If progress means better material conditions then, for the hunters, progress must necessarily mean more catch. Not only hunters and employees/administrators take part in the debate; people from all parts of society seem to have an opinion on the issue.

In 1916, a man signing himself as only 'Someone from Upernavik', the northernmost district, right next to Thule, published an article with the title 'Is the number of eider ducks decreasing?' (Upernavingmio 1916). The author most emphatically answers his own question in the negative. People who believe such nonsense are only worth a laugh, the anonymous author argues. In 1922, a young man named Anthon Kleist from Qeqertarsuaq in North Greenland responds to this article (Kleist 1922) without entering into a discussion about the numbers of birds. Instead, he gives a report of his experiences in Denmark. Kleist has been in Denmark since the autumn of 1919, where he is undergoing education. He is impressed with Denmark, where everything works so much better than in Greenland. Staying in Denmark has opened his eyes, and he wishes to share this knowledge with his compatriots (1922:42). What characterizes

Denmark, Kleist writes, is that in Denmark people are willing to comply with rules and regulations. In Denmark, the birds are protected during the breeding season and not a single gunshot is heard during the summer. Therefore, the birds are plenty, and they are much less shy than in Greenland, where even the seagulls flee at the slightest sound of human activity. In Denmark, they come in between the houses and sit on the roofs. Unfortunately, it is such that 'we Greenlanders are always having difficulties in adopting rules and regulations that would be to the benefit of our country' (1922:43). Kleist takes the opportunity to point out that down in Denmark they have also long since done away with the stupid practice that young people must always listen to the old, instead of the other way around – as on a ship, where everybody has to listen to the captain, no matter how old or young he is. This is what characterizes 'the developed societies on earth' (1922:44). Kleist is here echoing the contemporary discourse describing Denmark as a 'people of culture' and Greenland as a primitive people or 'people of nature'. Thus, the new ideas also come with a whole new set of social hierarchies. Kleist speaks with a self-confidence that might provoke some, especially the hunters.

Although he certainly shares Kleist's view, Jakob Asser Berthels, a catechist from Kangerluarsussuaq, south of Nuuk, begins his intervention (Berthels 1922) in a much more humble voice, probably aimed at not offending the people he wants to address. Being a catechist, Berthels knows that he may not be taken seriously by the full-time hunters. He therefore starts his article with a longer defence of his right to talk about the subject as he, like everyone else, is completely dependent on the products he can get from the sea and the land.[11] Likewise, he underscores that he speaks with the right of age, and he expresses the greatest respect for the hunters. It is both unfair and counterproductive when some people claim that the hunters have no real knowledge, he says. The hunters know very well what they are doing, and their voices should be listened to (1922:58).

After this long introduction, Berthels finally takes on answering the question of whether the number of eider ducks is decreasing. He does not deny that it may be true when the North Greenlanders talk about the vast flocks of common and Kingeider ducks that are so big that, even when watching them through a pair of binoculars, they cannot see any ending. Every year, the South Greenlanders who watch this migration from south to north from the month of March till the beginning of June say to one another: '*Avangnâ miteqartartorinássusia!*' – meaning 'Wow, what a lot of eider ducks they will have up there in North Greenland!' However, when the eider ducks return from their summer breeding camps up north, they are frightened and do not come near the settlements, and the kayaks cannot get near them. Berthels has himself experienced how it has become more and more difficult to hunt the eider ducks, and it is not because he is getting older, he claims! First and foremost, there are fewer and fewer young birds in the groups when they return. Some may say that this is because the birds go to other places inland – far from the settlements – so that their numbers seem to decline. However, Berthels does not buy into this explanation. The birds will have to go where they can feed. No, Berthels claims, the most important reason for the decline of eider ducks is that the shotgun is used to hunt them (1922:58). The young Greenlanders have become used to firearms, and they are of the opinion that firearms are the best weapons for hunting at all times and no matter what kind of animal.

Before, Berthels continues, people used to go egg-collecting at the breeding grounds of the eider ducks, and they used to come back with the kayaks full of eggs but with no birds, since they did not bring a gun. So, no birds were killed, and they could lay new eggs. Today, people bring their guns, and not only the eggs are removed, but the breeding birds are killed. Therefore, the North Greenlanders might right now be of the opinion that the eider ducks are abundant, but sometime in the future they will themselves come to miss going eider duck hunting. However, the South Greenlanders also share in the blame, since they also use the shotgun for hunting, instead of, like before, using the bird dart. When using the shotgun, far more birds are killed, and many go to waste because the hunter does not get hold of all of them. To promote the idea that the number of animals is limited, Bethels refers to a quotation from the Bible that the world will perish with all its riches (1922:59). In essence, Berthels is completely in line with the anthropologist Ann Fienup-Riordan in his analysis of how the view of the environment is embedded in belief, and it is important for him to demonstrate how the new view of the wild animals and their limited numbers is in good accordance with the Christian faith. This was also completely in line with the ideas of the day, that Christianity was the pathway to enlightenment, prosperity, and modernity for the people of Greenland (Thuesen 1988).

However, this connection between progress and regulations was strongly opposed by other debaters. In an article titled 'I do not consider the shotgun a bad weapon', Vilhelm Egede (1922) from the Qaqortoq district in South Greenland writes that he does not agree with the decision made by the North Greenlandic Council to delimit the distribution of free shotguns to young hunters (under the age of 20). Egede recalls how everyone is constantly talking about the progress of the Greenlanders. However, in his opinion, 'progress' must mean the elimination of need, and for that end the weapons of the previous generations (harpoons and rifles) are not good enough. When a man gets a more efficient weapon, he is able to take better care not only of himself, but also of those who are dependent on him. Ultimately, more efficient weapons will also benefit the next generations, ensuring them a better start in life. Therefore, the shotgun is a useful tool in addition to the harpoon and the rifle, and the more shotguns they get in North Greenland, the more they will realize this fact, he claims. In this line of thought, effective catching is the same as plenty of catch. The idea of overuse does not even seem to come to his mind. Or, it has to give way to something which to him is far more important, namely the hunters' sufficiency.

Colonialism and neo-colonialism versus local knowledge and sustainable, national democracy[12]

While, in most cases, it is difficult to determine whether the opposition to nature preservation has anything to do with a pre-colonial worldview, it is obvious that many perceive the restrictions as an external interference in Greenlandic conditions. The issue of nature preservation therefore becomes an issue in the Greenlandic elite's opposition to Danish colonialism.

In the debate book *Strejflys over Greenland* (Gleams of Light Over Greenland) written in Danish in 1929 while the author was on a long stay in Denmark, the Greenlandic Vice Dean Mathias Storch (1883–1957) tried to get through to the Danes the message that the Greenlanders had to have more power and influence in their own country (Thisted 2017). On the issue of nature management, Storch is first and foremost concerned with the catastrophic decline of the seal. Storch mentions many possible reasons, but interestingly, firearms is not one of them. By the 1900s, the Greenlanders were well aware that other people from outside Greenland were harvesting from the same resources, at other locations. Especially the Norwegians were often blamed for being the cause of the falling seal population. Storch also pursues this track. The Norwegians have driven whaling in the Greenland waters with modern gear and ships equipped with train-boilers. This has led to a dramatic decline in whales, which the killer whales, according to Storch, used to hunt in the open sea. Due to the lack of their usual prey, the killer whales now enter the fjords in search of seals, and the hunters have to compete with the killer whales for the few remaining seals (Storch 1930:104–105). The Norwegians' massacre of hooded seals on the sea ice east of Greenland also had the sad consequence that, where the ice was previously black because of the number of seals, it is now only the scraps that come drifting with the ice (1930:92). However, the biggest problem is, according to Storch, climate change. Also at this point in time, the temperatures had risen so that there was no solid ice in Disko Bay:

> And when there is no ice, the population has no opportunity either for sealing or fishing. This is comparable to unemployment in Denmark. But the Greenlanders are worse off than the unemployed. They receive unemployment benefit – for the Greenlanders there is no such thing.
>
> (1930:105)

This situation requires leadership, and the most pressing problem is the lack of communication between Danes and Greenlanders. The Greenlanders are badly prepared to speak up to Danish officials when they make poor decisions in relation to the catch, Storch argues (1930:84–85). Over and over again, the experienced hunters have warned against the new habit of establishing settlements on the outermost islands. Not only are the seals hunted prematurely, before they have settled in the fjords, they are also scared away by the smell of human activity. However, the Danish administration will not listen to this.

They assert that the Greenlanders have always moved around and settled where it suited them. According to Storch, this is a misunderstanding. Previously, the Greenlanders had understood that the seals needed a sanctuary. Otherwise, they did not stay at the coast.

With his emphasis on nature's cycle and the human responsibility not to disturb the balance, Storch's description is actually the one that comes closest to the notion of 'indigenous knowledge' in the Greenlandic sources investigated from this period. In Storch's opinion, this kind of knowledge, combined with the old hierarchy of age, is worth securing before the Danish administration totally undermines the authority of the experienced hunters. Thus, in Storch's description, it is the experienced hunters who work for more orderly conditions for the catch, while the Danish authorities are represented as supporting a headless, disorganized approach which, in pursuit of the immediate surplus for trade, overrides the consideration of the future. In this way, the Danes' accusations against the Greenlanders' irresponsible behaviour are reversed. Without saying it directly, Storch demonstrates how the Europeans largely contribute to creating the unregulated conditions which they so authoritatively attach to the '"nature" of the Greenlanders'.

This way of repositioning Danish authorities also served to produce Greenlandic unity. As emerged from the debate about eider ducks, the divisions – administrative and otherwise – between North and South Greenland resulted in mutual stereotyping and a lack of commonality. In the newspapers, authors from South Greenland draw a picture of North Greenland as a hunter's paradise, while the North Greenlanders see the South Greenlanders as spoiled by European goods due to the more regular ship supply from Europe. Storch moves the focus to the colonial administration and the lack of communication between Danes and Greenlanders. Thus, already at this point, 'traditional knowledge' is used as a political strategy. After 1953, the split between North and South was abrogated, and the opposition between Denmark and Greenland became the demarcation line in society until the opposition between the outlying districts and Nuuk took over this position with the implementation of Home Rule and Self-Government.

When power moved from Copenhagen to Nuuk, the concept of neocolonialism arrived in Greenland, and the Greenlandic and Danish political, administrative, and intellectual elites are often lumped together as 'the neocolonialists in Nuuk', as here in a comment on *Sermitsiaq/AG Online*:

> The neocolonialists in Nuuk do not know our nature. The guillemots we catch originate mainly from islands in the North East Atlantic, Svalbard, Jan Mayen and a number of other islands … We do not catch much of our own stock of the Northwest colonies as they overwinter in the waters south of Nova Scotia in Canadian waters. Depending on how it goes with the ice in the Davis Strait, they will arrive sometime between March and May on their way North. They come in large numbers. And it has been like this every spring in the last approximately 30 years where I have been on the

> West Ice hunting belugas and walruses. We who inhabit this part of the country observe millions of them by this time.
>
> (Fontaine 2017)

It is striking how similar this statement is to the article written by the anonymous man from Upernavik in 1916, discussed above. Nor does Fontaine advocate pre-modern ontology, but like the man from Upernavik, he claims to have a local knowledge. For them both, the core question is who are those in power to decide whether a resource is to be used or not – the local users or someone from the outside?

Through the last century, the number of full-time hunters has gradually declined, and today few households base their income entirely on this sort of activity (Rasmussen 2005). However, hunting and all other sorts of harvesting from nature (fishing for family use, berry picking, etc.) play an important role for most families as a leisure activity. In addition to the Greenlandic language, preference for Greenlandic food and outdoor life are the most important identifiers of being Greenlandic. Keeping the balance between these interests and the international agreements to which Greenland has committed itself in terms of ecological sustainability is an important issue for the country's politicians.

In 2014, an international report, *Arctic Biodiversity Assessment, Status and Trends in Arctic Biodiversity*, pointed out many areas where the Greenlandic nature is challenged (Meltofte *et al.* 2013). In the Danish media, the message was immediately reformulated and the Greenlanders were accused of preying on Greenlandic nature, with specific reference to the guillemots (Duus 2014a; Hannestad 2014). Such accusations hit Greenland hard; first, because Greenland wants to build up an image as a nation that can handle such issues, and second, because nature plays such an important role in Greenlandic self-identification – not to mention Greenland's brand. In a press release[13] with a reply to the report, the Premier of Greenland, Kim Kielsen (Siumut), referred to the aforementioned Act on Greenland Nature Conservation from 2003: 'It clearly states that the exploitation of our nature must take place on an ecologically sustainable basis in accordance with the precautionary principle and respect for human living conditions and the conservation of animal and plant life' (quoted in Duus 2014b).

In an interview in the newspaper *Sermitsiaq*, Kielsen expanded on this answer, inscribing Greenland into the community of modern, democratic nations, governed by rules and regulations, based on scientific exploration:

> It is our duty to ensure that the utilization of our living resources takes place on a sustainable basis. To ensure this we have in our stewardship of nature built up a system of biological advice and created a dialogue with interest organizations and the profession. Does it make us irresponsible? No, it makes us a democratic society where both users and decision makers help to take responsibility for how we use our nature.
>
> (Duus 2014b)

'How we use our nature' is obviously the punchline here. Kielsen is speaking on behalf of a 'we', the Greenlandic people, who have organized their society according to democratic principles for good governance and which, under international law, have the right to self-determine the use of their natural resources. In this line of thought, sustainability is not in the culture, but is accomplished in a dialogue between the triangle of biological advisers, users, and politicians in an international environment of rules, regulations, and expectations. In Kielsen's statement, this is the key to sustainability: the constant dialogue which constitutes a democratic society.

But Kielsen is running the same risk of being categorized as an outsider as the catechist Jakob Asser Berthels was when interfering in the debate about eider ducks a hundred years earlier. Therefore, Kielsen, just like Berthels, makes an effort to position himself as an insider, even though he is not a full-time hunter:

> Everyone who knows me knows that I love being out in nature, and hunting is a natural part of my life. Sitting in the office one day and going on a seal hunt the next day are not two incompatible pursuits. Does it make me exotic or make me a Nature Romantic? No. Does it make me special? No, it makes me quite ordinary. Here, at home. Nature is there to be used – with reason.
>
> (Duus 2014b)

In this part of the interview, Kielsen first ascribes to a discourse which might be associated with indigeneity. Then he rejects this very discourse through the terms 'exotic' and 'Nature Romantic'. The words are used as epithets, denoting an external perception. Kielsen distances himself from such positions. Obviously, he does not regard it as in his best interest to make himself a carrier of any such humanity's 'ancient origins' as promoted in the Brundtland Report. Instead, he defines himself as 'a pragmatist': 'Of political nature I am a pragmatist' (Duus 2014b). Thus, even though the word 'indigenous' is never mentioned, the discourse is negotiated and restructured around this opposition between external natural romanticism and internal pragmatism.

Sometimes, culture is indeed invoked in a way it was not before the introduction of the sustainability and indigeneity concepts: 'It is not in the Greenlandic culture to eradicate our wildlife – nor guillemots!' writes Minister for Fishing, Hunting, and Agriculture Nikolaj Jeremiassen in a response to a member of the Greenlandic Parliament who has questioned the government's decision not to follow the biologists' recommendation for an interim total ban on guillemot hunting (quoted in Nyvold 2016).[14] However, such arguments are rare – and when they occur, they are usually supported by other arguments. The main argument in Jeremiassen's statement does not concern culture, but independence and economy. Greenland must be self-sufficient with meat in order to limit the amount of imported food. This will improve the economy – a prerequisite for political independence, which is the stated future goal for the government's policy (Nyvold 2016).

And, every now and then, tradition is invoked as something worth sustaining rather than something contributing to sustainability. Two recent Coalition Agreements (Naalakkersuisut 2014, 2016) have put a renewed focus on traditional hunting: 'Traditional hunting must be protected as an occupation throughout Greenland, with emphasis on and protection of especially the use of traditional hunting methods.' Such a focus may also serve as something to refer to when speaking about the 'indigeneity' of Greenlanders. In 2016, two succeeding ministers in charge of hunting used the exact same formulation to reject the demand for an interim total conservation of guillemots: 'A total ban on guillemot hunting is expected to have a negative effect on eating habits, so that children and young people lose interest in the traditional eating.'[15]

Conclusion

Considering the place of indigeneity in international sustainability discourse, the concept seems to play a surprisingly insignificant role in internal Greenlandic political debates. As far as the environment is concerned, good governance is based on a dialogue between biological advisers, users, and politicians. In internal political discussions, Greenlanders rarely refer to indigeneity as a guarantor of a special sense of sustainability. The endless debates about firearms and fast-paced motorboats resulting in over-hunting, turmoil, and disturbance, not to mention the number of wounded animals that sink and are lost when shot by firearms, speak a language other than the discourse on indigenous sustainability. Basically, the Greenlanders consider themselves modern people due to the long process of modernization they have undergone during colonization and its aftermath. However, external expectations play a major role, and Greenlanders certainly share an outside admiration for the ancestors who, by their own power and technology, survived in the harsh Arctic environment.

Culture and tradition may be invoked, but autonomy is the fulcrum of Greenland's work with the sustainability concept. Above anything else, the initiatives to develop and incorporate the sustainability concept have aimed at shaping and branding Greenland as a nation state, ready to enter into international cooperation and agreements at the highest political level. In this context, indigeneity could be a hindrance rather than a help, and this is probably the main reason why the two concepts are not more closely interlinked in Greenlandic political discourse.

Notes

1 Agenda 21 is a non-binding action plan formulated at the Earth Summit UN conference on Environment and Development held in Brazil in 1992. The '21' refers to the twenty-first century (it is also the area code for Greater Rio de Janeiro).

2 According to Jørgen S. Søndergaard, senior adviser for the Self-Government, the term was first introduced in the Self-Government's Durability and Growth Plan from 2016 (Søndergaard 2017). This is incorrect, and must be based on a misunderstanding. Søndergaard's own list of applied sustainability glossaries (p. 121) registers

the term as used in (the Greenlandic translation of) Peter Nielsen's report from 2008.

3 Other terms for sustainability have been tried out and are still circulating, such as *ataavartoq* (lasting, persistent), *imminnut nammassinnaasoq* (self-supporting), or *imminut napatissinnaasoq* (self-containing).

4 For instance, 'sustainable cultural heritage' has recently been used to negotiate how cultural heritage can be renewed and kept alive and still continue in a form that will represent the connection with the past, as part of a current discussion on ownership of the material culture and the appropriateness of bringing elements from the national costume into commercial fashion and new designs (see NAPA 2017).

5 My translation from the Greenlandic version.

6 Rasmussen and Sand (1929). Kap York Thule Station's Laws of 1929, my translation from the Danish original.

7 Quoted in Sejersen (2003:127), my translation from the Danish version.

8 Quoted in Sejersen (2003:154), my translation from the Danish version.

9 Quoted in Sejersen (2003:133), my translation from the Danish version.

10 All quotes are my translation from the Greenlandic original.

11 Greenlandic catechists were not paid the same salary as a Danish teacher or auxiliary priest. The prerequisite was that the catechists also went hunting and were fairly self-sufficient with Greenlandic products.

12 Unless otherwise noted, quotes in this section are my translation from the Danish original.

13 Quotes are my translations from the Danish version.

14 My translation from the Danish version.

15 Minister for Hunting Karl-Kristian Kruse, in Hansen (2016), and Minister for Fisheries, Hunting and Agriculture Nikolaj Jeremiassen, in Nyvold (2016), my translation from the Danish version.

References

Berkes, F., Colding, F., and Folke, C. (2000). Rediscovery of traditional ecological knowledge as adaptive management. *Ecological Applications*, 10 (5), pp. 1251–1262.

Berthels, J.A. (1922). Akíssutínguaq. *Avangnâmioq* 8 August, pp. 58–59.

Bjørst, L.R. (2011). Arktiske diskurser og klimaforandringer i Grønland: Fire (post) humanistiske klimastudier. PhD dissertation, University of Copenhagen.

Burkett, M. (2013). Indigenous environmental knowledge and climate change adaptation. In: Abate, R.S. and Kronk, E.A. (eds), *Climate Change and Indigenous Peoples: The Search for Legal Remedies*. Camberley: Edward Elgar.

Dahl, J. (2012). *The Indigenous Space and Marginalized Peoples in the United Nations*. Basingstoke: Palgrave Macmillan.

Duus, S.D. (2014a). Forsker efter kritik: Husk at læse rapporten, *Sermitsiaq* 7 March. Available from: http://sermitsiaq.ag/forsker-kritik-husk-laese-rapporten (accessed 7 March 2014).

Duus, S.D. (2014b). Kim Kielsen afviser rovdrift-kritik. *Sermitsiaq* 2 March.

Egede, V. (1922). 'Sáko tingmiarsiut nikanartûtingilara', *Avangnâmioq* 7 July, pp. 49–50.

Fienup-Riordan, A. (1990). *Eskimo Essays*. New Brunswick, NJ: Rutgers University Press.

Fontaine, L. (2017). Commentary. *Sermitsiaq/AG Online* 21 January. Available from: http://sermitsiaq.ag/kl/appat-eqqissisimatinneqannginnerat-knapk-nuannaarutigineqarpoq (accessed 8 March 2018).

Hannestad, A. (2014). Grønland kritiseres for rovdrift i Arktis. *Politiken Internationalt* 1 March. Available from: http://politiken.dk/udland/art5504572/Grønland-kritiseres-for-rovdrift-i-Arktis (accessed 8 March 2018).

Hansen, N. (2016). Ingen planer om at totalfrede lomvier. *Sermitsiaq/AG Online* 11 February. Available from: http://sermitsiaq.ag/ingen-planer-totalfrede-lomvier (accessed 8 March 2018).

Kaspersen, S. (1922). Oqalugtualiaq Oqaloqatigîngneq. *Avangnâmioq* 12 December, pp. 89–91.

Kleist, A. (1922). Mitit agdlautigissaunerat pivdlugo. *Avangnâmioq* 6 June, pp. 42–44.

Lynge, F. (1923). 'Sigsuerniminínguit' [Small Observations]. *Avangnâmioq* 8 June, pp. 45–47.

Meltofte, H. *et al.* (2013). *Arctic Biodiversity Assessment: Status and Trends in Arctic Biodiversity*. Akureyri, Iceland: CAFF, Conservation of Arctic Flora and Fauna, Arctic Council.

Naalakkersuisut (2009). *2009-miit 2013-imut Naalakkersuisut suleqatiinniissaannut isumaqatigiissut* (Coalition Agreement 2009–2013). Signed by Inuit Ataqatigiit, Demokraatit, and Kattuseqatigiit Partiiat, 10 June 2009.

Naalakkersuisut (2013). *Unified Country – Unified People. Coalition Agreement 2013–2017*. Signed by Siumut, Atassut, and Partii Inuit, 26 March 2013.

Naalakkersuisut (2014). *Fellowship, Security, Development. Coalition Agreement 2014–2018*. Signed by Siumut, Demokraatit, and Atassut, 4 December 2014.

Naalakkersuisut (2016). *Equality, Security, Development. Coalition Agreement 2016–2018*. Signed by Siumut, Inuit Ataqatigiit, and Partii Naleraq, 27 October 2016.

NAPA (2017). *The Ice Crystal Princess Vol. II – Sustainable Heritage*. Available from: www.napa.gl/event/the-ice-crystal-princess-vol-2-sustainable-heritage/?lang=en (accessed 22 December 2017).

Nielsen, P. (2008). Arbejdet med bæredygtighed og globalisering i Grønlands Hjemmestyre. Unpublished report prepared for Greenland's Home Rule Government.

Nyvold, M. (2016). Jeremiassen: Vi udrydder ikke fangstdyrene. *Sermitsiaq/AG Online* 17 July. Available from: http://sermitsiaq.ag/jeremiassen-udrydder-ikke-fangstdyrene (accessed 8 March 2018).

Petersen, R. (1965). Some regulating factors in the hunting life of Greenlanders. *Folk*, 7, pp. 107–124.

Pisani, J.A.D. (2006). Sustainable development: historical roots of the concept. *Environmental Sciences*, 3 (2), pp. 83–96.

Rasmussen, R.O. (2005). *Analyse af fangererhvervet i Grønland*. Naalakkersuisut. Department for Fishery.

Rasmussen, K. and Sand, R. (1929). *Kap York Stationen Thule's Love./Inatsisit maligagssiat niuvertoqarfingme ~Umáname avangnardlerme (Kap York Stationen Thule) nunanilo tássunga atassunut*.

Sejersen, F. (2003). *Grønlands Naturforvaltning – ressourcer og fangstrettigheder*. København: Akademisk Forlag.

Sillitoe, P. (2007). Local science versus global science: approaches to indigenous knowledge in international development studies.In: *Environmental Anthropology and Ethnobiology*. New York: Berghahn Books.

Snively, G. and Corsiglia, J. (1998). Discovering indigenous science: implications for science education. Paper presented at the 1998 National Association of Research in Science Teaching, San Diego, CA, 19–22 April 1998.

Storch, M. (1930). *Strejflys over Grønland*. København: Levin & Munksgaards Forlag.

Søndergaard, J.S. (2017). Noget om aktuelle muligheder for anvendelsen af begrebet 'Bæredygtig udvikling' for et arktisk velfærdssamfund. *Tidsskriftet Grønland*, 2, pp. 118–124.

Thisted, K. (2013). Discourses of indigeneity: branding Greenland in the age of Self-Government and climate change. In: Sörlin, S. (ed.), *Science, Geopolitics and Culture in the Polar Region: Norden beyond Borders*. Farnharn: Ashgate, pp. 227–258.

Thisted, K. (2017). 'A place in the sun': historical perspectives on the debate on development and modernity in Greenland. In: Hansson, H. and Ryall, A. (eds), *Arctic Modernities: The Environmental, the Exotic and the Everyday*. Newcastle-upon-Tyne: Cambridge Scholars Press, pp. 312–344.

Thomsen, H. (1998). Ægte grønlændere og nye grønlændere – om forskellige opfattelser af grønlandskhed. *Den jyske Historiker*, 81, pp. 21–55.

Thuesen, S. (1988). *Fremad – Opad. Kampen for en moderne grønlandsk identitet*. København: Rhodos.

United Nations (1987). *Our Common Future*. Oxford: Oxford University Press.

Upernavingmio (1916). 'Merqit ikiligpat?'. *Avangnâmioq* 3 March, pp. 22–24.

Whyte, K. (2017). What do indigenous knowledges do for indigenous peoples?. In: Nelson, M.K. and Shilling, D. (eds), *Keepers of the Green World: Traditional Ecological Knowledge and Sustainability*. Available at: https://ssrn.com/abstract=2612715 (accessed 8 March 2018).

13 Sustaining Denmark, sustaining Greenland

Johanne Bruun

Early history and making of radioactive resources

The conceptual link between the notion of sustainability and the practice of extracting non-renewable resources is, at first sight, perhaps not an obvious one (see Jacobsen, this volume; Thisted, this volume). When the resource in question is the highly contentious uranium mineral with its affective ties to both humanitarian and environmental disasters, the link seems even more tenuous. In Greenland, however, the mining of rare earth elements, of which radioactive minerals are an unavoidable byproduct,[1] has become deeply entwined with Greenlandic nation-building and ideas about the economic sustainability of a future Greenlandic state (Nuttall 2013; Vestergaard 2015; Dingman 2014). Greenland's former Premier, Aleqa Hammond, for example, called the 2013 lifting of a 25-year moratorium on the mining of radioactive resources 'a great step towards independence' (quoted in Milne 2013; see also Mølgaard 2013). From their position within the deep interior of Greenland's mountains, uraniferous minerals are thus sustaining visions of an independent Greenlandic nation-state. Anticipation of the wealth and abundance that resource-based development will bring to the small island nation has thus placed a different object of sustainability front and centre; it is not only the environment that needs sustaining, but also Greenlandic society and its institutions (Tianen 2016). As emphasized in Greenland's most recent Oil and Mineral Strategy, the Government of Greenland (2014a) does not perceive sustainability as an exclusively environmentalist trope (see Bjørst, this volume; Gerhardt *et al.*, this volume). Sustainability is a complex which includes questions of infrastructure, education of the labour force, health, long-term community development, and not least the production of surplus value to sustain the state apparatus (Government of Greenland 2014a).

Sustaining Greenland requires capital, both social and economic, and it is common for official Greenlandic narratives to hinge this on the development of Greenland's extractive industries (Tianen 2016). Investment from foreign mining companies will, according to the Oil and Mineral Strategy, bring long-term benefits to Greenland by providing jobs and opportunities for professional development for the Greenlandic people (Committee for Greenlandic Mineral

Resources 2014; Government of Greenland 2014a). Furthermore, the mining of radioactive minerals in particular has been framed by leading political figures such as Aleqa Hammond (2013) and the recent Minister of Industry, Labour, Trade, Energy, and Foreign Affairs – Vittus Qujaukitsoq (2016) – as a form of mining which would ultimately benefit the environment. In defiance of the apparent paradox between the mining of non-renewable resources and the concept of sustainability, Hammond and Qujaukitsoq have presented nuclear energy as a 'clean' alternative to fossil fuels and, as such, as a means of mitigating climate change. Following their logic, mining Greenland's uranium would serve the dual purposes of sustaining a fragile Arctic environment while simultaneously economically sustaining a would-be independent Greenlandic state.

Such narratives of uranium mining as sustainable practice are, of course, both questionable and subject to much critique from multiple sources. Environmentalist organizations, such as Avataq, NOAH, and Greenpeace, question whether radioactive minerals can be mined in an ecologically safe manner (Jørgensen 2017; NOAH 2017). Concerned citizens raise questions regarding the social and cultural costs that large-scale mining might bring to the communities they are supposed to sustain, including harmful effects on activities such as hunting and fishing (Nuttall 2013; Hansen *et al.* 2016; Bjørst 2016; see also Lindqvist 2017). Invoking examples of exploitative mining in decolonized African states, opponents of uranium mining have questioned whether placing Greenland's rare earths in the hands of foreign companies runs the risk of replacing one postcolonial dependency with another.

Despite both national and international criticism, the geological imaginaries upon which a self-sustaining Greenlandic nation-state are so often based remain discursively linked to the economic sustainability of Greenland. As argued by Nuttall (2013, 2015), this link is not only discursive, but material as well. At the crux of these complicated narratives about the nation and its future sits the rock. At first glance, the rock appears grey, dull, and unassuming, as it is pictured in the government's promotional material (e.g. Government of Greenland 2014b). Yet the rock is the material foundation upon which a prosperous Greenlandic society is envisioned. Geological imaginaries of Greenland's future align visions of an independent Greenland with the rock – solid, stable, and durable. Firmly anchoring the state within the substrata in this manner rests on the extension of governmental logics to the underground beyond the spaces of human inhabitation (see Dodds and Nuttall 2016); an effort closely linked to the practices of modern science and a particular geological 'way of seeing' (Braun 2000).

Rock does not automatically emerge as a resource. Nor is it a given that the rock presents itself as the foundation of an economically sustainable nation state. As scholars of critical resource studies have long pointed out, resources are socio-material constructs, made and unmade (Bakker and Bridge 2006; Richardson and Weszkalnys 2014). Resource-making rests on an epistemological ordering of the geological world which involves the division of geological bodies into distinct categories in accordance with a set of valued qualities. The sciences of the earth play a central role in this practice. This division allows for the subsequent conversion of select geological bodies into economic and sub-economic objects (Bridge 2009a; 2009b; 2014; Klinger 2015). Such practices and the extractive imaginaries they inspire may fundamentally redefine the nature and character of a place. As argued by Dodds and Nuttall (2018; also in this volume), a critical Arctic geopolitics thus needs grounding in the practices through which its deep, material structures are made available for geopolitical consumption.

The imaginary (if not yet the reality) of an economically sustainable Greenlandic state founded on extraction is sustained by a complex network which includes scientists, politicians, and matter (Latour 1999). This network, however, is not merely the product of a relatively recent "rush for resources". The economization of Greenland's uraniferous minerals has a much longer history tied up in its own contentious politics as well as the (post)colonial relationship between Denmark and Greenland. The commodity status of Greenland's most famous site of possible uranium mining, the Kvanefjeld, is the product of practices which began in 1955 when its grey, rocky matter was first imagined as an object that could sustain a state and bring it into modernity – albeit the Danish state rather than the Greenlandic nation (Nielsen and Knudsen 2016). The remainder of this chapter tells a story of how the Kuannersuit/Kvanefjeld – a part of the mountain complex known by geologists as Ilímaussaq – was brought into the realm of political rationality as a resource.[2] As will become apparent, the productive fiction of Ilímaussaq which dominates current popular and political imaginaries of the mountain was not easily established. As such, the aim of subsequent discussions is to bring to light some of the work required to construct the kind of socio-material network needed to cast rock as a source of economic and national sustainability.

Claiming the Greenlandic "resource frontier"

Danish colonial imaginaries of Greenland as a resource frontier date back not only decades, but centuries. Throughout the colonial period,[3] non-renewable elements such as coal, lead, and most notably cryolite were mined and sent to Denmark for processing, consumption, and distribution (Berry 2012; Ries 2003). These extractive practices were dependent on the establishment of vast networks of geological knowledge extending deep into the Greenlandic subsoil. As Bridge and Fredriksen (2012:368) have noted, the production of knowledge is 'integral to the exercise of colonial power and the expansion of capital'. Practices of sustaining empire through the production of knowledge and subsequent "productive occupation" are well-documented by scholars writing on postcolonial resource environments. Knowledge construed as "superior" to local systems of meaning is used to legitimize colonial practices of dispossession while simultaneously communicating territorial sovereignty (see Scott 2008; Edney 1997; Livingstone 2010).

As noted, such processes of materially augmenting the Danish–Greenlandic colonial relationship through economic geology have a very long history. Greenland's position as a potential site of uranium extraction, however, was not established until the mid-1950s.[4] By that time, the previously tense political climate surrounding fissionable materials had begun to ease up following Eisenhower's 1953 *Atoms for Peace* address, and nuclear power gradually emerged as a marker of techno-scientific modernity and progress (Klinger 2015; Hecht 1997). At the same time, Denmark was eager to communicate territorial sovereignty over Greenland amid an uncomfortable American presence (Olesen 2013). During the Second World War, the US had taken control of Greenland to keep it from falling into the hands of the Nazi regime (Lidegaard 1997). When the war ended, the Cold War was already on the rise, and given Greenland's geostrategic position between the main industrial centres of the US and the USSR, the US maintained its presence on the island (Archer 1988; Heymann *et al.* 2010). The mining of Greenlandic uranium would thus sustain the Danish state in at least three ways: by communicating Danish territorial sovereignty over Greenland; raising Denmark's standing as a techno-scientifically advanced nation; and addressing post-war Danish energy deficiencies.

Once the ambition of a Danish nuclear energy infrastructure emerged, no time was wasted. In the spring of 1955, the Danish Government appointed an Atomic Energy Commission, and one of the first orders on its agenda was to initiate large-scale uranium prospecting of what was then considered Danish mountains (Koch 1958; Nielsen and Knudsen 2010). Wanting immediate action, the Commission approached the Greenland Geological Survey (Grønlands Geologiske Undersøgelse – GGU), the Danish state institution closest to the matter at hand, proposing a 1955 field expedition. The GGU, however, informed the Commission that in lieu of the lack of basic geological mapping of Greenland, such a rushed effort would be both foolish and futile and that no GGU geologists could be made available on such short notice (GGU 1955). Despite strong words of discouragement, the GGU reluctantly recommended the Ilímaussaq complex, alongside two other locales, as a likely site of uranium ore (Noe-Nygaard 1955). The Commission, however, was not dissuaded by the geologists' resistance to their assigned role in this Danish national project driven by a Copenhagen-based political elite. Instead, they approached the newly established Danish Defence Research Establishment – an institution which did not house geological experts, but which did provide basic training for conscripts of mixed scientific backgrounds. Seemingly against the odds, the Commission thus managed to piece together an expeditionary outfit consisting of 11 conscripts under the leadership of Lieutenant Colonel Vilhelm Valdemar Mouritzen.

GEOX 55

Mere weeks after the Commission had first approached the GGU, Mouritzen and his men set foot on the shores of Greenland's southwesterly coast. The only piece of information guiding their radiometric survey was three areas which the geologists had reluctantly recommended and marked on large-scale maps displaying very little detail of the field sites in question (Maag 1955). To these men, Ilímaussaq represented a fundamentally unknown geography, a landscape in excess of the flatness of their Danish homeland, and a frontier zone positioned at the very edge of the Kingdom. As such, their mission was to "plant the flag of civilization" on this unruly landscape by taking the first steps towards rendering it productive in the service of the Danish state (Seed 1995).

The GEOX 55 expedition, as it was subsequently dubbed, had been equipped like a military outfit in a temperate landscape, more so than a scientific field party in the north. Their base camp tents were ill-suited to the rocky terrain, their mattresses soaked up moisture from the ground below, and their clothing was camouflage, which made it difficult to conduct geographical sightings between points (Mouritzen 1955a). This, however, did not seem to impede Mouritzen's enthusiasm for what he perceived to be an honourable and important task (Mouritzen 1955b). Serving as impermanent markers of Danish territorial sovereignty, Mouritzen and the conscripts tirelessly carried their heavy equipment up and down mountains, day in and day out. Meticulously counting the "clicks" of the hand-held Geiger counters while manually keeping time with a stop watch, the conscripts mapped and measured the radiance of Ilímaussaq. Although the practice was time-consuming and tedious, the acoustic feedback of the Geiger counter worked not only as an indicator of radiation, but also worked to enthuse the fieldworkers and instil in them a sense of the thrill and excitement of the "hunt". The Geiger counters allowed the conscripts to enact an otherwise intangible quality of the mountain which they could never have sensed without this technological aid. An article published in the Danish newspaper *Berlingske Tidende* (1958) later reported how the fieldworkers testified to being driven by 'Geiger psychosis' and 'uranium fever' in their 'pioneering work'. While the conscripts never encountered uranium in any direct way, the affective qualities of this highly symbolic resource were ever present (Dodds and Nuttall 2018).

As Simon Naylor (2010) notes, fieldwork is a geographical performance. The performance of the GEOX conscripts served at least two purposes. A second article from *Berlingske Tidende* (1955), for example, reported on Danish popular enthusiasm at the sight of 'Danish men in Danish mountains … both amongst those who prioritise the economy, and those who put the national aspects first'. In addition to communicating Danish productive occupation in Greenland amid an uncomfortable American presence, the conscripts were simultaneously performing the value of a national reserve and the potential for this reserve to sustain the state economically.

An aerial experiment

Resistance, as the 1955 fieldwork amply demonstrated, did not only come from the geologists who continued to perceive the prospecting task as premature and ill-conceived (see Bondam 1957; Noe-Nygaard 1955). The mountains themselves were also pushing back against Danish desires for a self-sustaining energy infrastructure based on Greenlandic uranium. During the first year of fieldwork, the reach of the radiometric survey had been significantly impaired by the ruggedness of the terrain, which largely determined where the inexperienced fieldworkers, many of whom 'had never before seen a mountain, let alone climbed one' (Maag 1955), could physically go. Hence, in an effort to extend the horizontal reach of the survey, Mouritzen and the Atomic Energy Commission decided to dedicate the 1956 efforts to testing a technology that had not previously been used in Greenland – aerial scintillometry.

A scintillometer is a device for measuring radiation which mechanically counts the light flashes (scintillations) which occur as radioactive particles pass through a gamma-sensitive crystal. Capable of measuring radiation at a great distance, this device could be mounted to an aircraft, thus making it possible in principle to 'scan' the landscape from above and produce an automatic measure of radiation (Nuclear Chicago 1955). The aerial survey thus promised a speedy radiometric assay of Ilímaussaq which would cover vast stretches of otherwise impenetrable terrain in a matter of hours, rather than days. However, aerial scintillometry was not quite the 'easy fix' that the Atomic Energy Commission had perhaps hoped for.

In topographically complex areas of Ilímaussaq, unexpected manoeuvres of the heavy, amphibious aircraft that the expedition had at their disposal were often necessary, meaning that the detector could not be kept level. Even when the line of flight was stable, small irregularities in the landscape showed up on the automatic charts as changes in radiation. Finally, the highly sensitive device registered radiation from any direction, not only from the ground below, but also from nearby precipices (Vinther and Koch 1956; Mouritzen 1956a). Consequentially, much of Mouritzen's (1956a) detailed report on the aerial survey is dedicated to reconciling 'peculiar' results of the aerial survey with measurements recorded on the ground.

Despite all these difficulties and the seemingly dubious results of the aerial survey, Mouritzen's (1956a; 1956b) enthusiasm for the technology did not appear to waver. Befitting of the Atomic Energy Commission's emphasis on achieving hasty results, aerial scintillometry embodied values such as efficiency, speed, and utility rather than hard-earned accuracy in the Danish effort to extract wealth from the Greenlandic subterrain. Sustaining the promissory qualities of the national project, it would seem, was given priority over scientific rigidity.

Economizing the subterrain

While the division of the field sites into zones of red, green, and yellow was potentially highly affective, the radiometric charts which Mouritzen and his men produced were not straightforward markers of value. Since the maps revealed nothing about the geochemical or mineralogical composition of the rock or if uranium was present, high levels of radiation were not necessarily markers of a reserve with the potential to become a resource (Bondam 1956; Pauly 1955). Hence, while conducting their survey, Mouritzen and the conscripts had simultaneously engaged in geological sampling. This, however, did not sit well with the professionally trained geologists of the GGU who blatantly dismissed the samples, arguing that they had no geological value due to the amateur composition of Mouritzen's team (Pauly 1955). Extending the epistemological (and political) reach of the Danish state beyond the surface of the mountain to incorporate its deep, voluminous structures into the national project required an inward looking gaze which only trained geologists could provide (Braun 2000).

After two field seasons dominated by Mouritzen and his conscripts, the Atomic Energy Commission ended its arrangement with the Defence Research Establishment, effectively putting the GGU in charge of the prospecting task (Atomic Energy Commission 1956). Armed with locational devices, field notebooks, and colour pencils, the geologists set out to add depth, volume, and matter to the horizontal charts produced by the GEOX fieldworkers. Geology is a highly visual and theory laden practice. The trained geological eye, drawing on detailed knowledge of how to "read" the landscape in terms of its structures and formation, effectively penetrates the solid strata of the rock and brings into existence an image of the subterrain which no one could ever see (Rudwick 1976). Vertical sections mapping out the internal architecture of Ilímaussaq drew hard lines around suspected bodies of uraniferous rock (Bondam 1958), thus transforming '*terra incognita* into vertically organized goods to be exploited in the name of development, security and progress' (Klinger 2015:574).

Yet, embedded in the rock, the uranium had no value. To become a marker of economic sustainability, the uranium had to 'cut its ties with Greenland' (Sørensen 1966). Samples ranging from small specimens to giant boulders were extracted from Ilímaussaq and sent to Copenhagen for testing. In the laboratory, however, scientists soon found that the resistance embodied by the geologists' protests and the contours of the landscape was also manifest in the material structure of the rock itself. The uraniferous rock of Ilímaussaq constituted a geochemical anomaly which proved recalcitrant to standard practices used to extract uranium from ore (Nielsen 1981). Countless experiments eventually led to a method of extraction, yet the method was laborious, costly, and led to a considerable loss of the original uranium content of the rock (Sørensen 1966).

Despite the less than promising results of the geochemical tests, 1958 saw the first drilling programme at Ilímaussaq. The GGU geologists had initially objected to this project, which they perceived as premature from a scientific standpoint (Bondam 1957). Once again, however, their resistance to the national project had been overwritten by the politically influential Atomic Energy Commission (1957a; 1957b). During this fourth field season, thousands of metres of cylindrical rock cores were extracted and sent back to Copenhagen-based laboratories alongside tonnes of boulders blasted from the mountain (Bondam 1958). Notwithstanding this monumental effort, however, the geo-economic results remained less than promising. The refractive uranium of Ilímaussaq only existed in concentrations significantly lower than what was then considered viable for extraction (Nielsen 1981).

Conclusion

By the end of the 1958 drilling programme, scientists knew the Ilímaussaq uranium deposits to be sub-economic. Nevertheless, Danish prospecting continued for no less than 25 years before finally coming to a halt (Nielsen 1981). Yet, as Dodds and Nuttall (2018:143) write:

> projects do not need to undergo construction and operational phases and resources do not necessarily have to be extracted, but the idea of extraction and the hype surrounding it becomes part of a political economy concerned with the reproduction of speculation.

Likewise, Ilímaussaq did not need to be the site of any actual physical extraction of profit in order to sustain Danish resource dreams. Similarly, the *promise* of mineral riches is what sustains current visions of an independent Greenlandic state (Nuttall 2012). As this chapter has sought to draw out, this "promise" is not merely an artefact of what is "found" in nature or of political rhetoric. Rather, it requires considerable work to put in place a network of scientists, policy-makers, and matter with the force to sustain such political imaginaries.

Although the prospecting of Ilímaussaq began as Greenland was undergoing a process of formal decolonization, the economization of Ilímaussaq's uraniferous rock arguably bore the marks of a classical colonial relationship. Affirming the submissive relationship between centre and periphery, the prosperity and modernity that the rock was anticipated to bring was not intended to sustain Greenland, but rather to sustain its colonizer and the colonial relationship (see Scott 2008; Graugaard, this volume). Local involvement in the enactment of Ilímaussaq as a space of extraction was minimal and the scientific work and its planning was entirely in Danish hands.

The geologists who were employed at Ilímaussaq during the first years of prospecting described the complex as a seemingly endless frontier of new scientific discoveries – it was 'new land' brought within the reach of the Danish state through the means of science (Ellitsgaard-Rasmussen 1962; Noe-Nygaard 1957). Having been under intense scientific scrutiny for decades and part of an Inuit homeland for centuries, Ilímaussaq was (and is) hardly "new land" in any sense of the word. Yet it is still imagined and governed like a frontier (Nuttall 2013). The abstraction of uranium from rock is still key to its economization – to its solidity as foundation for financial and political independence for Greenland. Unlike the discourse of the 1950s, current uranium rhetoric does engage Ilímaussaq's other spatial qualities, including its use as farmland, hunting grounds, and a general space of habitation. However, when reading the Oil and Mineral Strategy, it is notable how most of the 12 points about sustainability are directed at sustaining the mining industry (Government of Greenland 2014a; Tianen 2016).

The Danish prospecting finally came to a halt when faced with growing popular resistance to nuclear power among both the Danish population and, more importantly, the increasingly organized Greenlandic political movements challenging Danish colonial rule. Opposing the mining of radioactive minerals was, according to Nielsen and Knudsen (2016), part of Greenland's project of liberation which culminated in the Home Rule Agreement of 1979. While resistance to the enrolment of Greenland's uranium into circuits of extractive capitalism was initially part of an anti-colonial move, it is perhaps not surprising that, now that Greenland controls its own subterrain, the reclaiming of the Ilímaussaq uranium to sustain itself rather than Denmark can be construed as an act of liberation. At the same time, however, this bears testament to the long-lasting durability of the historically rooted narrative of Ilímaussaq as a space of extraction.

As demonstrated by the early years of fieldwork, the "productive fiction" of the mountains of Ilímaussaq had proven challenging to establish (Klinger 2018). Institutional tensions and dissatisfaction among the geologists had hampered scientific collaboration, the steep and rugged terrain had limited the field-workers' movement through the landscape, and the uranium content of the rock was not only low, but also difficult to extract. Yet, once this narrative was in place, it proved even more difficult to do away with – even as the object of sustainability changed from the Danish to a would-be Greenlandic state.

The geological imaginary, largely unquestioned for decades, is part of the colonial heritage of Ilímaussaq. Speaking back against decades of 'hard science' in a landscape that has long borne the marks of extractive industries has proven difficult for local opponents. These people, who fear that they cannot sustain a way of life that rests on fishing and hunting in a highly polluted mining area, have found the public hearings ineffective and mostly symbolic (Nuttall 2013). Those in favour of uranium mining similarly employ the rhetoric of sustainability. As one proponent of the mine put it: 'You cannot live in a museum – you have the right to sustain your people' (Walsh 2017). As resource extraction remains emblematic of the sustenance and modernity of a future Greenland, opinions on uranium extraction continue to be highly divided. For now, the processes through which the extractive potential of the uraniferous rock of Ilímaussaq is normalized and rationalized continue.

Notes

1 Because of the geological coincidence between radioactive minerals and rare earth elements, uranium politics and rare earth politics are inextricably entwined (Klinger 2015).

2 A detailed analysis of the material, territorial, and institutional politics of the scientific practices presented in this chapter can be found in Bruun (2018). See also Nielsen and Knudsen (2013).

3 Gad (1984) defines the colonial period as the time from 1721 until the formal decolonization of Greenland in 1953. Notably, however, resource extraction did not formally take place until the seventeenth century (Nuttall 2013).

4 For an in-depth account of the politics surrounding Danish nuclear energy and the reasons why the Danish Government did not pursue Greenlandic uranium, see Nielsen and Knudsen (2010; 2013).

References

Archer, C. (1988). The United States defence areas in Greenland. *Cooperation and Conflict* 23, pp. 123–144.

Atomic Energy Commission (1956). Notat vedrørende organisationen af den kommende sommers eftersøgning af radioaktive stoffer på Grønland. Internal document, undated. Rigsarkivet, Copenhagen. AEK collection, Journalsager 1955–1960, Jn. No. 1957 12 28–15b 8, box 50.

Atomic Energy Commission (1957a). Minutes from meeting between GGU, Kryolitselskabet Øresund, and the Atomic Energy Commission. Document dated 31 October 1957. Rigsarkivet, Copenhagen. AEK collection, Journalsager 1955–1960, Jn. No. 1957 12 28-15b 8, box 50.

Atomic Energy Commission (1957b). Minutes from meeting between GGU, Kryolitselskabet Øresund, and the Atomic Energy Commission. Document dated 17 December 1957. Rigsarkivet, Copenhagen. AEK collection, Journalsager 1955–1960, Jn. No. 1957 12 28.15b 8, box 50.

Bakker, K. and Bridge, G. (2006). Material worlds? Resource geographies and the 'matter of nature'. *Progress in Human Geography* 30 (1), pp. 5–27.

Berlingske Tidende (1955). Fem hemmelige mænd paa Uran-færd. *Berlingske Tidende* 4 July, p. 9. Den Sorte Diamant, Copenhagen. Media archives, micro film.

Berlingske Tidende (1958). Uranbyen paa den gamle nordbomark. *Berlinske Tidende* 8th June 1958, pp. 13–14. Den Sorte Diamant, Copenhagen. Media archives, micro film.

Berry, D.A. (2012). Cryolite, the Canadian aluminium industry and the American occupation of Greenland during the Second World War. *The Polar Journal*, 2, pp. 219–235.

Bjørst, L.R. (2016). Saving or destroying the local community? Conflicting spatial storylines in the Greenlandic debate on uranium. *The Extractive Industries and Society*, 3 (1), pp. 34–40.

Bondam, J. (1956). Laboratorieprogram for undersøgelser af prøver, hjembragt af GEOX 56 (udkast). Declassified report from Geologisk Arbejdsudvalg for Radioaktive Råstoffer dated 31 July 1956. Rigsarkivet, Copenhagen. GGU collection, Journalsager 1946–1971, Jn. No. C6-16, box 4.

Bondam, J. (1957). Hvilken stilling skal GGU tage m.h.t. udviklingen i uransagen?. Internal GGU memo dated 11 November 1957. Rigsarkivet, Copenhagen. GGU collection, Journalsager 1946–1971, Jn. No. T12, box 58.

Bondam, J. (1958). Rapport vedrørende undersøgelserne på Kvanefjeldet i sommeren 1958. Internal GGU report, undated. GEUS internal collections, Copenhagen.

Braun, B. (2000). Producing vertical territory: geology and governmentality in late Victorian Canada. *Ecume*, 7 (7), pp. 8–46.

Bridge, G. (2009a). Material worlds: natural resources, resource geography and the material economy. *Geography Compass*, 3 (3), pp. 1217–1244.

Bridge, G. (2009b). The hole world: scales and spaces of extraction. *New Geographies* 2, pp. 43–48.

Bridge, G. (2014). Resource geographies II: the resource–state nexus. *Progress in Human Geography*, 38 (1), pp. 118–130.

Bridge, G. and Fredriksen, T. (2012). 'Order out of chaos': resources, hazards and the production of a tin-mining economy in northern Nigeria in the early twentieth century. *Environmental History*, 18, pp. 367–395.

Bruun, J.M. (2018) Enacting the substrata: Scientific practice and the political life of uraniferous rock in Cold War Greenland. *Extractive Industries and Society* 5 (1), pp. 28–35.

Committee for Greenlandic Mineral Resources to the Benefit of Society (2014). *To the Benefit of Greenland*, Expert report published by Ilisimatusarfik/University of Greenland and University of Copenhagen.

Dingman, E. (2014). Greenlandic independence: the dilemma of natural resource extraction. *Arctic Yearbook*, 2014, pp. 228–243.

Dodds, K. and Nuttall, M. (2016). *The Scramble for the Poles*. Cambridge: Polity Press.

Dodds, K. and Nuttall, M. (2018). Materialising Greenland within a critical Arctic geopolitics. In: Kristensen, K.S. and Rahbek-Clemmensen, J. (eds), *Greenland and the International Politics of a Changing Arctic: Postcolonial Paradiplomacy Between High and Low Politics*. London: Routledge, pp. 139–154.

Edney, M. (1997). *Mapping an Empire: The Geographical Construction of British India, 1765–1843*. Chicago: University of Chicago Press.

Ellitsgaard-Rasmussen, K. (1962). Geologien i Grønland. *Tidsskriftet Grønland*, 2, pp. 41–60.

Gad, F. (1984). *Grønland*. Copenhagen: Politikens Forlag.

GGU (1955). Letter from the GGU to the Danish Ministry of State, Grønlandsdepartementet. Correspondence dated 13 June 1955. Rigsarkivet, Copenhagen. GGU collection, Journalsager 1946–1971, Jn. No. T12, box 58.

Government of Greenland (2014a). *Greenland's Oil and Mineral Strategy 2014–2018*, dated 8 February 2014. Available from: http://naalakkersuisut.gl/~/media/Nanoq/Files/Publications/Raastof/ENG/Greenland%20oil%20and%20mineral%20strategy%202014-2018_ENG.pdf (accessed 8 March 2018).

Government of Greenland (2014b). *Our Mineral Resources: Creating Prosperity for Greenland – Greenland's Oil and Mineral Strategy 2014–2018*. Available from: http://naalakkersuisut.gl/~/media/Nanoq/Files/Publications/Raastof/ENG/Olie%20og%20mineralstrategi%20ENG.pdf (accessed 8 March 2018).

Hammond, A. (2013). Speech given at the 2013 Arctic Circle Assembly in Reykjavik. Available from: http://naalakkersuisut.gl/~/media/Nanoq/Files/Pressemeddelelser/ARCTIC%20CIRCLE%20presentation%20FINAL%20EN.pdf (accessed 8 March 2018).

Hecht, G. (1997). *The Radiance of France: Nuclear Power and National Identity After World War II*. Cambridge MA: MIT Press.

Heymann, M., Knudsen, H., Lolck, M.L., Nielsen, H., Nielsen, K.H., and Ries, K.J. (2010). Exploring Greenland: science and technology in Cold War settings. *Canadian Journal of the History of Science* 31 (2), pp. 11–42.

Jørgensen, T.J. (2017). Narsaq-resultater positivt for Kvanefjeldsprojekt. *Sermitsiaq*/AG. Available from: http://sermitsiaq.ag/narsaq-resultater-positivt-kvanefjeldsprojekt (accessed 13 April 2017).

Klinger, J.M. (2015). A historical geography of rare earth elements: from discovery to the atomic age. *The Extractive Industries and Society*, 2, pp. 572–580.

Klinger, J.M. (2018). *Rare Earth Frontiers: From Terrestrial Subsoils to Lunar Landscapes*. Ithaca, NY: Cornell University Press.

Koch, H.H. (1958). Atomenergikommissionen – baggrund og arbejde. *Nationaløkonomisk Tidsskrift*, 96, pp. 117–132.

Latour, B. (1999). *Pandora's Hope: Essays on the Reality of Science Studies*. Cambridge, MA: Harvard University Press.

Lidegaard, B. (1997). *I Kongens Navn: Henrik Kauffmann I Dansk Diplomati 1919–1958*. Copenhagen: Samlerens Forlag.

Lindqvist, A. (2017). Støv og vand kan sprede stråling. *Sermitsiaq*/AG. Available from: http://sermitsiaq.ag/stoev-vand-kan-sprede-straling (accessed 18 January 2017).

Livingstone, D. (2010). Landscapes of knowledge. In: Meusburger, P., Livingstone, D., and Jöns, H. (eds), *Geographies of Science*. Dordrecht: Springer, pp. 3–22.

Maag, M. (1955). Foreløbig rapport for Atomenergikommissionens arbejde i Sydgrønland i sommeren 1955. Internal GGU report. Rigsarkivet, Copenhagen. GGU collection, Journalsager 1946–1971, Jn. No. T12, box 59.

Hansen, A.M., Vanclay, F., Croal, P., and Skjervedal, A-S.H. (2016). Managing the social impacts of the rapidly-expanding extractive industries in Greenland. *The Extractive Industries and Society*, 3 (1), pp. 25–33.

Milne, R. (2013). Greenland prime minister eyes independence from Denmark. *Financial Times*. Available from: www.ft.com/content/4eb3e29c-416c-11e3-9073-00144feabdc0?mhq5j=e3 (accessed 30 October 2013).

Mouritzen, V.V. (1955a). GEOX 55. Report of 1955 fieldwork for the Atomic Energy Commission. Statens Naturhistoriske Museum, Copenhagen. Unsorted materials.

Mouritzen, V.V. (1955b). Personal diaries. Rigsarkivet, Copenhagen. AEK collection, Journalsager 1955–1960, Jn. No. 1957 12 28-15 b 8, box 50.

Mouritzen, V.V. (1956a). GEOX 56. Report of 1956 fieldwork for the Atomic Energy Commission. GEUS internal collections, Copenhagen.

Mouritzen, V.V. (1956b). Personal diaries. Rigsarkivet, Copenhagen. AEK collection, Journalsager 1955–1960, Jn. No. 1957 12 28-15 b 8, box 50.

Mølgaard, N. (2013). Nultolerancen over for uran er ophævet. *Sermitsiaq*.AG. Available from: http://sermitsiaq.ag/nultolerancen-uran-ophaevet (accessed 24 October 2013).

Naylor, S. (2010). Fieldwork and the geographical career: T. Griffith Taylor and the exploration of Australia. In: Naylor, S. and Ryan, J.R. (eds), *New Spaces of Exploration: Geographies of Discovery in the Twentieth Century*. London: I.B. Tauris, pp. 105–124.

Nielsen, B. (1981). Exploration history of the Kvanefjeld uranium deposit, Ilímaussaq intrusion, South Greenland. In: *ORGX, Uranium Exploration Case Histories*. Vienna: International Atomic Energy Agency.

Nielsen, H. and Knudsen, H. (2010). The troublesome life of peaceful atoms in Denmark. *History of Technology*, 26 (2), pp. 91–118.

Nielsen, H. and Knudsen, H. (2013). Too hot to handle: the controversial hunt for uranium in Greenland in the early Cold War. *Centaurus* 55, pp. 319–343.

Nielsen, H. and Knudsen, H. (2016). Cold atoms: the hunt for uranium in Greenland in the late Cold War and beyond. In: Doel, R.E., Harper, K.C., and Heymann, M. (eds), *Exploring Greenland: Cold War Science and Technology on Ice*. New York: Palgrave Macmillan, pp. 241–264.

NOAH (2017). Pressemeddelelse: Kuannersuit/Kvanefjeldsprojektet opfylder ikke Råstoflovens miljø- og klimakrav. Press release published jointly between Avataq, Urani? Naamik, NOAH Friends of the Earth Denmark, Det Økologiske Råd, VedvarendeEnergi, and Nuup Kangerluata Ikinngutai, dated 10 March 2017. Available from: https://noah.dk/index.php/pressemeddelelse/kvanefjeldsprojektet-opfylder-ikke-rastoflovens-miljo-og-klimakrav (accessed 8 March 2018).

Noe-Nygaard, A. (1955). Notat til departementschef H.H. Koch. Statement dated 7 May 1955. Statens Naturhistoriske Museum, Copenhagen. Arne Noe-Nygaard collection, folder marked 'Udvalget for geologiske arbejder med radioactive råstoffer 1949–1955'.

Noe-Nygaard, A. (1957). Letter from Noe-Nygaard to H.H. Koch dated 15 July 1957 Rigsarkivet, Copenhagen. AEK Collection, Journalsager 1955–1960, Jn. No. 1957 12 28-15 b 8, box 50.

Noe-Nygaard, A. and Ellitsgaard-Rasmussen, K. (1957). Kort rids af udviklingen omkring samarbejdet med Atomenergikommissionen. Internal GGU document, dated 26 November 1957. Rigsarkivet, Copenhagen. GGU collection, Journalsager 1946–1971, Jn. No. T12, box 58.

Nuclear Chicago (1955). Specifications: Geological surveys with radioactivity measuring equipment. Promotional material sent to the Danish Defence Research Establishment. Rigsarkivet, Copenhagen. AEK Collection, Journalsager 1955–1960, Jn. No. 1957 12 28-15 b 8, box 50.

Nuttall, M. (2012). Imagining and governing the Greenlandic resource frontier. *The Polar Journal*, 2 (1), pp. 113–124.

Nuttall, M. (2013). Zero-tolerance, uranium and Greenland's mining future. *The Polar Journal*, 3 (2), pp. 368–383.

Nuttall, M. (2015). Subsurface politics: Greenlandic discourses on extractive industries. In: Jensen, L.C. and Hønneland, G. (eds), *Handbook of the Politics of the Arctic*. Northampton, MA: Edward Elgar Publishing, pp. 105–127.

Olesen, T.B. (2013). Between facts and fiction: Greenland and the question of sovereignty 1945–1954. *New Global Studies*, 7 (2), pp. 117–128.

Pauly, H. (1955). Orientering om prøvematerialet fra Julianehaab 1955. Internal GGU memo, undated. Rigsarkivet, Copenhagen. GGU collection, Journalsager 1946–1971, Jn. No. C6-16, box 4.

Qujaukitsoq, V. (2016). Tale ved tiltrædelse af IAEA's konvention, 26 September 2016. Speech at IAEA Convention. Available from: http://naalakkersuisut.gl/~/media/Nanoq/Files/Publications/Erhverv/Tale%20IAEA/Tale%20ved%20tiltrden%20af%20IAEAs%20konvention%2026%20september%202016%20%20DK.pdf (accessed 26 September 2016).

Richardson, T. and Weszkalnys, G. (2014). Resource materialities. *Anthropological Quarterly*, 87 (1), pp. 5–30.

Ries, C.J. (2003). *Retten, Magten og Æren: Lauge Koch Sagen – en Strid om Grønlands Geologiske Udfoskning*. Copenhagen: Lindhart og Ringhof.

Rudwick, M. (1976). The emergence of a visual language for geological science 1760–1840. *History of Philosophy of Science*, 14, pp. 149–195.

Scott, H. (2008). Colonialism, landscape and the subterrain. *Geography Compass*, 2 (6), pp. 1853–1869.

Seed, P. (1995). *Ceremonies of Procession in Europe's Conquest of the New World, 1492–1640*. Cambridge: Cambridge University Press.

Sørensen, E. (1966). På spor af sjældne metaller i Sydgrønlands undergrund: V. Uran og Thorium – forekomst og udvinding. *Tidsskriftet Grønland*, 8, pp. 275–286.

Tianen, H. (2016). Contemplating governance for social sustainability in mining in Greenland. *Resource Policy*, 49, pp. 282–280.

Vestergaard, C. (2015). Greenland, Denmark and the pathway to uranium supplier status. *The Extractive Industries and Society*, 2, pp. 151–161.

Vinther, M. and Koch, S. (1956). Rapport over tjenesterejse til England 29–31 maj 1956. Internal report of the Atomic Energy Commission. Rigsarkivet, Copenhagen. GGU collection, Journalsager 1946–1971, Jn. No. T12m, box 58.

Walsh, M. (2017). 'You can't live in a museum': the battle for Greenland's uranium. *Guardian*. Available from: www.theguardian.com/environment/2017/jan/28/greenland-narsaq-uranium-mine-dividing-town (accessed 28 January 2017).

14 A new path in the last frontier state?

Transforming energy geographies of agency, sovereignty, and sustainability in Alaska

Victoria Herrmann

The strength of the path dependence of remote Alaskan villages over nearly six decades on diesel electricity generation seemed unending, until a decision by the Alaska State Legislature in 2008 to establish the Renewable Energy Fund (REF), administered by the Alaska Energy Authority. The stated goal of the fund was to help the state generate 50 per cent of its electricity from renewable sources by 2025 by awarding about $25 million yearly to utilities, independent power producers, local governments, and tribal councils (AEP 2010). The fund, buttressed by additional state incentives and Obama-era policies, created a path to redefine energy in Alaska away from diesel towards renewable energy innovation. More than that, the shift from diesel to renewable energy can be understood as a shift away from colonial transportation routes that lock-in money and sovereignty flows out of Native villages towards a sustainable, Native-empowering remote energy model. This chapter follows the transformation and offers a survey of the transition through the lens of path dependency and creation. It will argue that a combination of top-down and bottom-up catalysts opened a path to redefine the periphery–core relationship, empower Native Alaskan communities, and build community-based sustainability in electricity and heating generation in remote villages.

First, the chapter briefly examines the historic, colonial foundation of non-renewable energy path dependency and the triggers – both proximate and systemic – for path creation by (1) state policies within Alaska; (2) the Obama administration's Arctic and climate change policies; and (3) local, community-based ambitions for energy sovereignty. The body of the chapter will contend that the combination of these facilitators at three different scales created an opportunity to move away from diesel-generated power towards sustainable energy systems in remote Alaskan villages through investments in renewable research, development, and implementation. It uses path dependence theory and the rich scholarship on Arctic sustainable development and circumpolar postcolonial relations to understand the theoretical dimension and paradigm shift of such a transformation. In closing, the chapter considers the implications of the funding shortfall in Alaska due to lower-than-expected oil prices and the outlook on remote renewable energy penetration.

While this particular inquiry focuses on the 280 isolated villages in Alaska scattered across some 660,000 square miles, the heart of this chapter is an Arctic case study.[1] Northern citizens across the Arctic face exorbitantly high fuel prices, colonial legacies of built energy infrastructure, and the challenges of sustainable decision-making in a way that simultaneously empowers their sovereign status and provides the enviro-economic benefits of low-carbon development. Roughly half of the global Arctic's population live in rural remote communities, requiring off-grid electricity systems (Hoag 2016). In both North America and Europe, Arctic states are pursuing a quicker-than-expected transition to a less oil-dependent economy. The chapter provides an analysis of the shortcomings and successes along the path towards more sustainable, postcolonial community energy systems relevant to not only Alaska, but all Arctic spaces.

At the intersection of sustainability and sovereignty: a history of diesel path dependency

As the introduction of this collection notes, sustainability has been conceived as a concept that intervenes in discursive struggles over the future allocation of rights and resources. Sustainability is, at its core, a political struggle between competing visions of the future and the co-existing strategies needed to create those futures (Laclau and Mouffe 1985; Palonen 2006; Skinner 2002). In order to understand visions of the future, it is first important to understand the historical path that led to the current set of options for the future. In this sense, temporality becomes an important component to understanding postcolonial transitions in sustainable remote energy systems, in that sense-making of the past is considered alongside aspirations for the future, framed through the lens of what is transpiring in the present (Mead 1932; Ricoeur 1984). To understand the temporal frame of Arctic remote energy system development, the chapter relies on path dependence. Path dependence theory dictates that previous processes alter the costs/benefits of particular choices, and consequently narrow current decisions. Applied to energy technologies and investments, path dependence theory contends that all actions are governed by history, in that previous choices create self-reinforcement and lock-in regardless of their efficiency. Alongside the path dependence of the state of Alaska's economy on oil, remote Alaskan villages, primarily populated by Alaska Natives, became dependent on diesel fuel from the creation of remote systems dependent on diesel generators.

More than creating a lock-in to an unhealthy, expensive, and environmentally unsustainable fuel source, the path dependence on diesel and the history that created that reliance intersects with the colonial history of Alaska as a geography and its sovereignty as a Native homeland. Sovereignty for Indigenous peoples in the United States, inclusive of Alaska Natives, can be understood as the inherent right of a collective society to exercise self-determination and therefore express political authority over the ecosystems they occupy, the resources found within that geography, its cultures and heritage, and the lives of its members (Borrows 1997; Napoleon 2005; Whyte 2016a). Indigenous law understands sovereignty to mean 'the expression of regionally specific Indigenous legal orders that demand reciprocal relations in a community reliant upon the ecosystem, in order to shape environmental sustainability' (Brewer *et al.* 2017). This construction of sovereignty ties sustainability as a concept to its discursive use, specifically tied to a community's ability to maintain cohesion and ecological food, energy, and spiritual security in the face of environmental challenges. Sovereignty for Native communities thus exists at the nexus of sustainability and (post)colonialism – it is the right and responsibility to mend, cultivate, and expand social, cultural, environmental, and economic relationships damaged by colonialism. In reference to the reading strategy for this volume, this can be read as sovereignty should be made sustainable in relation to environmental – and by extension social, economic, and cultural – connections by Native communities. As a state, Alaska is unique in its legal framework for Alaska Natives because of the Alaska Native Claims Settlement Act (ANCSA), passed by the U.S. Congress in 1971 (Berardi 2005). The Act conveyed nearly one billion dollars and 44 million acres of land – the most tribal lands out of all US states – and divided it into 12 geographic regions of common heritage and interests. The sovereign status conferred on Native American reservations in the lower 48 does not exist in the Alaska context; instead, land provided by ANCSA is owned by Native-run corporations, with the intent being that all would benefit from resources on any given parcel of land (Resource Development Council n.d.). Twelve regional native corporations, which encompass 229 tribal groups, own most tribal land and their subsurface mineral rights, and as such rank as the largest private businesses in Alaska (Resource Development Council n.d.). According to a 2015 opinion piece advocating for the inclusion and elevation of Native corporations into any sustainable economic plan,

> There are approximately 120,000 shareholders who hold roughly 12% of the total land in Alaska. The concept driving ANCSA was that Native Corporations would utilize the money and land resources conveyed to them to engage in the capitalistic marketplace in order to provide profits to care for the social, cultural and economic benefit of their shareholders in perpetuity.
>
> (Godfrey 2015)

Alaska Native Corporations have done this by participating in a wide array of business sectors, including oil and gas exploration, development, and export. But while Alaska Native Corporations have benefited from the development and export of Arctic petroleum resources, for most of its history as a state Alaska Natives did not have energy sovereignty. Energy sovereignty can be understood as

> the right of conscious individuals, communities and peoples to make their own decisions on energy generation, distribution and consumption in a way

> that is appropriate with their ecological, social, economic and cultural circumstances, provided that these do not affect others negatively.
>
> (El Ecologista, Ecologistas en Acción Magazine 2014)

In short, if sovereignty refers to power, energy sovereignty refers to where power resides in energy affairs. At the local level of energy generation in remote Alaska villages, this power flows from refineries further south onto barges and aeroplanes and then out again – energy power lies entirely outside of the community because of the colonial path dependency on imported diesel fuel. Path dependency is created by the imprinting effects of initial conditions, exogenous contingencies, and self-reinforcing loops that lock actors into paths. In this combination, initial conditions are not given but are constructed by empowered actors who mobilize specific alignments to benefit themselves or their position. What is exogenous and what is endogenous is also not given, but depends on how those actors draw initial boundaries. Self-reinforcing mechanisms are not organic, but instead are cultivated by actors (Garud *et al.* 2010). In the mid-twentieth century, a series of 'policy-driven decisions by non-Indigenous governments' created a path dependency for Alaska Natives on diesel fuel for electricity, heating, and transportation (Brewer *et al.* 2017). Decisions to force, or at times incentivize, Indigenous communities to settle permanently at one village site rather than continue a long tradition of mobility between seasonal camps ushered in a new era of dependence on fossil fuels. As a replacement to moving from site to site depending on the subsistence season through traditional means, such as dog sled or canoes, most residents in remote Alaskan villages today use motorized vehicles in order to access remote subsistence locations (Brinkman *et al.* 2014). This requires villages to purchase diesel fuel from corporations that truck, fly, or barge in the resources (Gerlach *et al.* 2011; Szymoniak *et al.* 2010).

Breaking free: a path forward for renewable energy

Path dependence theory dictates that previous processes alter the costs/benefits of particular choices, and consequently narrow current decisions (Arthur 1994). Applied to Arctic community energy technologies and investments, path dependence theory contends that all actions are governed by history, in that previous choices create self-reinforcement and lock-in regardless of their efficiency. But just as there is path dependency for diesel, there is also path creation for alternative energy sources. Path creation can be understood as actors who attempt to shape an unfolding process in real time, with the omission that no one can fully determine the emergent ecology of socio-material entanglements. Garud and Karnøe (2001) argue that understanding path creation requires an ontological shift from the economic approach adopted in canonical path dependence theory to a sociological one. While lock-in is driven by forces like technological, economic, and political self-reinforcement, the initial conditions that support new path creation are constructed by reflective agents such as entrepreneurs, policy-makers, and newly empowered communities. Path dependencies serve to rob actors of agency as they find themselves pulled from one set of conditions to another, a phase shift wherein the parameters of engagement themselves change to offer the opportunity to create a new path.

Arguably the most monumental phase shifts come when there is an opening up of agency. Agency here is an emergent attribute. It is a capacity to do certain things and not others based on the specifically socio-material entanglements that ensue. When parameters of engagement change, so too do socio-material entanglements, so that an emergence of new nets of agency is part of unfolding actions around particular issues or events (Czarniawska 2008; Tsoukas 2008). Agency is not uniformly distributed within these nets because an actor's capacity to formulate new visions for the future and options to get there is still dependent on their specific, and at times previous, socio-material entanglements. Still, this shift in agency and entanglements, the interactions of actors and artefacts that constitute action nets within a new parameter for engagement, holds the immense potential to create a new path (Garud and Karnøe 2001).

Agency also implicates temporal elements in path creation (Tsoukas and Hatch 2001), and manifest in actors sensing the appropriate moment to strike to realize an option's value (Garud *et al.* 1997). This analysis centres on a moment in the recent history of the United States and the State of Alaska – 2009 to 2016. This time frame brought two events together – the creation of the Alaska REF and the Obama administration's suite of Alaska-specific initiatives for renewable energy – that changed the parameters of engagement for remote Native villages, which in turn created new nets of agency, and ultimately the creation of a new path. Public support is necessary to break the carbon lock-in (Safarzyńska and van den Bergh 2010; Schwoon 2006), which both President Obama had in his 2008 and 2012 elections, and the state of Alaska had in its creation of the REF. And while exogenous shocks (like the 2009 recession and the campaign for change) may lay the foundation for aggressive public support for clean energy by changing the cost–benefit ratio of new energy policies, Native Alaskan remote communities used the temporal agency created in this period to cultivate serendipity (Garud *et al.* 1997), as epitomized in Pasteur's dictum that 'fortune favors the prepared mind'. In doing so, they mobilized the path created for them at the state and national scale towards energy sovereignty. As a recent study of Fort Yukon, a remote Alaska Native village in the interior of the state, indicates,

> remote, rural Indigenous communities are not solely motivated by government policies that encourage decreased dependence on a transition away from nonrenewable energies. Rather, rural Indigenous communities implement alternative energy projects like this as a course of action towards their sustainable future development.
>
> (Brewer *et al.* 2017)

The reading strategy set forth by this volume's editors can be used to understand the path creation for renewable energy penetration into remote Alaskan village electrical grids. We can view the opening up of agency during the 2008–2015 time period as the foundation upon which localized energy is made sustainable in relation to Native agency by the federal (U.S) and state (Alaska) governments. The sections to follow attempt to evaluate 'what should be made sustainable in relation to what and how' at three different scales of actors: by the state of Alaska; by the Obama administration; and by Native communities in Alaska.

State-level policy: three bills, two legislatures, one Alaska

This first section explores how Alaskan remote village energy systems were made sustainable in relation to economic stability by the State Government of Alaska. That shift of parameters for remote energy systems in Alaska can be narrowed down to three key bills from 2008 to 2010: House Bill 152 in 2008, which created the Renewable Energy Grant Fund; House Bill 306 in 2009, which introduced a roadmap for Alaska's sustainable energy future; and Senate Bill 220 in 2010, which introduced a plan of action to achieve the goals set forth in HB 306. All three bills were instrumental in stimulating and coordinating government action on renewable energy. HB306 established an energy policy to guide the legislature, administration, utilities, conservation groups, and Alaskans towards the goal of providing more affordable, abundant, and reliable energy. The bill set a goal for Alaska to generate 50 per cent of its electricity through renewable resources by 2025, primarily through hydroelectric projects with contributions from wind, solar, geothermal, tidal, hydrokinetic, and biomass energy. But it was HB 152, the REF, that positioned Alaska as a national leader in funding for renewable energy. The Alaska REF provides benefits to Alaskans by assisting communities across the state to reduce and stabilize the cost of energy. The programme is designed to produce cost-effective renewable energy for heat and power to benefit Alaskans statewide. The programme also creates jobs, uses local energy resources, and keeps money in local economies. Established in 2008, the REF has appropriated $259 million for 287 qualifying projects, which has been matched with more than $152 million from other sources. The REF was extended by ten years in 2012, until 2023. The REF is managed by the Alaska Energy Authority (AEA) and provides public funding for the development of qualifying and competitively selected renewable energy projects in Alaska. The AEA estimates that renewable energy projects constructed with funding from the REF displaced 30 million gallons of diesel in 2016.

The importance of HB 152 lies not only in its catalyst to break financial and technical lock-in of diesel systems in remote communities; the Bill shifted the parameters for engagement by broadening the scope of which actors enjoyed agency in energy decisions, changed the interactions between actors, and in turn created the potential for a path the villages then took. The Bill itself was conceived and advocated for by a diverse group of stakeholders in Alaska, led by the Renewable Energy Alaska Project (REAP) rather than the State Legislature. REAP's mission is to increase the development of renewable energy and promote energy efficiency in Alaska through collaboration, education, training, and advocacy. It currently includes more than 710 organizational and contributing members representing small and large Alaska electric utilities, environmental groups, consumer groups, businesses, Alaska Native organizations, and municipal, state, and federal entities. Founded in 2004, REAP can be understood as a consensus-building organization (Callon 1986; Latour 1991), and, importantly, these deliberations included the Alaska Federation of Natives (AFN). In a 2012 message, then AFN President Julie Kitka noted that,

> AFN is very concerned about the high costs of energy to all Alaskans, especially those in rural Alaska. We all know that 'energy is the oxygen of the economy,' so it is no surprise with our high cost of energy, that we have underdeveloped local economies. In order to address our energy needs, AFN felt it was important to develop new and healthy forms of collaboration that can cross boundaries, including national, public–private, cross industry, business–nonprofit, and tribal entities.

Those entanglements from this net of agency made the REF possible, but they were not the only emerging constellations of actors to come together during this time frame. A similar parameter shift was occurring at the national level, with the Obama administration's push for renewable energy and initiatives during the US Arctic Council Chairmanship (2015–2017).

National-level policy: the Obama administration and America as an Arctic nation

Moving up a level, the second portion of this brief survey into the transition of Arctic energy landscapes focuses on how Alaskan remote village energy systems were made sustainable in relation to climate change by the US federal government of 2009–2016. President Obama took office during the most severe recession the US had witnessed in modern history. Within the first month of his tenure in the White House, President Obama signed the American Recovery and Reinvestment Act (ARRA) of 2009, on 17 February 2009. The ARRA was a wide-ranging effort to jumpstart the weakened economy through investment in infrastructure, energy, education, and tax cuts. The ARRA included a large number of funding opportunities and tax incentives to support investment in clean energy at the local level. These incentives were 'designed to strengthen the economy and to promote clean and renewable energy', and as such, the White House boasted that 'the Recovery Act made the largest single investment in clean energy in history, driving the deployment of clean energy, promoting energy efficiency, and supporting manufacturing' (White House 2016). The focus of this section is a particular moment six years after the passage of ARRA in President Obama's second term, when the Arctic moved from the periphery closer to the centre of American climate and energy policy-making. Still, it is important to begin with the early days of the Obama administration, showing its early commitment to path creation at the federal level for renewable energy.

The Obama administration made the development of smart, renewable microgrid technology and implementation a priority of the US Arctic Council Chairmanship (2015–2017). As part of President Obama's journey to Alaska in 2015, he visited the Arctic town of Kotzebue to talk about the promise of renewables in Alaska.

> We are the number-one producer of oil and gas. But we're transitioning away from energy that creates the carbon that's warming the planet and threatening our health and our environment, and we're going all in on clean, renewable energy sources like wind and solar. And Alaska has the natural resources to be a global leader in this effort,

the president said (MAHB Admin 2015). While there, President Obama convened a meeting with Alaskan Governor Walker, the Denali Commission, the AEA, and REAP to launch the Clean Energy Solutions for Remote Communities (CESRC). The goal of CESRC is to expand investment in climate solutions by identifying technical, financial, and logistical challenges and opportunities in clean energy innovation (White House 2015).

President Obama's 2015 commitments to clean energy in the north were followed by a number of financial and technical assistance programmes (US Department of Energy 2016a). The most recent of these came in February 2016, when Secretary of Energy Ernest Moniz travelled to the Alaskan remote village of Oscarville and announced another seven million dollars for the tribal energy programme for technical assistance and training for Native Alaskan and American Indian communities (US Department of Energy 2016b). Recipients of funds were chosen through a competitive process and received five weeks of intensive training with the Department of Energy, giving villages the knowledge, skills, and resources needed to implement successful strategic energy solutions, and ultimately to create a national network of regional, tribal energy experts who are able to provide technical energy assistance and information resources when an issue with the microgrid technology arises. And at the regional level under the tenure of the US Arctic Council Chairmanship, the Arctic Remote Energy Networks Academy (ARENA) programme seeks to establish professional, knowledge-sharing networks related to microgrids and integration of renewable energy resources for remote Arctic communities (Arctic Council Sustainable Development Working Group 2017).

The prioritization of knowledge, funding, and technology transfer directly to Native communities by the Obama administration provided a further catalyst for the emergence of agency as an attribute of remote communities in energy debates – in the present and in future sustainable pathways. Like the concept of sustainability, renewable energy transitions are fundamentally political. The

Obama administration strategically overfunded clean energy in Alaska when it was possible that the next administration would be hostile to sustainable energy policies in order to exploit positive reinforcement to lock-in its preferred energy policies for when a time came when his administration would no longer be in power. By changing the framework for engagement to include positive feedbacks for renewable energy and a widened net of agency, it partly ensured the continuation of its renewable energy legacy beyond the tenure of President Obama.

Local-level action: networks of agency in remote Alaska Native villages

The final scale studied is one that is often overlooked in the path creation of renewable energy – small-scale communities, and specifically here remote Alaska Native villages. Here, Alaskan remote village energy systems were made sustainable in relation to Native agency by Alaskan communities. As Howitt (2012:824) observes, 'sustainable Indigenous futures in communities and territories that are remote from mainstream markets and other institutional arrangements cannot arise from policy interventions' that rely on programmes that promote a trickle-down of wealth to local Indigenous communities for sustainable development. But what Howitt and other scholars miss is that the path creation in Alaska remote energy systems is a two-way exchange where both government and community have agency. Remote Native communities were not passive actors in this time frame; rather, they were empowered with agency and a vision for the future to pursue renewable energy sources as a means of further defining energy sovereignty on their terms. If agency is understood to be temporal, it requires aspirations for the future, sense-making of the past, and a conceptualization and strategy for what is transpiring in the present (Stewart 2011). As Stewart *et al.* (2011:3085) note, 'the transition to a low carbon economy provides potential opportunities for Indigenous communities living in remote areas'. These communities have a high carbon footprint due to 'a frequent reliance on diesel-powered electricity generators [and] fossil-fueled vehicles'. Some remote Indigenous communities are responding to this reliance on fossil fuels by pursuing innovative and sustainable approaches to meeting their energy needs, Alaska Natives included (see Johnson *et al.* 2016).

But more than an opportunity for electrical power transitions, making local energy systems sustainable is also a chance for political power transformations. Much like Whyte (2016a, 2016b) argues in his writings on Indigenous food sovereignty, renewable energy implementation is more than an aspiration of self-sufficiency, lowering costs, and raising public health standards. Making energy systems sustainable is relational to Alaska Native sovereignty – it is a strategy for communities to cultivate and repair social, cultural, and economic relationships that were damaged by colonialism. In a guide to restoring energy and food sovereignty in Native America, a publication of Honor the Earth (2009), the call to purse energy sovereignty through localized, low-carbon production is framed as a decision about a future path by breaking the cycle of dependency: 'We must decide whether we want to determine our own future or lease it out for royalties. In the end, developing food and energy sovereignty is a means to determine our own destiny.' Here, sustainability embodies the contours set forth by the editors of this volume. Implementing sustainable energy systems in Native communities is a political struggle between competition visions of the future – a vision of dependency, both on oil and on colonial systems of power, versus a vision of independence, wherein communities are able to create their own futures using renewable energy generation. Though the action to make an energy system sustainable transpires at the local level, sustainability, in the form of renewable energy, intervenes in the discursive struggle at the state level in Alaska over the future allocation of energy rights, resources, and Native sovereignty.

The village of Shaktoolik, Alaska is one of a number of remote Native villages to take advantage of the changing parameters for energy sovereignty by using sustainability to transform power, both political and electrical, relations. An Inupiat village of roughly 300 residents, Shaktoolik sits on the shore of the Norton Sound. In partnership with the AEA, the Alaska Village Electric Cooperative completed the construction of a 200kW wind farm in the village, with two 100kW Northern Power turbines, secondary heat loads, load controllers, and a new switchgear for the power house. The wind farm became operational in April 2012, with the estimated annual diesel fuel offset totalling 33,000 gallons at a fuel saving cost of $133,375 annually. This has cut utility bills by 50 per cent. In March 2015, an update to the wind farm began transferring heat to the community's water plant, which offsets up to 10,100 gallons of heating fuel every year with an annual saving of $20,034 (Alaska Tribal Health Consortium 2015). The stated objectives of this project 'were to displace diesel fuel and provide the community of Shaktoolik with a renewable, reliable, and cost-effective energy source' (Alaska Tribal Health Consortium 2015). This objective was a key aim for the community. But in interviews in Shaktoolik, residents and leaders spoke of the wind farm as a community-driven project. Eugene Asicksik, Shaktoolik's Mayor, noted in an interview that 'This community put in our own water and sewer, and we've got windmills with 200 kilowatts generating capacity ... If the village wants something, it has to come from the village. There needs to be blood, sweat, and tears.' In relation to the other actors involved, Eugene noted, 'The government comes in here and has preconceived ideas and plans and their own workers, with no community engagement. But there's always another way to solve a problem and get other benefits.'[2]

Alaska at a budget crossroad: the staying power of renewable energy path creation

There is, of course, one stark external shock not mentioned in this analysis. Globally, oil prices have fallen steeply since the final months of 2014. Alaska's state budget for 2015 was based on oil at $105 per barrel; the actual price of oil in March 2015 was just under $50 per barrel (Walker Mallott Administration 2015). With a 60 per cent decline in price, the state government faced a $3.5 billion shortfall in spending in 2016. Governor Bill Walker has proposed a cut in government spending of 5–8 per cent this year, with a total cut of 25 per cent over the next four years if prices stay low (Gutierrez 2015). As noted in the 2015 report *The Economy of the North* (Glomsrød *et al.* 2015),

> The backbone of the economy is the petroleum industry. However, the giant oil field of Prudhoe Bay on the North Slope is in the decline phase and uncertainty around future oil prices and international climate policies question the sustainability of petroleum income level.

While the report offers other sectors of growth as potential paths forward in Alaska, such as mining, the seafood industry, and tourism, it is still 'among the arctic regions that are most exposed to the green transition'. This is perhaps the next chapter to be written on the transformation of Alaska's renewable energy landscape at the nexus of postcolonial politics, sustainable futures, and energy sovereignty – a chapter that is still unfolding, and will continue to unfold for many years to come. Here, a new discursive struggle emerges over the future allocation of rights and resources that moves beyond energy sustainability, towards economic sustainability for the state in its entirety. Future scholars thus may take this as a point of departure for future scholarship to delve further into the political struggle Alaska currently finds itself entangled in over competing visions of the future and which co-existing strategies to pursue in order to create those futures (Laclau and Mouffe 1985; Palonen 2006; Skinner 2002). Expanding from energy sustainability and sovereignty, forthcoming research might explore how the state economy should be made sustainable in relation to budget necessities by any number of actors, including but not limited to the legislative branch of Alaska, the executive branch of Alaska, Native corporations, and so on.

But for now, with low state revenues in recent years, the AEA has been working with the Renewable Energy Fund Advisory Committee (REFAC) to adapt the programme to changing times. Recent years have seen additional emphasis placed on funding early stages of development that cannot easily be financed, and providing assistance to applicants and financing options to construct feasible projects. In 2017, it was announced that the fund would not seek new applications. In each state budget discussion, renewable energy support is on the chopping block (Hobson 2016).

Still, in spite of state funding shortfalls, the analysis of path creation above ultimately reveals two catalysts for renewable energy deployment in Alaska – the widened networking of agency and the multi-scalar policy at the national, state, and local levels. These two catalysts also hold the power to continue pushing the path creation of Alaskan remote renewables. At the national level, Senate Energy Bill S. 2012 currently being conferenced with the House includes multiple provisions to help state and tribal governments finance and deploy renewable and efficiency projects. Alaska's Senior Senator Lisa Murkowski is an advocate of the Bill, and Chris Rose has been a vocal supporter of language in the legislation clarifying that the Department of Energy can provide federal loan guarantees to states for projects that employ innovative technologies to reduce greenhouse gas emissions as something that would be particularly helpful for Alaska. A second key provision in the bill would reauthorize the Department of Energy's state energy programme, which provides federal funds for renewable and efficiency projects. Locally, Alaska Native Corporations have been weathering the state's economic hardships relatively well, and play a vital diversifying role in the state's economy by returning outside profits to Alaska. University of Alaska Professor Bob Poe has termed them

> the unsung economic heroes, holding twenty-one spots on the Alaska Business Monthly's Top 49er list, making up 69 percent of the Alaska jobs generated, 86 percent of the total jobs created, and 75 percent of the income produced by Alaska's top companies. Now, if only Alaska's Legislature could get its act together.

To Professor Poe, the ongoing success and growth of Alaska Native Corporations positions them as a 'critical component of ensuring a sustainable future for our state.' Combined with the net of agency created between 2009 and 2016 at the intersection of Native rights and energy sovereignty, Alaska Native Corporations could play a larger role in extending the path of renewable energy development in remote communities.

As the REAP executive director put it in a media interview, 'The grant era is over for the time being' (Koss 2016). While the reduction in funding for the AEA, the REF, and the Emerging Energy Technology Fund may risk losing momentum on renewable energy and a sustainable future, the foundations of an expanded network of agency and a path to break dependency have already been laid. 'You can keep momentum going by education, people being aware of the opportunities', Governor Bill Walker said in a 2016 interview after the annual renewable energy expo at the Chena Hot Springs Resort outside Fairbanks (Koss 2016). 'So we have to fix our budget, there's no question about that, and we will. It doesn't mean we stop being Alaskans, and stop doing what's happening here because we need to fix our budget' (Koss 2016). With the help of the national and local scale, the future of renewables in remote Alaskan communities may well still be a bright one.

Notes

1 This chapter is an expansion of a discussion on energy poverty and policy in the pan-Arctic region I began in March 2017 with a two-part article published by The Arctic Institute, titled 'What the forthcoming Paris Agreement rulebook means for Arctic climate change' and 'The geographies of energy poverty: where North and South intersect'. The series focused on Arctic impacts and opportunities from the 22nd Conference of the Parties of the United Nations Framework Convention on Climate

Change, commonly known as COP22, and a portion of this introduction is from there. Please see www.thearcticinstitute.org/paris-agreement-rulebook-arctic-climate-change and www.thearcticinstitute.org/geographies-energy-poverty-north-south-intersect.

2 Quote taken from an interview conducted by the author in Shaktoolik in September 2016 as part of a National Geographic Society-funded research project. For more information on this project, please see www.americaserodingedges.org.

References

Alaska Energy Policy (AEP) (2010). Available from: http://legisweb.state.wy.us/Interim-Committee/2011/Alaska%20State%20Energy%20Policy%20-%202010%20HB%20306.pdf (accessed 15 March 2018).

Arctic Council Sustainable Development Working Group (2017). Arctic Remote Energy Networks Academy. Available from: www.sdwg.org/wp-content/uploads/2017/03/ARENA-Brochure.pdf.

Alaska Tribal Health Consortium (2015). Rural Energy Initiative 2015 report on activities. Available from: https://anthc.org/wp-content/uploads/2015/12/2015-REI-Report-06.21.16.02-FINAL3_web.pdf (accessed 15 March 2018).

Arthur, W.B. (1994). *Increasing Returns and Path Dependence in the Economy*. Ann Arbor, MI: University of Michigan Press.

Berardi, G. (2005). Alaska Native Claims Settlement Act (ANCSA): whose settlement was it – an overview of salient issues. *Journal of Land Resources and Environment*, 25, pp. 131–139.

Borrows, J. (1997). Living between water and rocks: First Nations, environmental planning and democracy. *University Toronto Law Journal*, 47 (4), pp. 417–468.

Brewer, J.P., Vandever, S., and Johnson, J.T. (2017). Towards energy sovereignty: biomass as sustainability in interior Alaska. *Sustainability Science*, 13, pp. 417–429.

Brinkman, T., Maracle, K.T.B., Kelly, J., Vandyke, M., Firmin, A., and Springsteen, A. (2014) Impact of fuel costs on high-latitude subsistence activities. *Ecology and Society*, 19 (4), p. 18.

Callon, M. (1986). The sociology of an actor-network: the case of the electric vehicle. In: Callon, M., Law, J., and Rip, A. (eds), *Mapping the Dynamics of Science and Technology*. London: Macmillan Press, pp. 19–34.

Czarniawska, B. (2008). *A Theory of Organizing*. Northampton, MA: Edward Elgar.

Garud, R. and Karnøe, P. (2001). Path creation as a process of mindful deviation. In: Garud, R. and Karnøe, P. (eds), *Path Dependence and Path Creation*. Mahwah, NJ: Lawrence Earlbaum, pp. 1–38.

Garud, R., Nayyar, P., and Shapira, Z. (1997). Beating the odds: towards a theory of technological innovation. In: Garud, R., Nayyar, P., and Shapira, Z. (eds), *Technological Innovation: Oversights and Foresights*. Cambridge: Cambridge University Press, pp. 345–354.

Garud, R., Kumaraswamy, A., and Karnøe, P. (2010). Path dependence or path creation? *Journal of Management Studies*, 47 (4), pp. 760–774.

Gerlach, S.C., Loring, P.A., Turner, A., and Atkinson, D.E. (2011). Food systems, environmental change, and community needs in rural Alaska. In: Lovecraft, A.M. and Eicken, H. (eds), North by 2020. Fairbanks: University of Alaska Press.

Glomsrød, S., Duhaime, G., and Aslaksen, I. (2015). The economy of the north. Available from: www.ssb.no/natur-og-miljo/artikler-og-publikasjoner/the-economy-of-the-north-2015 (accessed 15 March 2018).

Godfrey, G. (2015). OP-ED: Alaska Native corporations are a uniquely Alaskan resource. *Alaska Business*. Available from: www.akbizmag.com/Alaskan-Native/OP-ED-Alaska-Native-Corporations-Are-A-Uniquely-Alaskan-Resource.

Gutierrez, A. (2015). Walker outlines plans for budget cuts. Available from: www.ktoo.org/2015/01/23/walker-outlines-plans-budget-cuts.

Hoag, H. (2016). Expanding renewable energy into remote communities is a team effort. Available from: www.opencanada.org/features/expanding-renewable-energy-remote-communities-team-effort (accessed 25 February 2017).

Hobson, M.K. (2016). Isolated Alaska villages struggle with energy, infrastructure problems. Available from: www.eenews.net/stories/1060032619.

Howitt, R. (2012). Sustainable Indigenous futures in remote Indigenous areas: relationships, processes and failed state approaches. *GeoJournal*, 77 (6), pp. 817–828.

Johnson, J.T., Howitt, R., Cajete, G., Berkes, F., Louis, R.P., and Kliskey, A. (2016). Weaving Indigenous and sustainability sciences to diversify our methods. *Sustainability Sciences*, 11 (1), pp. 1–11.

Koss, G. (2016). Cash-strapped Alaska looks to Senate energy bill for help. *E&E News*. Available from: www.eenews.net/stories/1060042704 (accessed 17 July 2017).

Laclau, E. and Mouffe, C. (1985). *Hegemony and Socialist Strategy: Towards a Radical Democratic Politics*. New York: Verso.

Latour, B. (1991). Technology is society made durable. In: Law, J. (ed.), *A Sociology of Monsters: Essays on Power, Technology and Domination*. London: Routledge, pp. 103–131.

MAHB Admin (2015). Transcript: President Obama's speech at Kotzebue Middle/High School. Millennium Alliance for Humanity and Biosphere, Stanford University. Available from: http://mahb.stanford.edu/library-item/transcript-president-obamas-speech-at-kotzebue-middlehigh-school (accessed 15 March 2018).

Mead, H. (1932). *The Philosophy of the Present*. LaSalle, IL: Open Court.

Napoleon, V. (2005). Aboriginal self-determination: individual self and collective selves. *Atlantis: A Women's Studies Journal*, 29 (2), pp. 31–46.

Palonen, K. (2006). Two concepts of politics: conceptual history and present controversies. *Distinktion: Scandinavian Journal of Social Theory*, 7 (1), pp. 11–25.

Resource Development Council (n.d.). Alaska Native corporations. Available from: www.akrdc.org/alaska-native-corporations.

Ricoeur, P. (1984). *Time and Narrative*. Chicago, IL: University of Chicago Press.

Safarzyńska, K. and van den Bergh, J.C. (2010). Evolutionary models in economics: a survey of methods and building blocks. *Journal of Evolutionary Economics*, 20 (3), pp. 329–373.

Schwoon, M., (2006). Simulating the adoption of fuel cell vehicles. *Journal of Evolutionary Economics*, 16 (4), pp. 435–472.

Skinner, Q. (2002). *Visions of Politics*. Cambridge: Cambridge University Press.

State of Alaska (2010). SB 220. 26th Legislature. The Alaska State Legislature. Available from: www.legis.state.ak.us/basis/get_bill.asp?bill=SB%20220&session=26 (accessed 15 March 2018).

State of Alaska (2012). 25th Legislature. The Alaska State Legislature. Available from: www.legis.state.ak.us/basis/get_fulltext.asp?session=25&bill=HB152 (accessed 15 March 2018).

State of Alaska (2014). HB 306. 28th Legislature. The Alaska State Legislature. Available from: www.akleg.gov/basis/Bill/Detail/28?Root=HB%20306 (accessed 15 March 2018).

State of Alaska (2017). Particulate matter: health impacts. Division of Air Quality: Air Non-Point & Mobile Sources. Available from: http://dec.alaska.gov/air/anpms/pm/pm_health.htm (accessed 17 July 2017).

Stewart, J., Anda, M., and Harper, R.J. (2011). Carbon management and opportunities for Indigenous Communities. In: UCLan Centre for Sustainable Development (CSD) & Confucius Institute International Conference, 19–21 May, Preston, UK.

Szymoniak, N., Ginny, F., Villalobos-Melendez, A., Justine, C., and Smith, M. (2010). Components of Alaska fuel costs: an analysis of the market factors and characteristics that influence rural fuel prices. Prepared for the Alaska State Legislature, Senate Finance Committee. Institute of Social and Economic Research, University of Alaska Anchorage, Anchorage.

Tsoukas, H. (2008). Towards the ecological ideal: notes for a complex understanding of complex organizations. In: Barry, D. and Hansen, H. (eds), *The Sage Handbook of New Approaches in Management and Organization*. New Delhi: Sage, pp. 195–198.

Tsoukas, H. and Hatch, M.J. (2001). Complex thinking, complex practice: the case for a narrative approach to organizational complexity. *Human Relations*, 54, pp. 979–1014.

US Department of Energy (2016a). Remote Alaskan communities energy efficiency competition. Office of Energy Efficiency and Renewable Energy. Available from: https://energy.gov/eere/remote-alaskan-communities-energy-efficiency-competition (accessed 15 March 2018).

US Department of Energy (2016b). Energy Department makes up to $7 million available for assistance to Indian tribes; releases Alaska solar prospecting report. Office of Energy Efficiency and Renewable Energy, Department of Energy. Available from: https://energy.gov/articles/energy-department-makes-7-million-available-assistance-indian-tribes-releases-alaska-solar (accessed 15 March 2018).

Walker Mallott Administration (2015). Doing what needs to be done: life at $50 a barrel'. Available from: https://gov.alaska.gov/Walker_media/documents/Governors-BudgetBook.pdf.

White House (2015). President Obama announces new investments to combat climate change and assist remote Alaskan communities. Fact sheet. The White House President Barack Obama Archive. Available from: https://obamawhitehouse.archives.gov/the-press-office/2015/09/02/fact-sheet-president-obama-announces-new-investments-combat-climate (accessed 15 March 2018).

White House (2016). The Recovery Act made the largest single investment in clean energy in history, driving the development of clean energy, promoting energy efficiency, and supporting manufacturing. Fact sheet. Office of the Press Secretary. Available from: https://obamawhitehouse.archives.gov/the-press-office/2016/02/25/fact-sheet-recovery-act-made-largest-single-investment-clean-energy (accessed 15 March 2018).

Whyte, K.P. (2016a). Indigenous food sovereignty, renewal and US settler colonialism. In: Rawlinson, M. and Caleb, W. (eds), *The Routledge Handbook of Food Ethics*. New York: Routledge, pp. 354–365.

Whyte K.P. (2016b). Indigenous experience, environmental justice and settler colonialism. In: Bannon, B. (ed.), *Nature and Experience: Phenomenology and the Environment*. Lanham, MD: Rowman and Littlefield, pp. 157–174.

15 Geo-assembling narratives of sustainability in Greenland

Klaus Dodds and Mark Nuttall

Introduction

In this chapter we consider the materializing of sustainability and how narratives about sustainability are geo-assembled. Our discussion is premised on the argument that space is not simply a stage on which practices of sustainability are enacted (Whitehead 2006). Materiality, objects, and networks of knowledge create multiple contexts for ideas about sustainability to emerge, circulate, play out, and make themselves felt in particular places and in and between particular kinds of assemblages. Sustainability narratives promoted by various human actors contribute forcefully to the production of sustainability in the Arctic (Gad *et al.*, this volume). But those narratives are also 'grounded' in material circumstances. Sustainability, therefore, is not an outcome but something rather more contingent, dynamic, and heterogeneous. We use the term *assemblage* to highlight not only those material circumstances but also to draw attention to the liveliness and possibility of sustainability.

In a Greenlandic context, socio-spatial relations, everyday things, and the more-than-human entities that constitute the world and bring it into being are enrolled and implicated in ideas and practices about sustainability projects and narratives about the past, present, and future (Nuttall 2017). In other words, there is a history and geography to narratives of sustainability as colonial administrators and Cold War planners and postcolonial Governments of Greenland have sought to mobilize ideas and implement plans for making environments and communities 'sustainable'. However, the colonization of the Arctic often generated profoundly unsustainable outcomes as living resources were overexploited, ecosystems damaged, and the lives of indigenous peoples disrupted.

Analytically, then, our approach is one that is grounded in a sense of sustainability assemblages being dynamic and contingent. We don't assume there is something self-evident about a referent object – Arctic sustainability or sustainability in the Arctic. Our focus is pragmatic and we chart how, first, the material environment of Greenland and, second, particular objects are central to narratives and practices of sustainability. When we think materially of Greenland and those accompanying narratives of sustainability, we highlight how sustainability acts upon Greenland geophysically but also geopolitically, as land, sea, air, and minerals become complicit in regional and global extractive and transport projects. For instance, the shipping container (which might not be first thought of as something typically associated with Arctic sustainability) and sea and air transport routes enable goods, objects, and people (including geosurvey equipment and scientists) to be put into motion along and around Greenland. Our Greenlandic examples serve to exemplify our approach, which is to interrogate how narratives of sustainability work through and on material environments and the objects and flows that are thought to make 'sustainable communities' in the Arctic possible.

Arctic sustainability is inherently multi-object, multi-scalar, and multi-sited. It makes little sense just to treat Arctic sustainability as if it applied to only one kind of locality, environment, resource, and/or biome. Arctic sustainability is also entangled and co-relational. The shipping route of the container and an extensive air route network bring people and their lived worlds into contact with other things. And how and where this is made possible is, on the one hand, dependent on grand schemes, past and current, but on the other hand, inescapably co-dependent on weather, ice, seasonality, sailing schedules, and the like. In remote parts of northern Greenland, weather and ice still disrupt and frustrate air schedules and travel by boat or ship. The town of Upernavik, for

Figure 15.1 Air Greenland serves the settlements of northern Greenland by a Bell helicopter, stationed at the Thule Air Base. Savissivik, Melville Bay, March 2015.

Source: © Mark Nuttall.

Figure 15.2 Vestlandia, the supply ship for the settlements in Northwest Greenland, arrives in Upernavik for the first time since spring ice break-up, 28 May 2017.

Source: © Mark Nuttall.

instance, is situated on a small island and the airport was constructed on its highest point by levelling the mountain top. Travelling to and from Upernavik by the Dash-8 service, for example, means being prepared to wait in either Ilulissat or Upernavik for several days because of a winter storm, a late spring or early summer snowfall, or the seemingly persistent low cloud cover and fog that obscures the runway and leads to cancelled flights in summer and autumn. Increasingly stormy weather and more fog during summer throws up considerable challenges for travelling to and from Upernavik by air (Nuttall in press). Communities in Greenland are thus stratified and prioritized in this economy of sustainability, with some enjoying greater connectivity than others, and Greenland is produced as an object of and for human sustainability.

Materializing narratives of sustainability

Sustainability is a complex area of enquiry and embraces not only discourse but also professions, disciplines, industry, and a host of objects, practices, and technologies that might attract the moniker 'sustainable' (Fondahl and Wilson

2017). Sustainability assemblages intersect with other assemblages involving actors, objects and practices associated with security, sovereignty, science, and stewardship. Fondahl and Wilson argue that, 'Where one is located – both geographically and socially – matters in terms of some of the most pressing issues of sustainability in the North today, be they social, economic or environmental' (Fondahl and Wilson 2017: 2). While we welcome the importance accorded to the 'where', it is not clear to us how much agency is accorded to 'place' beyond acknowledging that Arctic sustainability might mean different things to different people, depending on their location. As the contributions to this volume testify, the Arctic is many different locales and regions, comprising different levels and understandings of sustainability, many different experiences of the present and various ideas about the future – from political and industry discourses to indigenous articulations of sustainable livelihoods and sovereignty.

What Fondahl and Wilson's edited volume is better at, arguably, is recognizing that sustainability might have social, economic and environmental qualities. What is missing, though, is an accounting of how the material and intangible qualities of places and locations – that go beyond the human – might create, circulate, and regulate different sorts of sustainability. Or, for example, how past, present, and future visions and projects of and for sustainability leave their traces on human and non-human communities, entities, and ecosystems. How we account for such elemental things (or don't) has implications for narratives of Arctic sustainability.

It is not uncommon for Arctic states to argue that they wish to promote sustainability in the Arctic (Medby, this volume; Steinberg and Kristoffersen, this volume). However, Greenlandic examples and experiences seem particularly apt for discussions on the nature of contemporary narratives about Arctic sustainability because Greenlandic space is, in a specific sense, constituted by an abundance of states. Formal Danish sovereignty cohabitats – happily, it appears – with de facto US military sovereignty and ever-louder plans for future independent Greenlandic sovereignty (Gad 2014). As a self-governing territory of the Danish Realm, Greenland has assumed sharper international visibility and prominence in recent years as the effects of climate change and resource development, indigenous rights, conservation, sovereignty, and environmental and political security in the Global North attract and demand greater attention (Gad 2014; Nuttall 2017). Indeed, Greenland is often at the forefront of scientific and media reports about climate change and extractive industries. Ice is melting, hunting and fishing livelihoods are becoming increasingly precarious, and a number of places are in the process of being identified as having significant economic potential in terms of non-renewable resource development. The fate of the Greenlandic inland ice, on the one hand, and on the other hand the state of the country's resource development plans both continue to inform representations and impressions of the wider Arctic as a dynamic frontier for extractive projects (Nuttall 2017). Greenland looms large in debates about Arctic sustainability because of the confluence of interest in the prospect for further indigenous autonomy and governance, instability of the inland ice, and intensification of resource extraction projects. Mining has been identified by the Government of Greenland as integral to long-term plans for independence from the Kingdom of Denmark – geology and deep time play into discussions of Greenland's future (Nuttall 2017; see also Jacobsen, this volume; Bruun, this volume). Along with this come different scales and visions of sustainability. But all of this is unfolding alongside ongoing discussion about climate change and the consequences for the country and the wider world that follow with the mass loss of ancient ice.

In our reading of Arctic sustainability, Greenland is not an inert spatial platform for such conversations and speculations, but more akin to what we call a geo-assemblage; a space where topological and topographical relationships and networks play their part in assembling and mobilizing arguments and relationships about Arctic sustainability, not just within the political-territorial unit of Greenland, but also in the Kingdom of Denmark, the wider Arctic, and globally. The role of the material, as well as the intangible and more-than-human qualities of the Arctic deserves greater elaboration. It builds on a growing literature on assemblage and materialist thinking, some of which is directly pertinent to investigations on Arctic sustainability (for example, Depledge 2017).

Assemblage scholars draw attention to the manner in which things, objects, and bodies – including the human and more than human – are arranged, move, interact, and collide in time and space (Depledge 2017). An assemblage, as Deleuze and Parnet (2007) see it, is a multiplicity of numerous, diverse, but constituent, co-functioning parts. Deleuze and Guattari (1988) argue, however, that it is critical to see assemblages as constituted by both material entities (content) and non-material entities (expression); so there are material things such as people, animals, and objects in an assemblage, but non-material entities could be sentiments, statements, ideas, beliefs. Material and non-material entities move and interact across different temporal and spatial scales – histories of human action and human influence cannot be ignored, nor can the ways in which places are imagined, represented, and approached. Component parts of an assemblage, however, are all engaged in, entangled with, and affected by a complexity of processes that act to stabilize or destabilize them. In Greenland, for example, we can think of historical and contemporary encounters between indigenous Inuit and Europeans and North Americans as encounters between different kinds of assemblages. For example, explorers and whalers brought their own ideas about the Arctic and marine mammals, as well as different technologies, into Northwest Greenland and encountered indigenous Inuit who had their own ideas about place, space, and animals, as well as their own technologies (Hastrup 2015; Nuttall 1992); similarly, the massive mobilization of resources and people by the US military in the construction of Thule Air Base and the relocation of the area's indigenous inhabitants involved a clash of assemblages of scale with profound environmental effects and social and ecological rupture (e.g. Hastrup 2015; Gad 2017a; 2017b).

There is a diversity of approaches to thinking about assemblage and it is probably more productive to think of this body of work as a storehouse of approaches towards both the social and material. Due emphasis is given to the dynamic nature of assemblages of objects, actors, affects, and materials and their intersection with particular territories. Importantly, assemblage thinking does not presuppose any particular spatial unit. Where it has particular purchase for discussions on Arctic sustainability, though, is its attentiveness to how claims to knowledge about sustainability intersect with relationships and connections between social, technical, and natural actors. Sustainability assemblages, as Havice and Iles (2015) note with reference to aquaculture, involve additional disciplinary aspects such as defining what and where is sustainable. Rules, norms, and conventions play their part in shaping desired behaviour and promote evaluative mechanisms for judging sustainability in the Arctic and beyond. What counts as sustainable in an Arctic context or what Arctic sustainability makes manifest is also subject to negotiation and change. Taking assemblage thinking seriously implies that there is nothing static or assured about Arctic sustainability, or what it targets such as a fish stock, ice sheet, ecological resilience, architecture for modern living, or a mining project.

Narratives of sustainability might well then be transformed depending on how particular things and objects are targeted for sustainability. Living and working in Greenland is, for many of its residents, a profoundly elemental experience. Seasonal variations in lightness and darkness, the movement and migration of animals, fluctuations in weather, sea ice extent and thickness, in combination with long-term warming trends, combine to make everyday life entangled with relationships of earth, air, ice, and ocean. While residents in larger urban centres such as Nuuk might have much in common with the citizens of European or North American towns, our interest in geo-assemblage is premised on the recognition that Arctic sustainability is most profitably approached as a heterogeneous assemblage of human and non-human actors and materials organized and made manifest through apparatuses that bring those actors and materials together. Thus, Arctic sustainability is best approached as something where both 'Arctic' and 'sustainability' need to be interrogated and not assumed to be something simply reducible to a preformed location or reducible to authorized expertise, practices, and institutions.

Thus, our reading strategy of narratives of sustainability does not assume that there are straightforwardly coherent Arctic territories. Sustainability discourses and practices are inserted by human stakeholders in particular territories but they have a capacity to generate territorial units, including visions of territory that are more or less attentive to human and non-human communities and aerial, terrestrial, marine and subterranean components of those territories. In the Greenlandic context, this appears to us to be particularly important as the territory has been imagined and re-imagined in multiple ways, which in turn reveal colonial, Cold War, and postcolonial legacies and visions of (un)sustainability (Dodds and Nuttall 2018).

Colonial, Cold War, and postcolonial Greenland

Danish Minister for Equality and Nordic Co-operation Karen Ellemann told the Arctic Frontiers conference in Tromsø in 2017 that,

> The consequences of the global climate changes are obvious in the Arctic Region – a region that only contributes minutely to the reasons for these changes. It is in the best interest of the Arctic states, with support from the international community, to find the right balance between economic development and the protection of the environment. That includes respecting the way of life of the indigenous population.
>
> (Cited in *Copenhagen Post* 2017)

This view is not untypical of Arctic state officials, where sustainability is often prescribed as a policy-challenge regarding 'balance' – balancing the competing demands for social and economic development and environmental protection (see Keil, this volume; Wilson Rowe, this volume).

Within that framing of sustainability, two things strike as worthy of elaboration. First, the interests of indigenous communities are located as if to suggest that ongoing struggles over land, resources, and rights can also be subject to a so-called balancing act. The explicit assumption is made that Arctic states will be the ones to steer communities and ecologies towards more sustainable futures (see Medby, this volume; Steinberg and Kristoffersen, this volume). It usually necessitates on the part of government ministers (on behalf of settler colonial states) that the 'ways of life' of indigenous communities (rather than, say, northern settler residents) are respected in the process of achieving this balance. Land claims in parts of the Arctic have been settled and governments have recognized indigenous rights and they are protected constitutionally. Northern settler residents are seldom represented by organizations that emphasize rights and consultation. Icelandic fishers and Norwegian whalers may be represented by occupational organizations, but indigenous peoples' organizations such as the Inuit Circumpolar Council (ICC) work to represent indigenous residents in broader areas of human rights and have become key players in Arctic politics, playing a major role in the Arctic Council and other international fora. So what we might have, as a consequence, are competing narratives of sustainability with very different ideas of what 'balance' might entail.

Second, the minister's referencing of 'global climate change' acts as a proxy to highlight how the Arctic is undergoing material change. Material changes to the Greenlandic ice sheet have become integral to narratives of Arctic sustainability, so that melting ice is considered indicative of unsustainability. For much of Greenlandic–Danish history, the material qualities of Greenland have largely been taken for granted. The colonization of Greenland by Danish administrators, missionaries, and traders from the eighteenth century onwards was economically opportunistic but also culturally expansionist. Making Greenland a 'sustainable' part of the Kingdom of Denmark has involved a wholescale intervention in Greenlandic society from building institutions of governance to developing a public media, transforming the economy to commercial fisheries and other large-scale industries, and, most infamously, dispatching groups of Greenlandic children to schools in Denmark (Rud 2017).

Greenland's material qualities and strategic location also enabled, however, a series of international negotiations with other parties over a number of decades. To end uncertainty over sovereignty following Robert Peary's activities in Northwest Greenland (Petersen 1996), Denmark negotiated with the US a territorial settlement in the former Danish West Indies in return for US acquiescence over Danish control of Greenland in the aftermath of the First World War (Olsen *et al.* 2017). Later, Denmark – supported by Greenlandic politicians – won an international judgment against Norway over East Greenland in the 1930s (Cavell 2008; Hudson 1933), the Danish case for which was supported by extensive geological mapping carried out over the previous decade. So when we consider narratives of sustainability, it is worth asking what actually is the referent object of sustainability – coastlines, ice sheets, waters, and communities have all been implicated and targeted.

During the Second World War, the Danish colonial administrators and a diplomat, technically gone rogue, allowed the US armed forces to maintain a military presence in Greenland for continued American recognition of Danish sovereignty (Lidegaard 1996). Making Denmark a 'sustainable' member of NATO during the Cold War entailed developing and maintaining a secretive military relationship with the US, involving locating nuclear weapons on the island. A defence agreement signed in 1951 confirmed formal recognition of the American military presence in Greenland and unlimited freedom for US armed forces to roam over much of Greenland's territory. The strategic value of the Greenlandic ice sheet was integral to the agreement and, hence, to making Danish sovereignty over Greenland sustainable. A US-sponsored Project Iceworm and construction of Camp Century, some 240 km east of Thule Air Base, in the late 1950s represents the zenith of the strategic value ascribed to the inland ice as such in the late 1950s (Dodds 2017; Nielsen and Nielsen 2017). The Danish Government was never told that the project involved plans for establishing a system of railroads to move American missiles around under the Greenlandic ice sheet (Camp Fistclench had been constructed a couple of years earlier to the south, and served in part to test techniques later employed at Camp Century). US Army documents released by Denmark in 1997 include details about the extensive complex planned for Project Iceworm, which were sketched out specifically in a report on the *Strategic Value of the Greenland Icecap*. What sustained this particular form of bilateral security relationship in the Arctic was persistent deception and misdirection, a fact not lost on later academic and media accounts of that relationship (Dodds and Jensen 2019).

These activities might not have been framed at the time by Danish governments as examples of 'sustainability' but they were sustaining of Danish sovereignty over Greenland, and deflecting criticism from the USA that Denmark (a fellow NATO member state) was not investing enough in military capability.

Danish academic Lars Jensen has been at the forefront of interrogating colonial and postcolonial histories and geographies of the Kingdom of Denmark (2012). What makes narratives of sustainability possible is the not just material environments such as the Greenlandic ice sheet, but also the role of knowledge production and critical infrastructures. Science's role in the colonial and Cold War Arctic occupies a deeply ambivalent role given funding sources (Arctic militaries were substantial funders (van Dongen 2016)) and the manner in which the environmental and physical sciences as well as social sciences provided expert knowledge on the region's peoples and ecologies (Petterson 2012).

Even before the Cold War was declared over by 1991, however, we find evidence of challenge and change to those Cold War narratives of sustainability. One notable area was indigenous governance and autonomy (Dahl 2012). Across the North American and Nordic Arctic, indigenous communities and organizations campaigned for greater cultural recognition, access to natural resources, and engagement with political decision-making. In Greenland, this process culminated in 1979 with the introduction of Home Rule after a referendum supported the establishment of the Greenlandic parliament and the granting of sovereign powers over most domestic issues, including education, fisheries, the economy, and health (Skydsbjerg 1999). By 1985, the Greenlandic people had left the EEC, a decision made based on Greenlandic ideas regarding what was 'sustainable' and what was not in relation to the realization of Greenlandic self-determination. In that case, the withdrawal from the EEC was based largely on an evaluation of the Common Fisheries Strategy and the potential harm it might do to the Greenlandic economy and to fishing households and communities (Rebhan 2016). Further devolution in 2009, in the form of Self Rule and a greater degree of self-government, resulted in the momentous decision to allow the Government of Greenland to enjoy sovereign rights over the subsurface resources of the country (something Denmark denied Greenland during negotiations for Home Rule). While Denmark contributes to this day an annual block grant, the Government of Greenland has become a far more active and assertive participant in areas such as resource governance (Nuttall 2017), and is also demanding a greater say in some areas of foreign policy and international relations (Gad 2016a; Kristensen 2005).

Hence, the introduction of Home Rule and Self-Government in effect supports Greenland in continuing to be an indispensable space for the sustainability of Danish Arctic practices. When Danes think of the Arctic, they would typically think of Greenland. When the Danish government called for an Arctic Ocean conference in May 2008, they selected Ilulissat because it was a place wider global audiences had come to know as a place-based metaphor for climate change and anxieties over an imperilled Greenlandic ice sheet (Dodds and Nuttall 2018; Bjørst, this volume). Without Greenland, there would be no *Kingdom of Denmark Strategy for the Arctic* (Nuttall 2014; Rahbek-Clemmensen 2018) and there would have been no need to submit extensive materials to the UN Commission on the Limits of the Continental Shelf in lieu of extending Denmark's sovereign rights over parts of the Arctic Ocean seabed (Strandsbjerg

2012). Greenland acts as the precondition for wider Danish discussions about its role as an 'Arctic great power' (Gad 2016b). As part of that discussion of Denmark's great power capabilities, it has been suggested that it be made sustainable by recruiting local Greenlandic communities to participate in search and rescue and maritime surveillance activities (Sejersen 2015; Strandsbjerg 2012).

The five Arctic Ocean coastal states have reinforced their special geographical relationship. While Greenland played host to that first meeting in Ilulissat, a second meeting was organized in Canada in 2010. While the seabed was a powerful material marker of their relationship with one another, in more recent years the fate of the high seas of the central Arctic Ocean has provided further incentives. In 2014, Nuuk was host to a meeting on potential fishing activity in the central Arctic Ocean, which later led to the 2015 Oslo Agreement (Declaration Concerning the Prevention of Unregulated High Seas Fishing in the Central Arctic Ocean) by the coastal states to prohibit their vessels from fishing in the region until a regional fisheries agreement is in place. This occurred in consultation with extra-territorial actors such as China, South Korea, and the European Union. While those negotiations led to a legally-binding international accord in late 2017, the net result has been to reinforce, according to some Danish observers (Rosamond 2015), a view of the Kingdom of Denmark as a 'middle power' with a vested interest in the governance of the Arctic Ocean. On the one hand, Greenland's geographical qualities are clearly critical to this in terms of identification of the Kingdom of Denmark as a coastal state with specific sovereign rights. On the other hand, the quality of the postcolonial relation between Copenhagen and Nuuk is critical to the very existence of a kingdom including Greenland (Gad 2016a).

Objects and flows of Arctic sustainability

Greenland is not just held together as an entity by territorial imaginations of sovereignty. The material environments of Greenland are also shaped by the interactions of minute objects and mundane flows. Narratives of sustainability work with different scales, objects, actors, and geographical areas. Sometimes sustainability can make itself felt in more everyday, even banal, objects. In the case of a shipping container, the sustainability here is arguably more localized and co-dependent on shipping schedules, sea conditions, electronic tracking of cargo, and port facilities. In the example of a shipping vessel or plane, the sustainability in question appears to pivot on the infrastructural capabilities of the Danish Kingdom to stretch, exercise, and sustain its sovereign rights to offshore maritime areas and airspace. Transport links to Denmark and inter-island connectivity across Greenland are not just crucial in stitching together sustaining narratives of Denmark and Greenland. They are also crucial in stitching together Greenland as an entity to sustain in the first place: it might be that military strategy and narratives of sovereignty produce Greenland as an island, but in terms of everyday livelihood, it is rather an archipelago consisting of dozens of separate settlements, connected mainly by air and, depending on season, water or ice. To make sustainability 'work' in both individual settlements and the Greenlandic archipelago requires the harvesting and managing of natural and social actors and environments, and those sustainability assemblages also have the capacity to break down if circumstances alter. In other words, there is nothing immutable about those expressions of sustainability.

Shipping containers are often to be found neatly stacked in Nuuk harbour. Royal Arctic Line A/S (RAL) reminds its audiences that the company serves to provide 'a lifeline to the Greenlandic Society [en livline til det grønlandske samfund]'. The company works via 13 harbours and ports throughout Greenland. If the container – and the ship that brings it – represents a 'lifeline' to Greenlandic society (especially to those smaller, remote communities in the far northwest, where sea ice closes the shipping lanes for several months of the year), then it does so because the Government of Greenland awarded RAL (created in 1993) an exclusive concession to transport sea cargo between and within Greenland. RAL's provenance lies with Royal Greenland Trade and earlier iterations of Danish–Greenlandic trading relationships, but the Home Rule Government established RAL as an A/S – i.e. a stock-based corporation – to bring in the private funding and expertise deemed necessary to implement containerization of both the transatlantic and the domestic trade and supply. Now, the shipping container and the networks that sustain its mobility are crucial to the country. Without those supply chains, the monopolies that facilitate their sustainability for their operators, and to say nothing of subsidies, remoter communities in Greenland would be forced to adapt in ways that might be narrated to be 'unsustainable'.

If we follow the 'lifeline' (i.e. logistical networks), via the container, we find evidence of how the materiality of Greenlandic society in the most remote communities – along with the idea of sustainable livelihoods – is tied to attempts by government and corporate actors to generate a habitable geography by developing modern infrastructure. Communities up and down the coastlines of Greenland are not only tied to those infrastructural flows, but the container itself becomes a material marker of sustainability. Essential supplies, such as foods, fuel, and clothing, or equipment needed for hunting and fishing such as guns, ammunition, outboard engines and new boats, building materials, or parts for machinery or new classroom furniture for village schools, are brought in by ship. As in almost any part of the world, other kinds of goods, objects, and materials circulate within a complexity of networks of production and distribution. They cross oceans and reach Greenland because people have ordered household appliances, computers, toys, dog food, cars, snowmobiles, and an array of other consumer goods. When important things – an engine, a freezer – break down, months before the next ship arrives, it becomes clear how sustaining livelihood when cut off from global circulation demands creativity (Gad 2017b; Hastrup 2015). As Birtchnell *et al.* (2015) put it, containerization reshapes societies, economies, and geographies. New stabilities and securities are accompanied by new instabilities and insecurities. And the supply chain takes on a life of its own, prioritizing some flows and destinations over others.

Without the shipping container, it would be impossible (or expensive if flown by air) to move everyday objects such as household possessions and strategic supplies such as food and energy. However, the durability and extent of the container 'lifeline' depends on the entangled relations between human factors (such as shipping schedules, ship capacity, container facilities, computer programmes, and transport costs) and non-human agents and forces such as weather, sea ice, and ocean conditions that affect the ability of the ship to traverse such maritime spaces. Sustaining community life in Greenland entails a complex process of keeping goods in motion (Birtchnell *et al.* 2015). Recently, RAL has discovered this can be difficult when computer programmes lose track of containers (Schultz-Nielsen 2017). On top of any IT or logistical challenges, capricious weather and the presence and movement of sea ice can add further to the material expressions of sustainability. Mobility and immobility are integral to how Arctic sustainability gets understood and experienced.

Depending on an array of factors and forces, there may well be different thresholds of and for sustainability. In Greenland, for example, ports vary in size and capability so some areas of the country are better served than others. In Nuuk, a new container terminal has recently been inaugurated for the capital in order to allow for further container traffic. An estimated 300,000 cubic metres of rock was blasted from the shoreline and the harbour area dredged of subterranean material. New infrastructure is planned, including warehousing and administrative buildings, and a quay has been extended in order to handle the expected boost to shipping container traffic. In Upernavik in Northwest Greenland, the changing material condition of sea ice means the supply ship can arrive with much-needed freight a few weeks earlier than in previous years (by late winter and early spring, supplies run low in the stores in the town of Upernavik and the villages of the district). The open-water season around the town is now from around early/mid-May to early November, although sea ice can remain in the bays, channels, and fjords, often blocking access to and from the villages until mid-June. The flip side of this accessibility for the supply ships, however, is that as sea ice is forming later in the winter it makes hunting and fishing by dog sledge more precarious when the ice is not solid (Nuttall in press). Midway between Nuuk and Upernavik, the practice of automobiles crossing the sea ice from Ilulissat to settlements on Disko Island is long gone. Meanwhile, travel between communities by open boat is tenuous as the ice which lingers during the early weeks of open water hinders mobility. In Qaanaaq, further to the north, the lack of a pier means that standard size containers are out of the question; boxes and barrels must be ferried ashore on barges and zodiacs. So the absence and presence of both sea ice and port facilities in some parts of Greenland can have important consequences for how sustainability looks and feels for communities, depending upon their interests. And the 'lifeline' provided by RAL turns out to be a bit more complicated depending on where you are in Greenland and what the supply ship network represents.

If we look to the other end of the flow of containers, RAL's ships all depart from Aalborg, a provincial town in Denmark. In fact, the vast majority of goods imported to and exported from Greenland pass through the city's 'Greenlandic harbour'. For good reason: through centuries of Danish supplies, Greenlanders have acquired a taste for Danish brands of foods and other products, and most people migrating in and out of Greenland have Denmark as the country of departure and destination (Gad 2016a). But once in place, the flow of goods has meant that major processing and storage facilities for the largest Greenlandic seafood exporters are placed in and around Aalborg, rather than in places with cheaper labour (or in Greenland, for that matter). Recently, however, RAL has followed the general impetus of Greenland government policies pursuing a diversification of the island's dependency on the outside world. Specifically, the company has entered into collaboration with Icelandic shipping giant (relatively, when seen from Nuuk) Eimskip. In a few years, Reykjavik in Iceland will be the hub for containers going in and out of Greenland. Initially, most of the goods will probably still flow to and from Denmark – alas, now via Eimskip's Danish feeder port, Aarhus (Dollerup-Scheibel 2017). But exchanging the provincial/imperial bonds of Aalborg with the global connections of Reykjavik promises future sustainability of an independent Greenland (Gad 2016a).

Sustainability thresholds can and do take on a variety of shapes and feelings. Some of the material infrastructure facilitating flows across as well as in and out of Greenland is also a legacy of US military operations there. From the Second World War onwards, the US played a pivotal role in establishing the first airports on the island. So while RAL might make claims that their shipping networks provides a 'lifeline' to remoter Greenlandic communities, the US military provided its own 'lifeline' both to the Danish authorities, to certain Greenlandic settlements, and to the integration of Greenland into one national community (Gad 2017a).

If you were to take to the air, courtesy of Air Greenland, at the Second World War-era airport of Kangerlussuaq on Greenland's west coast, it is hard not to be reminded of Greenland's importance as an air bridge sustaining North American and European commercial and geopolitical interests. The physical geography and location of this vast island was pivotal to sustaining a Cold War geopolitical strategy, which witnessed airport construction, submarine patrolling, and geophysical investigation of the ice sheet for scientific–military–strategic purposes. As Cold War historians of the physical and environmental sciences have observed, Greenland attracted a great deal of interest both above and below the ice sheet. North American continental defence depended on its long-term sustainability, on physical and environmental knowledge about weather, ice, and sea conditions (Dodds and Nuttall 2018; van Dongen 2015).

Kangerlussuaq, seasonally supplemented by Narsarsuaq in the South, remains the only airport for transatlantic aeroplanes:[1] This is where Greenlanders meet; not in Nuuk, but en route to Copenhagen. Ever since the introduction of Home Rule, the Government of Greenland has invested in new airstrips in ever-smaller towns along the coast (Taagholt and Steenfos 2012) to hook up with this transatlantic infrastructure designed to make Greenland serve US military purposes rather than the people of Greenland. Lately, the parliament of Greenland decided to extend two airstrips to allow transatlantic planes to come directly to the preferred destinations of the Danish migrant workforce and the Greenlandic elite (the capital, Nuuk) and of the tourists (Ilulissat with its impressive icefjord). Even if the decision is meant to facilitate the development of a sustainable Greenlandic national economy, economic experts doubt the sustainability of the business case (Christensen 2017). Adding to unsustainability is the fact that Kangerlussuaq – in the middle of nowhere – might need to be kept open and renovated, adding a huge sum to the investment plans, simply to sustain the flow of aeroplanes: the US Air Force carefully picked a spot where no Greenlanders lived, so the Danish colonial authorities did not have to fear fraternization with soldiers, and where the topography of Greenland produces a record regularity of 364 days a year of flying-friendly conditions (Taagholt and Steenfos 2012). Nuuk, to the contrary, enjoys a coastal climate with frequent fogs and heavy winds.

The US military presence – the nuclear umbrella 'taking care' of the basics of territorial defence as part of the defence of the North American continent – also provides the background for Danish military activities. A handful of military personnel patrol the uninhabited Northeast Greenland by dog sledge to sustain Danish sovereignty. The Danish naval patrol vessels, which are employed in search and rescue operations, and fisheries control, as well as scientific survey work, should likewise be interpreted as helping to make Denmark's presence in the Arctic 'sustainable' and indicative of Copenhagen's commitment to be not only an Arctic Ocean coastal state but also an 'Arctic middle power'. The Arctic sustainability assemblage in this case involves policy documents (e.g. *Kingdom of Denmark Strategy for the Arctic 2011–2020* and sea charts), ships, sea, military personnel, infrastructure, and legal rights afforded to the Kingdom of Denmark as a coastal state. So the interplay between ships being tied up in harbours and ports in Greenland and then things moving onshore (either in the form of containers, machinery, and equipment) helps to remind us that sustainability is not rooted in a single place, but rather it involves topographical and topological relationships between human agents, objects, and environments linking many places, or linking and connecting to different narratives of sustainability. Even air-links to Denmark are also vulnerable to disruption and curtailment. As if to illustrate this point well, we can both relate experiences of being delayed by many hours, even days, because Air Greenland's only Airbus was grounded not by weather but by a combination of crew shortage, travellers who have wandered away from the airport terminal, and schedule disruptions to and from Nuuk. While not unique to the Arctic region per se, air delays are an integral part of making northern life time-consuming, precarious, and expensive.

Conclusion

It is now 30 years since the Brundtland Report (*Our Common Future*) was published by the UN World Commission on Environment and Development. There is still a great deal to do regarding 'sustainable development' and the report betrays its academic and policy vintage. When thinking about sustainability in an Arctic context, we are reminded that for all the references to 'future generations and their needs', there are plenty of stories still to be told about past and present sustainability and unsustainability. The habitability of the Arctic is changing, and elemental forces such as heat are playing their part to ensure that geophysical (as much as the geopolitical) terrain of sustainability is dynamic and lively.

However, thinking in sustainability terms, the Arctic is far from a homogeneous space. The flows sustaining a variety of human life forms in the Arctic are dependent on infrastructure established and driven by outsiders, and resource use (think fisheries, or aboriginal subsistence whaling) is determined and regulated to a great extent by international governance arrangements. As we have pointed out with reference to northern Greenland, communities are stratified and prioritized in this economy of sustainability, with some enjoying greater connectivity than others (and this is evident in the digital divide, not just in air and shipping schedules). Over the decades, there has been debate about whether some smaller communities should be amalgamated with others, or even people simply moved to larger, more urban centres like Nuuk to ensure efficiencies and economies of scale (e.g. Nuttall 1992). Making Greenland sustainable is an ongoing project, but one which is not in the gift of one agent alone (such as the Government of Greenland).

The Government of Greenland and hunting communities throughout the country have had to confront the disruptive activities of non-government organizations such as Greenpeace (see Gerhardt *et al.*, this volume; Nuttall 1992), fluctuations in global commodity prices (Graugaard, this volume), the vagaries of weather and environmental conditions, and address sociocultural anxieties over the possibility of large-scale immigration, pollution, and accusations that local communities were not being properly consulted about major resource development projects (Nuttall 2017; Sejersen, this volume). While for most Greenlandic political leaders mining has been positioned as elemental to plans for future independence, dissenting voices argue in relation to most specific mining and resource extraction projects that they are incompatible with a sustainable Greenland. While sustainability narratives might appear coherent and under some form of human control, they can nonetheless be either overwhelmed by non-human factors such as air, ice, and water and/or disrupted by attempts by stakeholders such as governments, commercial companies, and others to manipulate ecologies and non-human communities. For example, the accidental or deliberate introduction of non-native species by human intervention in the Arctic has often been complicit in generating unsustainable outcomes, as non-native animals and plants end up disrupting those ecologies. But as this chapter has shown, the people and government of Greenland cannot even take their island for granted: its identity and coherence as well as the lives lived on it rely on strategies devised by outsiders in the past and on current flows across the oceans.

Thinking about the political terrain of Greenland demands that one thinks beyond the figurative. Near the statue of Hans Egede (the first Danish–Norwegian missionary to arrive in Greenland in 1721), close to the old harbour, there are ample reminders of how colonial and Cold War histories and geographies have made their mark on this island. For over 300 years, Greenland has been 'harvested' by some agent or other – taking resources such as fish, ice, and minerals out of the country and extracting information about ice and climate often in the name of sustaining the Danish Realm and later in the name of NATO and the defence of the West. Could we even say that human and non-human populations and habitats in Greenland have 'suffered' for the sustainability of others? In short, when we cite narratives of sustainability, we need to recognize that those stories enrol objects and environments in the work that they do. If we see this enrolment as assemblage-like (or geo-assemblage in our terms), we not only retain an attentiveness of the contingency of those narratives but we also alert ourselves to the possibility that there are human and more-than-human qualities to account for, from shipping containers to sea ice.

Note

1 Since the establishment of a separate airport in Qaanaaq, Thule, the only active US base, only sparsely facilitates civilian traffic – contrary to the decades immediately following its establishment where it was instrumental to the integration of the distant population in the nascent Greenlandic nation (Gad 2017a). Another, smaller, airport left behind by the Americans in Kulusuk, and an airstrip constructed by an oil exploration company in Nerlerit Inaat (Taagholt and Steenfos 2012), both on the east coast, allow 'local' traffic to Iceland. Indeed, for years, citizens of Ittoqqortoormiit had to go abroad to Reykjavik first to go to any other destination – signifying the fragility of Greenland as an assemblage.

Bibliography

Birtchnell, T., Savitzky S., and Urry J. (2015). Moving cargos. In: Birtchnell, T., Savitzky, S., and Urry J. (eds), *Cargomobilities: Moving Materials in a Global Age*. London: Routledge, pp. 1–16.

Cavell, J. (2008). Historical evidence and the eastern Greenland case. *Arctic*, 61 (4), pp. 433–441.

Christensen, A.M. (2017). Grønland udfordret trods stærkt fiskeri. Danmarks Nationalbank. Report 13.

Copenhagen Post. (2017). Denmark part of Arctic conference for a sustainable future. Available from: http://cphpost.dk/news/denmark-part-of-arctic-conference-for-a-sustainable-future.html (accessed 24 January 2017).

Dahl, J. (2012) *The Indigenous Space and Marginalized Peoples in the United Nations*. Basingstoke: Palgrave Macmillan.

Deleuze, G. and Guattari, F. (1988). *A Thousand Plateaus: Capitalism and schizophrenia*. London: Bloomsbury.

Deleuze, G. and Parnet, C. (2007). *Dialogues II*. New York: Columbia University Press.

Depledge, D. (2017). *Britain and the Arctic*. London: Palgrave.

Dodds, K. (2017). Fissure. Available from: https://culanth.org/fieldsights/1237-fissure (accessed 24 October 2017).

Dodds, K. and Jensen, R. (2019). Documenting Greenland: popular geopolitics and film. In: Stenport, A., Mackenzie, S., and Kaganovsky, L. (eds), *Arctic Cinemas and the Documentary Ethos*. Bloomington, IN: University of Indiana Press.

Dodds, K. and Nuttall, M. (2018). Materialising Greenland within a critical Arctic geopolitics. In: Kristensen, K.S. and Rahbek-Clemmensen, J. (eds), *Greenland and the International Politics of a Changing Arctic*. London: Routledge.

Dollerup-Scheibel, M. (2017) Århus Havn bliver ny Grønlandshavn. Available from: http://sermitsiaq.ag/aarhus-havn-ny-groenlandshavn (accessed 12 December 2017).

Farish, M. (2010). *The Contours of America's Cold War*. Minneapolis, MN: University of Minnesota Press.

Fondahl, G. and Wilson, G. (eds) (2017). *Northern Sustainabilities: Understanding and Addressing Change in the Circumpolar World*. Berlin: Springer.

Gad, U.P. (2014). Greenland: a post-colonial nation-state in the making. *Conflict and Co-operation*, 49 (1), pp. 98–118.

Gad, U.P. (2016a). *National Identity Politics and Postcolonial Sovereignty Games: Greenland in the Margins of Europe*. Copenhagen: Museum Tusculanum Publishers.

Gad, U.P. (2016b). Hvad Danmark skal forstå om Rigsfællesskabet. *Tidsskriftet Grønland*, 64 (3), pp. 156–159.

Gad, U.P. (2017a). Pituffik i Praksis: Nationale reskaleringer i Avanersuaq. *Tidsskriftet Grønland*, 65 (2), pp. 151–169.

Gad, U.P. (2017b). Sex, Løgn og Landingsbaner: Thule set relationelt. *Tidsskriftet Grønland*, 65 (3), pp. 216–241.

Hastrup, K. (2015) *Thule på Tidens Rand.* København. København: Lindhardt and Ringhof.

Havice, E. and Iles, A. (2015) Shaping the global aquaculture sustainability assemblage: revealing the rule-making behind the rules. *Geoforum*, 58, pp. 27–37.

Hudson, M.O. (1933). An important judgment of the World Court. *American Bar Association Journal*, 19 (7), pp. 423–425.

Jensen, L. (2015) Greenland, Arctic Orientalism and the search for definitions of a contemporary postcolonial geography. *KULT – Postkolonial Temaserie*, 12, pp. 139–153.

Kristensen, K.S. (2005). Negotiating base rights for missile defence: the case of Thule air base in Greenland. In: Heurlin, B. and Rynning, S. (eds), *Missile Defense: International, Regional and National Implications*. London: Routledge.

Kristensen, K.S. and Rahbek-Clemmensen J. (2018). *Greenland and the International Politics of a Changing Artic*. London: Routledge.

Lidegaard, B. (1996). *I kongens navn: Henrik Kauffmann i dansk diplomati 1919–1958*. København: Samleren.

Nielsen, H. and Nielsen, K.H. (2017). *Camp Century: Koldkrigsbyen under Grønlands Inlandsis*. Aarhus: Aarhus Universitetsforlag.

Nuttall, M. (1992) *Arctic Homeland: Kinship, Community and Development in Northwest Greenland*. Toronto: University of Toronto Press.

Nuttall, M. (2014). Territory, security and sovereignty: the Kingdom of Denmark's Arctic strategy. In: Murray, R. and Dey Nuttall, A. (eds), *International Relations and the Arctic: Understanding Policy and Governance*. New York: Cambria Press.

Nuttall, M. (2017). *Climate, Society and Subsurface Politics in Greenland: Under the Great Ice*. London, Routledge.

Nuttall, M. (in press). Places of memory, anticipation and agitation in Northwest Greenland. In: Pratt, K.L. and Heyes, S.A. (eds), *Language, Memory and Landscape: Experiences from the Boreal Forest to the Tundra*. Calgary: University of Calgary Press.

Olsen, P.E., Brimnes, N., Hernæs, P., Pedersen, M.V., and Gøbel, E. (2017). *Danmark og kolonierne: Vestindien – st. Croix, st. Thomas & st. Jan*. Denmark: Gads Forlag.

Petersen, K. (1996). Handelsstationen i Thule. *Tidsskriftet Grønland*, 44 (6), pp. 222–250.

Petterson, C. (2012) Colonialism, racism and exceptionalism. In: Loftsdottir, K. and Jensen, L. (eds), *Whiteness and Postcolonialism in the Nordic Region*. London: Routledge.
Rahbek-Clemmesen, J. (2017). The Arctic turn: how did the High North become a foreign and security policy priority for Denmark? In: Kristensen, K.S. and Rahbek-Clemmensen, J. (eds), *Greenland and the International Politics of a Changing Arctic: Post-colonial Paradiplomacy between High and Low Politics*. Oxford: Routledge, pp. 54–69.
Rebhan, C. (2016). *North Atlantic Euroscepticism: The Rejection of EU Membership in the Faroe Islands and Greenland*. Tórshavn: Froðskapur Faroe University Press.
Rosamond, A.B. (2015). The Kingdom of Denmark and the Arctic. In: Jensen, L.C. and Hönneland, G. (eds), *Handbook of the Politics of the Arctic*. Cheltenham: Edward Elgar, pp. 501–516.
Rud, S. (2017). *Colonialism in Greenland: Tradition, Governance and Legacy*. London: Palgrave.
Schultz-Nielsen, J. (2017) RAL knækker fejlkurve. Available from: http://sermitsiaq.ag/ral-knaekker-fejlkurve (accessed 12 December 2017).
Sejersen, F. (2015). *Rethinking Greenland and the Arctic in the Era of Climate Change*. London: Routledge.
Skydsbjerg, H. (1999). *Grønland. 20 år med hjemmestyre*. Nuuk: Forlaget Atuagkat.
Strandsbjerg, J. (2012). Cartopolitics, geopolitics and boundaries in the Arctic. *Geopolitics*, 17 (4), pp. 818–842.
Taagholt, J. and Steenfos H.P (2012). *Grønlands teknologi historie*. Gyldendal.
van Dongen, J. (2015). *Cold War Science and the Transatlantic Circulation of Knowledge*. Leiden: Brill.
Whitehead, M. (2006). *Spaces of Sustainability*. London: Routledge.

16 Conclusion

Sustainability, reconfiguring identity, space, and time

Ulrik Pram Gad and Jeppe Strandsbjerg

We embarked on this research project with the ambition to establish a new way of understanding sustainability – a way that would provide an alternative, and maybe a constructive provocation, to most common use and understanding of what sustainability means. From this perspective, sustainability is intrinsically neither good nor bad, but a concept that has effects in political discourse. We wanted to establish a framework to capture this. In terms of bringing together a varied group of insightful scholars with distinct research profiles we, of course, got much more than we asked for. Each of the chapters convey fascinating insights into life and politics in the Arctic. In the following, however, we focus on the specific contributions they make to our overall analysis of the politics of sustainability in the Arctic. Before we turn to extracting these pearls from the chapters, we briefly recap the basics of our analytical framework.

The basic idea conveyed by the concept of sustainability is that things are connected – so that to sustain something valuable, you need to sustain an environment upon which it depends. In this sense, the idea of sustainability structures narratives about the future directions of societies and ecologies. And differences over what are the most important things to sustain remains key to the struggles over rights and resources in the Arctic. Moreover, the intensified integration of the region with the rest of the world fuels these struggles.

The preceding chapters have confirmed – not surprisingly – that the complexities of the Arctic in many ways cloud our neat schematic framework for analysing and understanding the politics effects and content of the concept of sustainability. The constellations of struggles of rights and resources vary between different postcolonial configurations. Yet there are tendencies and patterned variations stemming from peculiarities of the Arctic shared across the region, and from the way in which the conception of the Arctic as one shared region is increasingly institutionalized. In the following, we will deal with the relations between sustainability and, in turn, identity, space, and time.

As our point of departure, we discuss what happens to identity, space, and time separately, as they have come into focus for analytical questions developed in our introductory chapter: What is to be sustained? In relation to what? How? However, empirical narratives consist of knitting together identity, space, and time. So our initial analytical distinction is to break down as we approach the end of this concluding chapter, in which we condense our take on how sustainability works as a political concept in the Arctic; how sustainability reconfigures struggles over rights and resources.

Reconfiguring identities: communities and states, individuals and corporations, systems

When something is pointed out as important enough to be sustained, certain identities are empowered, others rearticulated or marginalized. This is why the first step in the reading strategy laid out in the introductory chapter was: 'What is it that should be sustained?' Sometimes, upon close inspection, the referent object of sustainability may indeed be one specific identity. Either way, claims to political legitimacy are mobilized by relating sustainability to specific identities in specific ways, and struggles over rights and resources are hence reconfigured.

What exactly deserves to be sustained is, indeed, a central part of ongoing identity negotiation in any society. One instructive example of this is produced by Becker Jacobsen's analysis of what it is that Greenlandic fisheries should sustain. When mining debates in Nunavut and Greenland produce very different sustainability narratives, it is – as is clear from Jacobsen's chapter – a symptom of very different identity discourses posing Nunavummiut primarily as members of Inuit communities worthy of being sustained and Kalalliit primarily as citizens of a prospective Greenlandic state in search of ways to make its national economy sustainable. As in all cases of minerals extraction, what is sustained is obviously not the non-renewable resource, but human societies (Lempinen 2017:40–42; Mikkelsen and Langhelle 2008:20). The question, then, is: What kind of human societies?

In the contributions to this volume, a general picture emerges of the nation state coming out on top. In their chapter, Gerhardt, Kristoffersen, and Stuvøy present three very different yet parallel examples of this core observation: resistance to Greenpeace's visions of global sustainability reproduces three distinct types of states – but they all prioritize sovereignty. The prominence of the state as a preferred identity formation is visible in different ways in other case studies. First, other actors look to the authoritative centre for affirmation of and support for their stories, as is particularly clear in Herrmann's chapter. Second, as Jacobsen sums up nicely in his chapter, 'sovereignty is closely connected to the question of who gets to decide what to sustain'. This comes out very clearly when claims to sovereignty are implicitly and/or explicitly challenged. In Svalbard, both Russia and Norway subsidize extractive industries, which, left to their own devices, would be economically unsustainable (Grydehøj 2013). In parallel, Bruun explicates how extractive projects designed to make Danish imperialism economically and societally sustainable served, first, as a point of condensation for anti-colonial resistance – but as control over mineral affairs moved to Nuuk, the very same projects were put in the service of a sustainable northern economy. Jacobsen describes how, vice versa, a recent extractive project devised to the same purpose spurred anxieties in the imperial metropole over how to stay in control over sovereignty. Third, the very complexity of a holistic approach to sustainability – the promise that all good things can be combined, if we just do this right – becomes an argument for relying on the managerial expertise embodied in the state (cf. Dryzek 2013:174; Hajer 1995). Arctic states each tell their stories of how they are particularly well equipped to take up such responsibility in part as a way of writing-back against those who would wish to 'globalize' the Arctic. Steinberg and Kristoffersen's chapter details this in their analysis of a Norwegian comprehensive maritime policy, while Gerhardt, Kristoffersen, and Stuvøy compare Greenlandic, Norwegian, and Russian reactions to Greenpeace protests against oil extraction. Likewise, Medby can conclude her chapter reporting how, according to Icelandic, Canadian, and Norwegian officials a variety of referent objects in the Arctic are 'to be maintained through time. However, importantly, "to be maintained" implies an Arctic that is passive, static, necessitating the active intervention of a "stewarding" state.'

Each Arctic state comes to this political-theological position in its own way. As documented by Medby, sustainability helps to underpin national identity narratives; Iceland's unique independence and elemental ties to nature prepares it, according to the narrative, to be part of the making and remaking of the geographical and human futures of the Arctic. Canada's responsibility for the Arctic springs from its colonial experience with northern stewardship, which belies a more troubling history of settler colonialism. Norway, as also explicated by Steinberg and Kristoffersen in their chapter, brings together histories of polar exploration, maritime life, and advanced technology in a special capacity for securing the sustainable co-habitation of all kinds of marine life and commercial enterprise in oil, gas, and shipping sectors. In a detailed analysis of Greenlandic mining discourses before and after the devolution of authority over subsoil resources, it appears that opposition to extractive projects – when authority in these matters rested in Copenhagen – retracted when politicians in Nuuk took charge (Schriver 2013). In effect, an argument has won which claims that 'since we Greenlanders have for millennia been dependent on natural resources without compromising them, extraction under our supervision will be environmentally sustainable'. In the perspective of all Arctic states and peoples, some humans are better stewards of the Arctic than others. Representatives of different states and peoples, however, differ over just which humans.

Among the sustainability narratives surveyed in this volume, a couple of surprises occurred: when carefully reading regulatory texts and policy statements regarding current and future maritime traffic on the Northern Sea Route, Keil found that next to the usually expected marine environment, the lives and safety of seamen appear as one of the referent objects of sustainability. Individual lives (of men, overwhelmingly often from countries not routinely associated with the Arctic such as the Philippines) need to be sustained as part of making polar navigation safe and sustainable. At an altogether different scale, the political stability of international relations is also a valid referent object of sustainable shipping. Maritime commerce is often put in relation to peace and cooperation – conflict should not rock the boats; more boats should not create conflict. However, these peculiar referent objects (individual lives, international cooperation) most often appear in conjunction with – and often merely as preconditions for – economic development instead of sustainability goals in their own right. To be economically sustainable, the Northern Sea Route both depends on and contributes to maritime safety and international stability. However, in discourse on Arctic shipping, the bottom line is the bottom line: to be sustainable for business it is a simple equation: no profitability, no traffic.[1] Keil's analysis presents a nice example of just how many identities may be organized by a constellation of sustainability narratives. And how the ultimate referent object of sustainability might, in some narratives and constellations, be the identity or smooth operation of complex systems of interdependence: the global economic order or the international society of states.

Reconfiguring spaces: sectors, zones, and scales

Sustainability is a relational concept: it establishes the dependence of the continued existence of its referent object in its relation to a specified environment. This is why the second step in our reading strategy asks in relation to what environment the referent object of sustainability should be sustained. The specific relation and the constitution of the specific environment often amount to a reconfiguration of the space inhabited by the referent object. In other words, sustainability attaches itself to some places, ecologies, and communities, while others are excluded, in both cases to mobilize political legitimacy.

The Brundtland report introduced a metaphorics of sustainability having three 'dimensions', which later morphed into a notion of sustainability relying on three 'pillars': environment, economy, society. However, if what was dimensions of one thing or pillars upholding an integrated phenomenon are rather apprehended as 'sectors', a particular project or practice can legitimately be measured against (only) the sectoral sustainability preferred by its promoters. 'Environmental sustainability', 'economic sustainability', 'social sustainability', and more – 'cultural sustainability', etc. – may, correspondingly, be managed more or less separately, as Jacobsen relays in his analysis of a mining project in Nunavut, with more or less due respect to their interconnectedness. More important, the sectors can be prioritized, as discussed, e.g. by Bjørst and Becker Jacobsen in their chapters.[2] Again, sectoral separation does not necessarily mean a denial of the overall relational and holistic impulses of the concept of sustainability. Rather, it inserts the possibility of a strategic choice by particular stakeholders of how to stitch relevant environments together. As one example, Bjørst describes how in a national discourse three sectoral sustainability projects – sustainable economy, sustainable community, and sustainable polity – are co-arranged in a way, which, incidentally, crowds out concerns about a sustainable climate. Cutting up sustainability in sectors can work as a way of avoiding holistic and global responsibilities. But sectorization may, conversely, also be appropriated as a way of seeking legitimacy by expanding responsibility. Becker

Jacobsen's analysis shows how actors may take responsibility for extending sustainability narratives beyond the sector of which they have traditionally been in charge. In her case, no one questions that large-scale fisheries conglomerates seek to optimize the environmental sustainability of their operation and their contribution to the sustainability of the national economy. However, the jury is still out on the question of whether their much-heralded care for the sustainability of local communities and traditional small-scale fishing amounts to real 'corporate social responsibility' or a more pernicious form of 'sustainability-washing'.

At the beginning of this book, we introduced the concept of scale as a heuristic of space-making. On the one hand, sustainability is often articulated to preconceived societal scales; a geographically defined 'local' community or an institutionalized territorial entity, most prominently the nation. On the other hand, the environmental aspect of sustainability introduces distinct scales: a regional ecosystem, a migratory stock, or a global climate. In effect, prioritizing a distinct scale of sustainability often entails a priority of one dimension or sector of sustainability.

In several of the chapters, the national scale takes precedence – and along with it, claims to politico-economic sustainability. Norway asserts itself as a guarantor of sustainability. On their side, Greenland regularly declares its ambitions to be independent (Gad 2016). This obviously has the implication that the national scale becomes somewhat crowded, as sustainability narratives need to juggle both the present Danish state and a prospective Greenlandic one. But such a priority given to a Greenlandic proto-state also has implications for what kind of room there is for sustainability concerns at other scales. Social sustainability sometimes piggybacks the ambitions for a sustainable national economy in Greenland, but often this aspect either implicitly or explicitly ends up subjugating the local scale – in contrast to Jacobsen's rendition of mining debates in Nunavut, where the local scale seems to be awarded priority focusing on the local community and particular ecosystems as the relevant environments. The priority of these small-scale communities and ecosystems as pivotal to sustainability is, of course, legitimized by the central place of Indigenous peoples in particularly Canadian discourses of the north, and the specific way these peoples are seen as defined by their relation to nature and culturally defined communities.

If Indigenous narratives can be said to challenge the national scale from below, the global scale challenges – albeit with little success – from above. Sustainability at the global scale – specifically concerning climate change – is less important to national-scale sustainability ideas and practices in Norway, Russia, and Greenland (see Wilson Rowe; Gerhardt, Kristoffersen, and Stuvøy; and Bjørst, in this volume). Gerhardt, Kristoffersen, and Stuvøy note how 'while Greenpeace's vision of a world beyond oil is controversial, it appears that their geographical imaginary of such a world [relegating nation states to second place] is even more so'. Consequently, it appears that the national scale is taking precedence in Arctic sustainability narratives. This means that the state – as an abstract denominator for territorial identities – has come to occupy a central discursive position and as such confirmed its continuing importance in the Arctic – although not unchallenged.

Beyond the effect of any particular construction and priority of this or that scale lies the dynamics of consecutive, coordinated rescaling integral to management in the name of sustainability. As formulated by Sejersen in his chapter, 'scalings of sustainability may indeed frame the way the social world emerges. Understandings of what constitutes the social world in sustainability discourses are also carving out the political manoeuvre rooms and the subject positions of people.' Notably, the polyvalence of the concept of sustainability ensures that scaling is not a procedure conducted once and for all: the scalar environment of a project or practice may contract or expand flexibly according to the needs of the project and its managers and promoters. Similar scalar divisions of labour characterize projects to drill for oil: The Arctic scale warrants plans for adaptation to climate change following from CO_2 emissions, whereas mitigation is referred to the global scale, even if the oil consumed is produced in the Arctic (see Bjørst, this volume; Kristoffersen and Langhelle 2017: Lempinen 2017:114).

An alternative to scale as a way of organizing space is zoning: Wilson Rowe, in her chapter, explains how a central element in the Russian approach consists of designating separate spaces: environmental protection over here, (sustainable) development over there. Zoning works to legitimate extraction by separating it from environmental protection. In doing that it works to exclude other perspectives, other sustainabilities, and other referent objects at other scales. As the promotion of a mining project in the Northeast Greenland national park, detailed in Jacobsen's chapter, shows, this approach comes with the flexibility that the borders between zones of preservation and extraction can always be moved if convenient. The borders are simply rolled up, like a carpet, and simply re-laid over physical and administrative space. Rosing *et al.* (2014) advocates that zoning should be employed in Greenland as a general planning device, prompting de Rosa to suggest interpreting mining sites as 'sacrifice zones' (2014:57–59; see Skorstad *et al.* 2017:20). But borders are also capable of 'rippling', so even when they are moved, altered, or eradicated, they continue to make their presence felt on bodies and things occupying landscapes, seascapes, and icescapes: Inughuit hunters and families were relocated to Qaanaaq in 1953 to make room for the expanding air base at Thule, sustaining US nuclear strategy. A series of court cases some 40 years later established that the relocation was far from voluntary, confirming to the Inughuit that their fate is dependent on greater outside forces. Even if the 'Defense Area' was reduced in 2003 so that they are now allowed to camp at the old Uummannaq/Dundas settlement, the old border is still effectively cutting their hunting grounds in two, contributing to making it difficult to sustain a household from hunting (Gad 2017; Hastrup 2015). Cutting out a distinct space to facilitate the environmental depreciation 'necessary' for development might be painful, but delimiting a distinct zone may simultaneously serve as a reminder of the connection between past and future identities.

In conclusion to the discussion of space, it is important to remember how Dodds and Nuttall point out that materiality in the Arctic can also be rather insistent in its resistance against certain sustainability projects. Relatedly, Bruun points out that once hard labour has succeeded in framing something as a 'resource', it may be very difficult to eradicate from the public imagination ideas about how sustainability works. Indeed, the equally material networks and technologies are exactly established to overcome such resistance. But even when some actors and narratives work hard to establish the precedence of a specific scale or a distinct territorial container, flows and networks in the Arctic often reach across formal borders of both states and regions. Finally, forbidding material configurations may somewhat paradoxically also serve as an enabler for 'sustainability', by deferring potentially unsustainable activities indefinitely. Keil concludes her chapter's analysis of how the Northern Sea Route is envisioned:

> One explanation as to why stakeholders generally perceive shipping as (possibly) sustainable is precisely because there are still so many obstacles to Arctic shipping becoming a large-scale, long-term activity. In this interpretation, the obstacles are a means to render shipping levels in the Arctic sustainable by generally keeping them low in spatial and temporal extent.

In other words, the very materiality of the Arctic in the form of sea ice and unpredictable weather joins hands with political uncertainty with regard to Russia to dampen down interest in boosting Arctic shipping potential.

Reconfiguring temporalities: futures, presents, pasts

Sustainability is a temporal concept: it insists on the possible and desirable durability of its referent object and its relation to some environment. Often, the specific way in which durability and value are inscribed involves a broader alignment of pasts, presents, and futures – particularly when sustainability narratives articulate competing and supplementing concepts with different temporalities, e.g. development or indigeneity. This is the reason for posing the third question of our reading strategy: 'How should sustainability come about?' Not because we do not know in advance the temporality of 'sustainability' but because sustainability narratives invariably involve more concepts. Sustainability narratives need to organize time to be able to credibly claim the environments which makes their preferred referent objects sustainable.

Most prominently, sustainability is a future-making device. Specifically, sustainability narratives often involve, as Sejersen observes in his chapter, a 'future social world … creatively made to talk to the present in a way that underpins a contemporary political discourse emphasizing the emergency of change and reform'. Sustainability might refer to pasts, sustainable or not, but in political use it generally labels a task involving an unsustainable present, naturally pointing out a course for the future which, if upheld, will place us in an unsustainable future – but along with this bad future, a preferred, sustainable alternative future is envisioned.[3] As such, 'sustainability-speak' often asks us to change our direction now, or it seeks to assure us that responsible authorities already have their hands on the steering wheel (as the etymological roots of the word 'governance' suggests). This is not a unique feature to sustainability; other concepts articulate good and bad futures in combination to redistribute authority, responsibility, and choice as they participate in moral economies of judgement and risk management about the past and present.

However, one intriguing aspect of sustainability talk is the way it is simultaneously, on the one hand, a managerial concept valuing technical and scientific expertise, while, on the other hand, it remains systematically open to new claims, which can and does call into question specific sustainability projects. The concept envisions the creation of a sustainable future as a top-down process. Nevertheless, it invites bottom-up participation not just in the form of engagement but even in the form of corrections and interventions.[4] Sejersen, in his chapter, points attention to one important way of regaining managerial control over a future-making process opened up by infusing a project with the concept of sustainability. Cutting the future into distinct phases allows the production of a single, seemingly coherent narrative of a future to exclude certain actors from having responsibility for others, and others from having a legitimate voice.

Zooming out, sustainability reconnects pasts, presents, and futures. As noted by Thisted in her chapter, global sustainability discourse holds a special place for Indigenous peoples in the Arctic and elsewhere. Having survived in a sustained relation to the natural resources where they live, important strands of global discourse have presented Indigenous peoples as exemplars in the face of the self-destructive environmental problems of industrialism. Moreover, the institutionalization of the Arctic region (epitomized by the intergovernmental forum of the Arctic Council) has – as an answer to prolonged territorial submission to Southern metropoles – involved a particularly prominent position for Indigenous peoples' organizations as so-called permanent participants. Nevertheless, as Graugaard and Thisted show in their chapters, in certain postcolonial configurations, sustainability and indigeneity refuse to be aligned. Even if their accounts of the normative effects of articulating sustainability are quite different, they both conclude that sustainability in the Greenlandic case connects to modernity rather than to indigeneity. Thisted dissects how Greenlandic identity discourse is national rather than Indigenous; sustainability, rather than an Indigenous heritage, comes as one of the responsibilities a state in the making must take upon itself to be a respectable member of the international community. Graugaard distils how Danish efforts for making imperialism sustainable disrupted Indigenous ways of life; ways which would in today's world have counted as environmentally sustainable. Moreover, this consecution of imperial projects produced criteria for both indigeneity and sustainability, which are irreconcilable, in effect confining Greenlanders to an indefinite colonial status. In a parallel account, Bruun, in her chapter, lays out the pains taken by Danish authorities to inscribe promissory value, potentiality, modernity, sovereignty, and hope for the sustainability of the Danish welfare state – in the form of uranium – in the rocks of Kuannersuit in Southern Greenland. And how the efforts inscribed now set the parameters for the value, potentiality, modernity, sovereignty, and hope for the sustainability of a future *Greenlandic* welfare state. The sustainability projects of the colonial past have produced the contemporary conditions of possibility for the sustainability projects of the postcolonial future.

The agency of a concept: substantive and processual effects of sustainability in the Arctic

As a political concept, sustainability came to the fore referring to humanity as a single identity and the globe as a single ecosphere. This volume has focused on the effects of sustainability entering struggles and negotiations preoccupied with human collectives and natural environments defined by a specific regional scale (the Arctic) and by each of the postcolonial governance arrangements and ambitions at national and territorial scales across the Arctic. We have done so by asking how identity, space, and time have been reconfigured by the concept of sustainability. Our survey shows that sustainability has important substantial effects pointing in one direction, but indeed also processual effects, which might point in different directions. In other words, the concept not only reconfigures struggles over rights and resources, but also the playing field on which these struggles are taking place.

In the contributions to this volume, there is a clear trend in revealing how nation states and their concerns with economic development take on a privileged position. As such, sustainability turns to reconfirm states and markets as central identities for the future development of the Arctic. Related, there is a sense in which sustainability simply has become a licence to exploit (Gad *et al.* 2017:20; Humrich 2017). The concept reconfigures rights and resources particularly by making 'preservation or production' less of a choice between two mutually exclusive options, more of a technical problematique. It insists on exploitation of natural resources, but promises to make it gentle. It insists on exploitation of human resources, but promises that it will be in the service of a greater good. Related, as an effect of the Arctic sustainability narratives, we see privilege ascribed to national spaces and a territorial order.

Inserting 'sustainability' in Arctic discourses has served to bypass apparently unilateral claims for the preservation or protection of the natural environment. This is the case within Arctic shipping (see Keil's chapter), and within the nexus of oil and state interests, where sustainability lends itself to a 'managerial ontology of natural capital in which a state's natural resources become valued as capital stock' (Steinberg and Kristoffersen, this volume). This way of approaching nature reactivates the roots of 'sustainability' in eighteenth-century *Hausvaterliteratur*, introduced in the introductory chapter. The same image is confirmed from an environmentalist position: where sustainable development is seen as a means to economic growth achieved 'with the requisite safety protocols being followed', sustainability discourse appears 'increasingly vacuous' (Gerhardt, Kristoffersen and Stuvøy, this volume).

At the same time, however, the narratives analysed in this volume also show that relieving 'sustainability' of its environmental commitment does not imply that it is without substantial effects. As Medby notes, sustainability is not meaningless; Arctic environments and natures, lives, and cultures are indeed to be maintained through time. Graugaard distils how, historically, colonial projects constructed their relation to not only the natural resources of a colonialized territory but also the human societies colonialized along the lines of the *Hausvaterliteratur*, valuing – under various conditions – Indigenous populations as capital stock. Sejersen in his chapter presents a particularly instructive analysis of what sustainability can do when inserted – not necessarily explicitly as a word, but as a conceptual logic – in a current neo-liberalized discourse of what society is: society needs to be sustained; society is operationalized as national economy; relegating individuals to be counted as per their productivity rather than as citizens taking part in decision-making or community members taking part in sustaining or developing cultural traditions. All of which stands at odds with the more radical potential of the discourse on limits to growth discourse in the 1970s and 1980s, which was explicitly critical of capitalism and its expansionist logic (Dryzek 2013:27–51).

These discussions highlight one very important message concerning the political nature of sustainability. However, rather than criticizing the concept for being only a licence to exploit – or simply 'sustainable economic growth' as it was critically described by Griffiths and Young (1989) 30 years ago – we learn more about the concept by regarding it as having a more subtle and nuanced character. It plays a key role in 'normalizing' the Arctic – i.e. reaffirming the position of central actors – but it also inevitably ties in an environment to occupy a central place in these developments and, at the same time, allow critical and alternative voices.

So on the one hand, including more and more aspects under the auspices of sustainability has a potentially depoliticizing effect, given that capital and states stand ready to employ their governmental resources and coordinate relevant technical expertise to manage our way to sustainability. On the other hand, as Keil notes in her chapter, the large number of referent objects is indicative of the inherent conflict potential in sustainability discourse. A corollary of the holistic promise of the concept is that it is difficult for a discourse centred on sustainability to close itself off against new claims promoting yet another referent object as worthy of being sustained. Herrmann, in her chapter, analyses how Indigenous communities have seized the moment created by the state of Alaska and US federal incentive structures to create sustainabilities for themselves beyond that of energy security. And in Sejersen's chapter there are small instances where individual citizens in public hearings insist on rearticulating how the authorities employ 'sustainability' to legitimize the establishment of a huge aluminium smelter. In other words; all critiques of sustainability discourses can be formulated as internal critiques. In that sense, sustainability is not just an essentially contested concept (Gallie 1956; Jacobs 1999); it is an inherently contestable concept. It might try to close itself off from politics by reverting into technical expertise, but it can hardly insulate itself from accepting responsibility for new claims.

The openness to new claims generated by the holistic tendencies of the concept does not mean that the politics of sustainability necessarily takes the form of a conflict between opposing actors with well-defined interests. Becker Jacobsen lays out in her chapter how 'these conflicts may actually run through many individuals that are seemingly perfectly able to recognize and value competing referent objects at the same time'. As recent work on climate change scepticism suggests, voting publics are also adept at ignoring scientific expertise and technical advice if it does not match their feelings and sentiments about a particular topic or issue (McCrea *et al.* 2015). States and multinational corporations are not the only stakeholders that have an interest in expanding sustainability concerns beyond the natural environment; ordinary people may too, as a way of insisting on both having a cake and eating it too – or at least resisting the demand of analysts, pollsters, and policy-makers to prioritize incompatible goals.

Finally, 'sustainability' has become a battle cry of and for inclusive processes. Both Wilson Rowe's and Thisted's chapters include stories to the effect that sustainability is not the label of a substantial configuration of potentially opposing values, but rather the name of and for a certain kind of process rather than outcome. According to Russian policies, development is only 'sustainable' if it is achieved through the 'effective participation of the affected voices, which would arguably speak on behalf of competing interests and issues to be balanced' (Wilson Rowe, this volume). Sustainability, in the Greenlandic interventions analysed by Thisted, must be 'based on a dialogue'. In the abstract, this could be read as an argument for a corporatist style of government, where all interests are represented and taken into account in order to avoid explicit conflict. Alternatively, when found in the Arctic, it also puts into play a specific history of the concept of sustainability in this region, institutionalized via the Sustainable Development Working Group of the Arctic Council (cf. Kristoffersen and Langhelle 2017:30–1). In this governance process, sustainability has become a way of insisting that the Arctic is not just pristine nature but also involves people living there with a right to collective development and a desire to plan and promote their visions for the future.

The message from the Arctic to the global conversation on sustainability

For thinking further about the general lessons that can be derived from Arctic discourses on sustainability, it is useful to remind ourselves of the point made by both Dryzek (2013:159ff) and Kates *et al.* (2005:20) who describe the global stature of sustainability and sustainable development as a result of a process which is wide and open. Many voices and interests can commit to and articulate their version of sustainability, even if they do not agree on the details. Diversity is part of what moves us as a global community forward towards both finding out what sustainability is and might be, and implementing it. Our volume confirms that in the Arctic sustainability has become a magnetic concept, attracting actors and attaching itself to many discourses and practices. But just because sustainability attracts, it does not mean that all those parties caught up in this gravity field are united by a single goal or vision for the future. The attraction may stem from its very openness towards claims to take into account various identities and valued environments.

Part of the magnetism of sustainability is that it comes with a promise that 'we can do this'; we can and should sustain multiple, related referent objects. This holism, however, makes it inherently vulnerable to challenges, if one scale is prioritized over others, or if a referent object is excluded. Challengers need not win, but the specific function of the concept of sustainability is to open up discourse to challenges. When challenges in the name of a greater good or a marginalized identity are dismissed, the dismissal comes through the force of a referent object or scale, which is prioritized by a specific constellation of actors (often including the state) – not because the concept of sustainability as such closes off. Future struggles will not revolve around the question of whether we should strive for sustainability. Rather they will be a contest between competing narratives about what it is that should be sustained.

Notes

1 The most recent introduction of the UN Sustainable Development Goals into the work of the Arctic Council by the Finnish Chairmanship has invited the explication of similar narratives:

> If the Artic is to be developed in sustainable manner, we need responsible companies to play a role in turning risks into opportunities and support the economic and sustainable development of the region. Opportunities are ripe. … Done right, there is the opportunity for creating a unique business model for sustainable business in the Arctic.
>
> (Kingo 2017)

2 Petrov *et al.* (2017:66) call for more research into how governance structures can counteract that such 'trade-off' arguments undermine the holistic ambition of the sustainable development discourse. See also Lempinen's analysis of how lasting social development is supposed to outweigh environmental degradation in evaluations of energy extractive projects in the Arctic (Lempinen 2017:120–122).

3 On the distinctly bifurcated structure of policy narratives, see Gad (2010:70–88; 2017).

4 Kates *et al.* (2005) insist that openness and experimentation are what drives the discourse on sustainable development. In Dryzek's analysis (2013) a distinct feature of the 'sustainable development' discourse is its civil society basis – in contrast to the technocratic 'ecological modernization' discourse (see Hajer 1995).

References

Bjørst, L.R. (2008): *En anden verden. Fordomme og stereotyper om Grønland og Arktis.* Copenhagen: Forlaget BIOS.

Bjørst, L.R. (2017): Arctic resource dilemmas: tolerance talk and the mining of Greenland's uranium. In: Thomsen, R.C. and Bjørst, L.R. (eds), *Heritage and change in the Arctic: resources for the present, and the future.* Aalborg: Aalborg Universitetsforlag.

de Rosa, M. (2014): Mining in Greenland: the science–policy network. Thesis. Available at: https://brage.bibsys.no/xmlui/bitstream/id/248908/Masteroppgave%20De%20Rosa.pdf (accessed 13 September 2018).

Dryzek, J. (2013): *The politics of the earth: environmental discourses*, 3rd edn. Oxford: Oxford University Press.

Gad. U.P. (2010): *(How) can they become like us? Danish identity politics and the conflicts of 'Muslim relations'*. PhD dissertation, Department of Political Science, University of Copenhagen.

Gad, U.P. (2016): *National Identity Politics and Postcolonial Sovereignty Games: Greenland, Denmark, and the EU*. Copenhagen: Museum Tusculanum Press.

Gad, U.P. (2017): Sex, Løgn og Landingsbaner: Thule set relationelt. *Grønland* 65(3):216–241.

Gad, U.P., Jakobsen, U., and Strandsbjerg, J. (2017): Politics of sustainability in the Arctic: a research agenda. In: Fondahl, G. and Wilson, G.N. (eds), *Northern Sustainabilities: Understanding and Addressing Change in the Circumpolar World*. New York: Springer, pp. 13–24.

Gallie, W.B. (1956): Essentially contested concepts. *Proceedings of the Aristotelian Society*, 56:167–198.

Griffiths, F. and Young, O.R. (1989): *Sustainable development and the Arctic: impressions of the co-chairs, second session, Ilulissat and Nuuk, Greenland 20–24 April*. Working Group on Arctic International Relations, reports and papers 1989-1.

Grydehøj, A. (2013): Informal diplomacy in Norway's Svalbard policy: intersection of local community development and international relations. *Global Change, Peace & Security* 26(1):41–54.

Hajer, M. (1995): *The politics of environmental discourse: ecological modernization and the policy process*. Oxford: Oxford University Press.

Hastrup K. (2015): *Thule på Tidens Rand*. Copenhagen: Lindhardt & Ringhof.

Humrich, C. (2017): Sustainable development in Arctic international environmental cooperation and the governance of hydrocarbon related activities. In: Pelaudeix, C. and Basse, E.M. (eds), *The governance of arctic offshore oil and gas (global governance)*. London: Routledge.

Jacobs, M. (1999): Sustainable development as a contested concept. In: Dobson, A. (ed.), *Fairness and futurity: essays on environmental sustainability and social justice*. Oxford: Oxford University Press, pp. 21–45.

Jacobsen, M. (2015): The power of collective identity narration: Greenland's way to a more autonomous foreign policy. *Arctic Yearbook* 2015:102–118.

Kates, R.W., Parris, T.M., and Leiserowitz, A.A. (2005): What is sustainable development? Goals, indicators, values and practice. *Environment: Science and Policy for Sustainable Development* 47(3):8–21.

Kingo, L. (2017): The Sustainable Development Goals and why they matter for the Arctic. Presentation at the Conference on SDGs in the Arctic, Ministry of Foreign Affairs, Copenhagen. Available at: http://um.dk/en/foreign-policy/the-arctic/the-sdgs-in-the-arctic/speeches/lise-kingo (accessed 13 September 2018).

Kristoffersen, B. and Langhelle, O. (2017): *Sustainable development as a global-Arctic matter: imaginaries and controversies*. In: Keil, K. and Knecht, S. (eds), *Governing Arctic change: global perspectives*. London: Palgrave Macmillan, pp. 21–42.

Lempinen, H. (2017): *The elusive social: remapping the soci(et)al in the Arctic energyscape*. Thesis. Available at: http://lauda.ulapland.fi/handle/10024/62884 (accessed 13 September 2018).

McCrea, R., Leviston, Z., and Walker, I.A. (2015): Climate change skepticism and voting behavior: what causes what? *Environment and Behavior* 48(10):1309–1334.

Medby, I.A. (2014): Arctic state, Arctic nation? Arctic national identity among the post-Cold War generation in Norway. *Polar Geography* 37(3).

Mikkelsen, A. and Langhelle, O. (2008): *Arctic oil and gas: sustainability at risk?*. New York: Routledge.

Petrov, A.N., BurnSilver, S., Chapin III, F.S., Fondahl, G., Graybill, J., Keil, K., Nilsson, A.E. Riedlsperger, R., and Schweitzer, P. (2017): *Arctic sustainability research: past, present and future*. London: Routledge.

Poppel, B. (2018): Sustainable development and sustainability in Arctic political discourses. In: Nuttall, M., Christensen, T., and Siegert, M. (eds), *Handbook of the Polar Regions*. London: Routledge.

Rosing, M., *et al.* (2014): *To the Benefit of Greenland.* University of Greenland/University of Copenhagen. Available from: http://greenlandperspective.ku.dk/this_is_greenland_perspective/background/report-papers/To_the_benefit_of_Greenland.pdf (accessed 13 September 2018).

Schriver, N. (2014): Enig om målet, men hvad med vejen? – en analyse af Grønlands debat om udvinding af råstoffer i 2012, *Tidsskriftet Grønland* 2:121–132.

Skorstad, B., Brigt, D., and Bay-Larsen, I. (2017): Governing complexity: theories, perspectives and methodology for the study of sustainable development and mining in the Arctic. In: Dale, B., Bay-Larsen, I., and Skorstad, B. (eds), *The will to drill: mining in arctic communities*. New York: Springer, pp. 13–32.

Steinberg, P.E., Tasch, J., and Gerhardt, H. (2015): *Contesting the Arctic: rethinking politics in the circumpolar North.* London: I.B. Tauris.

Swyngedouw, E. (2007): Impossible 'sustainability' and the postpolitical condition. In: Krueger, R. and Gibbs, D. (eds), *The sustainable development paradox*. New York: Guilford Press, pp. 13–40.

Wilson, E. (2007): Arctic unity, Arctic difference: mapping the reach of northern discourses. *Polar Record* 43(2):125–133.

Index

For Product Safety Concerns and Information please contact our EU
representative GPSR@taylorandfrancis.com
Taylor & Francis Verlag GmbH, Kaufingerstraße 24, 80331 München, Germany

www.ingramcontent.com/pod-product-compliance
Ingram Content Group UK Ltd.
Pitfield, Milton Keynes, MK11 3LW, UK
UKHW022030061126
477531UK00003B/147
* 9 7 8 1 1 3 8 4 9 1 8 3 0 *